# Environmental Science and Engineering

## Environmental Science





For further volumes:
http://www.springer.com/series/3234

Reinhard A. Klenke · Irene Ring
Andreas Kranz · Niels Jepsen
Felix Rauschmayer · Klaus Henle
Editors

# Human-Wildlife Conflicts in Europe

## Fisheries and Fish-eating Vertebrates as a Model Case

Springer

*Editors*
Reinhard A. Klenke
Klaus Henle
Department of Conservation Biology
UFZ—Helmholtz Centre for Environmental Research
Leipzig
Germany

Irene Ring
Department of Economics
UFZ—Helmholtz Centre for Environmental Research
Leipzig
Germany

Andreas Kranz
Institute of Wildlife Biology and Game Management
University of Natural Resources and Life Sciences
Vienna
Austria

Niels Jepsen
Section of Freshwater Fisheries Ecology
National Institute of Aquatic Resources
Technical University of Denmark
Silkeborg
Denmark

Felix Rauschmayer
Department of Environmental Politics
UFZ—Helmholtz Centre for Environmental Research
Leipzig
Germany

ISSN 1431-6250
ISBN 978-3-642-43109-8 ISBN 978-3-540-34789-7 (eBook)
DOI 10.1007/978-3-540-34789-7
Springer Heidelberg New York Dordrecht London



Printed on acid-free paper

Springer is part of Springer Science+Business Media (www.springer.com)

# Foreword

The background of this book and the FRAP project concept is presented in great detail elsewhere in this book, so I will concentrate on the contents of the book and the policy that lay behind it.

The book analyses three animal species groups—cormorants, otters and seals—that by their ecological interaction with human activities cause challenges of human/wildlife conflict reconciliation. The book further focuses on conflicts between the three fish-eating species groups and fisheries and aquaculture activities. The species studied are slowly increasing in number and distribution in parts of Europe, as a consequence of changes in human attitude and successful conservation measures, which also augment the need for reconciliation activities. In the words of the introductory policy brief of the book "we need ecologically effective, economically efficient, and socially acceptable means" to manage these conflicts and reconcile them. This book fulfils these challenges using a case-by-case approach, because reconciliation measures differ greatly among conflicts and countries, and it is shown that there is simply no basic solution to all problems encountered even at the species level.

The book gives an excellent overview of the conflicts at a European level, and proposes a framework for the development of conflict reconciliation action plans and analyses of the conflicts at a local, national or European level. Diet and damage assessment is a central theme, because knowledge of the amount of fish consumed is an important argument in the reconciliation process. Superimposed on that information is the analysis of regional economics, policy and stakeholder positions and of course the ecological mitigation measures. Another "module" screens viability and management of the target species and how this knowledge can be used in monitoring and modelling.

At the end of the book there are numerous conclusions, recommendations and consequences for the reconciliation process.

Finally, there is a section on the evaluation of policy instruments and conclusions and recommendations regarding the different conflict management strategies and how participatory decision strategies should be designed.

Suffice it to say, that this book is indispensible for anyone interested in human/ wildlife interaction problems and how these problems may and should be tackled to bring about reconciliation between various stakeholders and the target species. It will help wildlife managers and other decision makers, scientists and laymen alike to design an appropriate approach to this participatory process, and thereby mitigate the problems that will arise as a consequence of expanding animal species that compete with humans for the fishes caught or farmed in European waters.

It has been a privilege to write a foreword and recommend this book, not only because of the concrete virtues of the book but also because I was given the task to give advice on the FRAP process at an early stage. I can only congratulate the FRAP team on this formidable accomplishment, where numerous obstacles would have to be tackled and equally numerous other problems be solved.

Dr. Lennart Nyman
Royal Swedish Academy of Agriculture and Forestry (KSLA)
Tranebergs strand 35
16740, Bromma
Sweden

# Contents

# Contributors

**Peter Alberti** Department of Ecological Modelling, Helmholtz Centre for Environmental Research, Permoserstraße 15, 04318 Leipzig, Germany;
*Present Address*: Department of Building Services and Power Engineering, University of Applied Sciences, Altonaer Straße 25, 99085 Erfurt, Germany, e-mail: alberti@fh-erfurt.de

**Paula Antunes** Centre for Environmental and Sustainability Research, New University of Lisbon, Campus da Caparica 2829-516 Caparica, Portugal, e-mail: mpa@fct.unl.pt

**Karl Bruckmeier** School of Global Studies Human Ecology, University of Gothenburg, Box 700, 40530 Gothenburg, Sweden, e-mail: karl.bruckmeier@globalstudies.gu.se

**Karin Frank** Department of Ecological Modelling, Helmholtz Centre for Environmental Research, Permoserstraße 15, 04318 Leipzig, Germany, e-mail: karin.frank@ufz.de

**Dália Freitas** Departamento de Biologia Animal Faculdade de Ciências, Centro de Biologia Ambiental, Universidade de Lisboa, Bloco C2–5° Piso, Campo Grande 1749-016 Lisboa, Portugal, e-mail: difreitas@fc.ul.pt

**Jorge Gomes** Centre for Environmental and Sustainability Research, New University of Lisbon, Campus da Caparica 2829-516 Caparica, Portugal, e-mail: jvg@fct.unl.pt

**Bernd Gruber** Department of Conservation Biology, Helmholtz Centre for Environmental Research, Permoserstraße 15, 04318 Leipzig, Germany;
*Present Address*: Spatial Analysis and Ecological Modelling, University of Canberra, Bruce, ACT 2601, Australia, e-mail: bernd.gruber@canberra.edu.au

**Randi Habighorst** Department of Environmental and Planning Law, Helmholtz Centre for Environmental Research, Permoserstraße 15, 04318 Leipzig, Germany; *Present Address*: Zur Krumke 11, 49326 Melle, Germany, e-mail: randi.habighorst@gmx.de

**Klaus Henle** Department of Conservation Biology, Helmholtz Centre for Environmental Research, Permoserstraße 15, 04318 Leipzig, Germany, e-mail: klaus.henle@ufz.de

**Niels Jepsen** Section of Freshwater Fisheries Ecology, National Institute of Aquatic Resources, Technical University of Denmark, Vejlsøvej 39, 8600 Silkeborg, Denmark, e-mail: nj@aqua.dtu.dk

**Reinhard A. Klenke** Department of Conservation Biology, Helmholtz Centre for Environmental Research, Permoserstraße 15, 04318 Leipzig, Germany, e-mail: reinhard.klenke@ufz.de

**Andreas Kranz** Institute of Wildlife Biology and Game Management, University of Natural Resources and Life Sciences, Gregor Mendel Str. 33, 1180 Vienna, Austria;
*Present Address*: alka-kranz, Consulting Engineers, Wildlife Ecology and Nature Conservation, Am Waldgrund 25, 8044 Graz, Austria, e-mail: andreas.kranz@aon.at

**Luísa Madruga** Centre for Environmental and Sustainability Research, New University of Lisbon, Campus da Caparica 2829-516 Caparica, Portugal, e-mail: lmcm@fct.unl.pt

**Jaroslav Myšiak** Department of Economics, Helmholtz Centre for Environmental Research, Permoserstraße 15, 04318 Leipzig, Germany;
*Present Address*: Fondazione Eni Enrico Mattei and Euro-Mediterranean Center for Climate Change, Castello 5252, 30122 Venezia, Italy, e-mail: jaroslav.mysiak@cmcc.it

**Carsten Neßhöver** Department of Conservation Biology, Helmholtz Centre for Environmental Research, Permoserstraße 15, 04318 Leipzig, Germany, e-mail: carsten.nesshoever@ufz.de

**Lennart Nyman** Royal Swedish Academy of Agriculture and Forestry (KSLA), World Wide Fund for Nature, Ulriksdals Slott, 170 81 Solna, Sweden;
*Present Address*: Tranebergs strand 35, SE 167 40 Bromma, Sweden, e-mail: lennart.nyman@manandwater.com

**Thomas Olesen** Institute for Fisheries Management and Coastal Community Development, Willemoesvej 5, 9850 Hirtshals, Denmark;
*Present Address*: Udviklingskonsulent Regional Udvikling Region Midtjylland, Tingvej 15, 8800 Viborg, Denmark, e-mail: c/o dw@ifm.aau.dk

**Lukáš Poledník** Institute of Wildlife Biology and Game Management, University of Natural Resources and Life Sciences, Gregor Mendel Str. 33, 1180 Vienna, Austria;
*Present Address*: ALKA Wildlife, o.p.s., Lidéřovice 62, 38001 Dačice, Czech Republic, e-mail: lukas.polednik@alkawildlife.eu

**Kateřina Poledníková** Institute of Wildlife Biology and Game Management, University of Natural Resources and Life Sciences, Gregor Mendel Str. 33, 1180 Vienna, Austria;
*Present Address*: ALKA Wildlife, o.p.s., Lidéřovice 62, 38001 Dačice, Czech Republic, e-mail: katerina.polednikova@alkawildlife.eu

**Felix Rauschmayer** Department of Environmental Politics, Helmholtz Centre for Environmental Research, Permoserstraße 15, 04318 Leipzig, Germany, e-mail: felix.rauschmayer@ufz.de

**Irene Ring** Department of Economics, Helmholtz Centre for Environmental Research, Permoserstraße 15, 04318 Leipzig, Germany, e-mail: irene.ring@ufz.de

**Teresa Sales-Luís** Departamento de Biologia Animal Faculdade de Ciências, Centro de Biologia Ambiental, Universidade de Lisboa, Bloco C2–5° Piso, Campo Grande 1749-016 Lisboa, Portugal, e-mail: tsl@fc.ul.pt

**Rui Santos** Centre for Environmental and Sustainability Research, New University of Lisbon, Campus da Caparica 2829-516 Caparica, Portugal, e-mail: rfs@fct.unl.pt

**Margarida Santos-Reis** Departamento de Biologia Animal Faculdade de Ciências, Centro de Biologia Ambiental, Universidade de Lisboa, Bloco C2–5° Piso, Campo Grande 1749-016 Lisboa, Portugal, e-mail: mmreis@fc.ul.pt

**Kathleen Schwerdtner Máñez** Department of Economics, Helmholtz Centre for Environmental Research, Permoserstraße 15, 04318 Leipzig, Germany;
*Present Address*: Department of Social Sciences, Leibniz Center for Tropical Marine Ecology, Fahrenheitstr. 6, 28359 Bremen, Germany, e-mail: kathleen.schwerdtner@zmt-bremen.de

**Jukka Similä** Environmental Policy Centre, Finnish Environment Institute (SYKE), Box 140, 00251 Helsinki, Finland, e-mail: jukka.simila@ymparisto.fi

**Leila Suvantola** Department of Law, University of Eastern Finland Joensuu campus, P.Box 111, 80101 Joensuu, Finland;
*Present Address*: Heikinpohjantie 22, 57100 Savonlinna, Finland, e-mail: leila.suvantola@suursaimaa.com

**Riku Varjopuro** Environmental Policy Centre, Finnish Environment Institute (SYKE), Box 140, 00251 Helsinki, Finland, e-mail: riku.varjopuro@ymparisto.fi

**Vivien Weiss** Department of Urban and Environmental Sociology, Helmholtz Centre for Environmental Research, Permoserstraße 15, 04318 Leipzig, Germany, e-mail: vivien.weiss@ufz.de

**Håkan Westerberg** Department of Research and Development, Swedish Board of Fisheries, Box 423, 40126 Gothenburg;
*Present Address*: Påskvägen 32, 426 53 Västra Frölunda, Sweden, e-mail: hakan.westerberg@slu.se

**Douglas Clyde Kongshøj Wilson** Professor MSO Fisheries Sociology, Department of Development and Planning, Innovative Fisheries Management, An Aalborg University Research Centre, Nybrogade 14, 9000 Aalborg, Denmark, e-mail: dw@ifm.aau.dk

**Heidi Wittmer** Department of Environmental Politics, Helmholtz Centre for Environmental Research, Permoserstraße 15, 04318 Leipzig, Germany, e-mail: heidi.wittmer@ufz.de

# Policy Brief

## Meeting the Challenges of Human-Wildlife Conflict Reconciliation

**Klaus Henle, Andreas Kranz, Reinhard A. Klenke and Irene Ring**

Conflicts arising from the competition of humans and wildlife for biological resources are as old as humankind. Changes in civil society's attitudes towards wildlife and the success of conservation management have resulted in wildlife prospering again and returning to areas from where they had disappeared and even spreading to new habitats. This is reigniting old conflicts between humans and wildlife.

To reconcile such conflicts, we need ecologically effective, economically efficient, and socially acceptable means to manage human-wildlife conflicts. It is an arduous task that requires time, commitment, and knowledge. It is most successful if management and policy have adequate tools in place well before a conflict becomes virulent.

K. Henle (✉) · R. A. Klenke
Department of Conservation Biology,
UFZ—Helmholtz Centre for Environmental Research,
Permoserstraße 15, 04318 Leipzig, Germany
e-mail: klaus.henle@ufz.de

A. Kranz
Institute of Wildlife Biology and Game Management,
University of Natural Resources and Life Sciences,
Gregor Mendel Street 33, 1180 Vienna, Austria
e-mail: andreas.kranz@aon.at

R. A. Klenke
e-mail: reinhard.klenke@ufz.de

I. Ring
Department of Economics,
UFZ—Helmholtz Centre for Environmental Research,
Permoserstraße 15, 04318 Leipzig, Germany
e-mail: irene.ring@ufz.de

R. A. Klenke et al. (eds.), *Human-Wildlife Conflicts in Europe*,
Environmental Science and Engineering, DOI: 10.1007/978-3-540-34789-7_1,

Human-wildlife conflicts often differ strongly, depending on the species concerned and the regionally specific social contexts for a single species. They can range from no conflict to escalation at local, regional, or international level. Attempts to reconcile conflicts are usually developed on a case-by-case approach. As a consequence, reconciliation activities differ greatly among conflicts and among countries, both in approaches followed and in their successes and failures.

Against this background, researchers from the natural and social sciences from nine European countries have joined in a project to develop a generic framework for the reconciliation of human-wildlife conflicts using fisheries and fish-eating vertebrates (grey seal, Eurasian otter, and great cormorant) as model cases.

This policy brief summarizes important messages learned in the development of the generic framework. Above all, successful conflict management requires interdisciplinary and participatory approaches, among these an identification of the ecological, socio-economic, and cultural factors that play a key role in the conflict. We recommend using a conflict manager, who coordinates these activities. Such a person must be accepted by all stakeholders.

In the past, assessments of conflicts focused on the consumption of contested resources and almost always neglected landscape factors. However, the presence of conflicting wildlife and the potential for conflicts is not evenly distributed across the landscape or the sea. For example, the impact of cormorants on fish depends on the distance to major breeding colonies. While diet studies are comparably straightforward in well-delimited environments, such as inland fish ponds, they pose major challenges in open systems, such as coastal areas or the open sea.

Conflict perceptions by stakeholders can differ immensely from country to country, even in the presence of comparable policy instruments (e.g., damage compensation schemes). Similarly, the perceptions of the same conflict can vary widely among stakeholders depending on their specific interests in the conflict. A systematic description of the facts, values, and interests of the different stakeholder groups is essential for successful conflict reconciliation. It is important to realize that EU state aid rules inhibited in some countries the application of policy instruments that are effectively used for conflict resolution in other countries. Structural funds provided by the EU could be better used to reduce the conflicts and are under-utilized in most countries.

Classic ecological mitigation strategies in wildlife management, such as lethal and fertility control, wildlife translocation, or repellents, usually work only under restricted conditions. Typical conflict species tend to be highly adaptable, skilful, and clever and thereby counteract the efficacy of the chosen mitigation strategy. Moreover, any manipulation of the wildlife species to reduce its impact on the competed resource can have adverse effects on the species. Thus, monitoring must be implemented as an integral part of human-wildlife conflict management. Modeling the effects of management alternatives on the viability of the wildlife species can greatly help to evaluate alternative management options.

Single instruments are rarely adequate to solve conflicts. Rather, a combination of different instruments is usually asked for and their selection must be based on the key factors identified in the assessment of the ecological and socio-economic basis of the conflict. Suitable instruments must be ecologically effective, economically efficient, and socially acceptable. They help to distribute, more equally, the benefits and costs among various stakeholder groups. In addition, one must take into account that civil-society action is an essential ingredient of socially acceptable conflict management. Participatory approaches are particularly asked for when there is a need to shift from species conservation to species management, when new actors emerge in the conflict, or when the conflict escalates due to environmental change or changing human and/or animal behavior.

In summary, if human-wildlife conflict reconciliation strategies are to work, they must avoid simplified views and use truly interdisciplinary approaches instead, involving all relevant stakeholders, engaging some coordinator (conflict manager), and being based on sound scientific principles. Reconciliation takes time to achieve, is a permanent process, and needs research that combines different governmental levels and ecological scales from local to international. Reconciliation approaches are most successful if they are already in place before a conflict becomes salient.

**Acknowledgments** This work was financed by the EU 5th Framework Program (5th FP) Project "FRAP" (Development of a procedural Framework for Action Plans to Reconcile conflicts between large vertebrate conservation and the use of biological resources: fisheries and fish-eating vertebrates as a model case), contract number EVK 2-CT-2002-00142-FRAP.

# Introduction

**Klaus Henle, Irene Ring, Reinhard A. Klenke, Andreas Kranz, Niels Jepsen and Felix Rauschmayer**

Wildlife captures the imagination of humans. The image of wildlife and human-wildlife conflicts differs among people. Therefore, it is essential in a book that addresses human-wildlife conflicts to first clarify what we understand by "wildlife" and "human-wildlife conflicts". In the broadest sense, all wild,

K. Henle (✉) · R. A. Klenke
Department of Conservation Biology,
UFZ—Helmholtz Centre for Environmental Research,
Permoserstraße 15, 04318 Leipzig, Germany
e-mail: klaus.henle@ufz.de

I. Ring
Department of Economics,
UFZ—Helmholtz Centre for Environmental Research,
Permoserstraße 15, 04318 Leipzig, Germany
e-mail: irene.ring@ufz.de

R. A. Klenke
e-mail: reinhard.klenke@ufz.de

A. Kranz
Institute of Wildlife Biology and Game Management,
University of Natural Resources and Life Sciences,
Gregor Mendel Street 33, 1180 Vienna, Austria
e-mail: andreas.kranz@aon.at

N. Jepsen
Section of Freshwater Fisheries Ecology, National Institute of Aquatic Resources,
Technical University of Denmark, Vejlsøvej 39, 8600 Silkeborg, Denmark
e-mail: nj@aqua.dtu.dk

F. Rauschmayer
Department of Environmental Politics,
UFZ—Helmholtz Centre for Environmental Research,
Permoserstraße 15, 04318 Leipzig, Germany
e-mail: felix.rauschmayer@ufz.de

R. A. Klenke et al. (eds.), *Human-Wildlife Conflicts in Europe*,
Environmental Science and Engineering, DOI: 10.1007/978-3-540-34789-7_2,

undomesticated animals and plants belong to wildlife but most frequently the term refers to medium to large-sized terrestrial vertebrates that are hunted by humans as a resource, trophy, or because they compete with humans for food or space. This is particularly the case when talking about human-wildlife conflicts and we use the word wildlife in this sense throughout this book.

Wildlife is a conspicuous component of the diversity of life, which in our modern days usually is called biodiversity (Wilson 1988; Reaka-Kudla et al. 1997). Throughout evolution humans have had a wide and complex interrelationship with biodiversity as our species *Homo sapiens* is an integral part of the global ecosystem. Humans depend on the "*ecosystem services*" provided by animals, plants, microorganisms, and genes and their natural products (MEA 2005). We use biodiversity for food, medicine, and construction material, and we compete with some species for space or natural resources. Growing human populations, expanding to almost every corner of the earth, and growing individual resource demands exert increasing pressures on biodiversity and natural resources. Species loss and habitat destruction have been highlighted for several decades and even further back (e.g., Kleinschmidt 1937; Leopold 1949; Carson 1962). The scale and potential consequences of this loss has led to action to combat it, notably the Convention on Biological Diversity (http://www.cbd.int) and the commitment of the European Union to protect and restore habitats and to halt the loss of biodiversity by 2010 (EC 2006).

Wild animal species have always played an important role in human exploitation of nature, both as resources and as competitors for food and space. Hence, the need for and ways to their protection are particular.

Given their importance for humans, wildlife figured prominently in human imagination and considerably contributed to the shaping of human culture as witnessed by the oldest documents of art, the elaborate cave drawings of our ancestors. Despite increasing detachment from nature, these influences still continue in modern times, albeit often less consciously, for example, in the widespread use of wildlife species names for human products, characters, and esteems. Notwithstanding the positive values attached to these names, human attitudes to wildlife have often been at odds (Brown 2009).

For thousands of years humans responded to damage caused by wildlife species by protecting their resources and by prosecuting these competitors (Conover 2002). With human population growth and modern societies demanding ever increasing amounts of resources, wildlife dwindled and regionally went extinct. Concomitantly, an increasing part of society was no longer affected by wildlife causing damage. The traditional dualistic view of "good" and "bad" species lost significance and was increasingly replaced by an ecological, and in some sectors of society even romantic, view emerging after World War II (Callicot 1998; Kruuk 2002; Konold 2004). People increasingly developed interest in preserving wildlife species, because they enjoyed their presence and because they identified a set of positive values associated with such species. Last but not the least the modern mass media like cinema and broadcast have brought imaginations about wildlife even to citizens in the largest cities. Who in Europe was not impressed by

Bernhard Grzimeck's "Serengeti shall not die" or Sir David F. Attenborough's amazing series on BBC? Both television series were only representatives for a whole genre that has shaped the perception of nature by people throughout the last decades.

Decision makers and politicians—on behalf of a majority of people not directly affected by wildlife damage—adopted legal and institutional frameworks in order to protect wildlife and prevent it from going extinct. In Europe, the Habitats Directive and the Birds Directive are the most relevant legal regulations in this area, and national laws have to conform to these Directives. These changes in civil society's attitudes towards wildlife and the success of conservation management have enabled some wildlife populations to prosper again and return to areas from where they had disappeared and even spread to new habitats (e.g., Enserink and Vogel 2006).

When wildlife returns, divergent beliefs and interests trigger and exacerbate conflicts among humans (White et al. 2009). While society at large enjoys the presence of wildlife, rural resource holders often have to carry the burden of wildlife damage unless the conflicts are well managed. The consequence is a polarization of the society into urban versus rural residents and local development versus national conservation interests (Conover 2002) that can lead to severe conflict among stakeholder groups. Thus, the need to reconcile such conflicts is the logical consequence of the change of fundamental paradigms in human-wildlife relationships from a purely utilitarian view to a perspective that includes non-utilitarian values of nature and wildlife. If reconciliation fails, either local resource holders loose, wildlife looses since their fate is still in the hands of the local habitants or both of them loose. The challenge in wildlife management is to convert such a *loose–loose* scenario into a *win–win* situation (Woodroffe et al. 2005).

Human-wildlife conflicts encompass two main aspects. The first, more material-based aspect deals with the conflicting "interests" of humans and wildlife competing for the same resources. The second aspect refers to intra-societal or stakeholder relations; protectors of wildlife species, in particular larger vertebrates, are in conflict with the human competitors for the resource who do not want to accept the damage caused by wildlife.

Various species and types of resources can be involved in human-wildlife conflicts depending on conditions and human perspectives. Typical "conflict species" are large carnivores, fish eaters, raptors (eagles etc.), large herbivores, bears, beavers, large owls, and some corvids (e.g., Common raven *Corvus corax* etc.) (Woodroffe et al. 2005). Typical resources involved in human-wildlife conflicts are livestock and game species, farmed and wild fish, but also crops and fruits (Conover 2002).

In this book we do not cover abundant, non-protected species causing damage to crops or forests, such as wild boar (*Sus scrofa*), deers, rodents, and European starling (*Sturnus vulgaris*). These species are not involved in the type of conflict that is the topic of this book. There is a widely-shared consensus that they may be managed and controlled or even eradicated in the most effective way or by means, which conform to a general ethics of animal rights (Caughley and Sinclair 1994). Rare and protected species suffering from habitat destruction (e.g., Baier et al. 2006) or any kind of human resource exploitation, such as by-catch of marine turtles, birds, or mammals (Piatt and Nettleship 1987; Lutcavage et al. 1997; Read et al. 2006), are also not the target of this book, although they also frequently cause conflicts between stakeholders. Nevertheless, some of the principles developed in this book may be applied to such conflicts as well.

So, what is this book about? The book is dedicated to the reconciliation of conflicts rising from the protection of species and the use of biological resources by humans. Next to presenting a number of illustrative case studies, the major objective of this book is to provide a generic framework for human-wildlife conflict reconciliation. In this way, our objective clearly is to move beyond a case-by-case approach. For successful human-wildlife conflict reconciliation, integrative biodiversity research is required, involving interdisciplinary and applied approaches (Jentsch et al. 2003). Conflicts evolve between people: between individual actors or stakeholder groups that hold different views, values, and interests. Hence, for successful biodiversity conflict reconciliation, (1) ecologists and social scientists need to closely collaborate and (2) societal actors and stakeholders need to be involved in participatory research. In short, the human dimensions of wildlife management become essential, especially when dealing with human-wildlife conflicts (Ring 2009).

Thus, this book is about damages caused by protected vertebrates and their management causing considerable conflicts within society. It is about key features of typical conflict species, about economics, attitudes, and positive or negative emotions. Such, often strong, emotions arise when animal species are regarded as good or bad: charismatic and cute or blood thirsty beasts. Such conflict-species are often adaptable and elusive and therefore difficult to manage.

In North America there is a long-standing tradition in the human dimensions research related to wildlife management, and dealing with human-wildlife conflicts (e.g., Arner and Dubose 1982; Hygnstrom et al. 1994; Hadidian et al. 1997; Conover 2002; Treves et al. 2006; Brown 2009; Unsworth and Petersen undated). Within this tradition, the U.S. Fish and Wildlife Service was established as a governmental authority to deal with damage and conflicts due to wildlife. It was also here that the focus in wildlife management gradually shifted from wildlife itself towards better considering the human dimensions: Wildlife management is people management (Leopold 1933; Maehr et al. 2001; Westley and Miller 2003; Fascione et al. 2004; Manfredo et al. 2009). It is therefore crucial to focus on humans, their behavior and attitudes, in order to reconcile such conflicts (Conover 2002).

In Europe, there is no central authority dealing with the management of damage and the reconciliation of conflicts caused by wildlife. As in most other parts of the world, human-wildlife conflicts are usually addressed in a case-by-case approach, which may differ considerably among and within countries. Generic approaches based on experience from other, similar conflicts have not yet been developed though recently White et al. (2009) developed a framework for assessing and understanding human-wildlife conflicts. Lessons learned from case studies and generic approaches could greatly improve our capacity to reconcile conflicts because conflict reconciliation is an arduous task that requires time, commitment, and knowledge (Conover 2002). It is most successful, if management and policy have adequate tools in place well before a conflict becomes virulent, as was well illustrated by the highly publicized story of the "problem" brown bear (*Ursus arctos*) Bruno in Germany (Anonymus 2006; Enserink and Vogel 2006).

Against this background, researchers from the natural and social sciences from nine European countries joined in a project to develop a generic framework for the reconciliation of human-wildlife conflicts consistent across national boundaries. The project's acronym was named FRAP, standing for Framework for Biodiversity Reconciliation Action Plans. FRAP was funded by the EU and used fisheries and larger fish-eating vertebrates (Baltic grey seal *Halichoerus grypus*, Eurasian otter *Lutra lutra*, and great cormorant *Phalacrocorax carbo sinensis*) as model cases to evaluate and illustrate successful (and less successful) approaches for conflict reconciliation (http://www.frap-project.net). As for many other human-wildlife conflicts the relationships between fisheries and the conservation of these vertebrates differ strongly across Europe and among species, ranging from no conflict at all to escalations on a local, regional, or even international level. This is well illustrated by the case studies presented in this book.

In the first part of the book we present case studies of human-wildlife conflicts in Europe and the various practical approaches used for conflict reconciliation. The case studies focus mainly on fish-eating vertebrates and fisheries, with a chapter on the golden eagle (*Aquila chrysaetos*) and reindeer (*Rangifer tarandus*) husbandry broadening this scope. As far as possible, the presentation of the cases follows the structure of the generic framework for conflict reconciliation, as presented in the book's second part. Thus, our framework is illustrated with examples from the conflicts between the conservation of seals, otters, respectively cormorants and fisheries. We evaluated existing information and studied the conflicts for regions that differ either in the ecological basis of the conflict or in the use of socio-economic mitigation strategies. The major regional comparisons are Denmark versus Italy for cormorants, Central Europe versus Portugal for otters and Finland versus Sweden for grey seals. The case studies conclude with lessons learned, be it from failures or success stories, and recommendations for improved conflict reconciliation.

In the second part of the book we draw on the experience from the case studies and the multi-disciplinary background of the FRAP team. We present a generic framework for the development of reconciliation action plans that adequately considers the human dimensions of such conflicts and the need for participatory research. The generic framework provides guidelines on how to analyze and assess the ecological and socio-economic basis of conflicts. It continues with evaluating and developing successful mitigation strategies, including technical mitigation, monitoring and population viability modeling, and mixes of policy instruments. It concludes with the design of participatory decision strategies and recommendations for effective stakeholder interactions.

While this generic framework was written from an EU perspective, many of its principles can be directly applied to other European countries not involved in FRAP or with little modifications to human-wildlife conflicts in other parts of the world as well. When the framework is transferred beyond the EU context, national regulations, institutional and cultural differences will certainly become more relevant (Ring 2009). In low-income regions, where poorer people are affected by human-wildlife conflicts, sustainable livelihood issues need to be properly integrated into conflict management strategies (Woodroffe et al. 2005; Johannesen 2007). Transfering the framework to wildlife endangering human life itself [such as wolf (*Canis lupus*) and brown bear (*Ursus arctos*)] might need a more explicit recognition of the emotional aspects of the conflict by psychological or anthropological research.

Last, but not least we hope that our integrative approach to biodiversity conflict reconciliation helps to improve our capacity to reconcile such conflicts in order to facilitate a sustained and acceptable coexistence of humans and wildlife.

**Acknowledgments** This work was financed by the EU 5th Framework Program (5th FP) Project "FRAP" (Development of a procedural Framework for Action Plans to Reconcile conflicts between large vertebrate conservation and the use of biological resources: fisheries and fish-eating vertebrates as a model case), contract number EVK 2-CT-2002-00142-FRAP.

## References

Anonymus (2006) Godspeed Bruno: brown bear meets a tragic end. www.spiegel.de/international/0,1518,423629,00.html. Accessed 01 Dec 2008

Arner DH, Dubose JS (1982) The impact of the beaver on the environment and economics in southeastern United States. Trans Int Congr Game Biol 14:241–247

Baier H, Erdmann F, Holz R, Waterstraat A (2006) Freiraum und naturschutz. Die Wirkungen von Störungen und Zerschneidungen in der Landschaft, Springer, Berlin

Brown PJ (2009) Introduction: perspectives on the past and future of human dimensions of fish and wildlife. In: Manfredo MJ, Vaske JJ, Brown PJ, Decker DJ, Duke EA (eds) Wildlife and society—the science of human dimensions. Island Press, Washington, pp 1–13

Callicott JB (1998) The great new wilderness debate. University of Georgia Press, Athens. http://www.ugapress.uga.edu/082031983X.html. Accessed 18 Feb 2009

Carson RL (1962) Silent spring. Riverside Press, Cambridge

Caughley G, Sinclair ARE (1994) Wildlife ecology and management. Blackwell Science, Cambridge

Conover M (2002) Resolving human-wildlife conflicts: the science of wildlife damage management. Lewis Publishers (CRC Press LLC), Boca Raton

EC (European Commission) (2006) Communication from the commission: halting the loss of biodiversity by 2010—and beyond. Sustaining ecosystem services for human well-being. COM/2006/216 final, Brussels, 22 May 2006. http://ec.europa.eu/environment/nature/biodiversity/comm2006/bap_2006.htm. Accessed 14 Dec 2008

Enserink M, Vogel G (2006) Wildlife conservation: the carnivore comeback. Science 314(5800):746–749. doi:10.1126/science.314.5800.746

Fascione N, Delach A, Smith M (2004) People and predators—from conflict to coexistence. Island Press, Washington

Hadidian J, Hodge G, Grandy J (eds) (1997) Wild neighbors: the humane approach to living with wildlife. The Humane Society of the United States and Fulcrum Publishing, Washington

Hygnstrom SE, Timm RM, Larson GE (eds) (1994) Prevention and control of wildlife damage. Univ Nebr Coop, Extension, USDA-APHIS-Animal damage control, and great plains agricultural council wildlife committee. Nebraska Cooperative Extension, University of Nebraska, Lincoln, vol 2. p 822. http://digitalcommons.unl.edu/icwdmhandbook/ or http://icwdm.org/handbook/index.asp#om. Accessed 14 Dec 2008

Jentsch A, Wittmer H, Jax K, Ring I, Henle K (2003) Biodiversity. Emerging issues for linking natural and social sciences. GAIA 12(2):121–128

Johannesen AB (2007) Protected area, wildlife conservation, and local welfare. Ecol Econ 62:126–135

Kleinschmidt O (1937) Die singvögel der heimat. Quelle & Meyer, Leipzig

Konold W (2004) Traditionen und trends im naturschutz. Ber ANL 28:5–15

Kruuk H (2002) Hunter and hunted. Relationships between carnivores and people. Cambridge University Press, Cambridge

Leopold AA (1933) Game management. Charles Scribner's, New York

Leopold AA (1949) A sand county almanac. Oxford University Press, Oxford

Lutcavage ME, Plotkin P, Witherington B, Lutz PL (1997) Human impact on sea turtle survival. In: Lutz PL, Musick JA (eds) The biology of sea turtles. CRC Press, Boca Raton, pp 387–409

MEA (Millennium Ecosystem Assessment) (2005) Ecosystems and human well-being: synthesis. Island Press, Washington

Manfredo MJ, Vaske JJ, Brown PJ, Decker DJ, Duke EA (2009) Wildlife and society—the science of human dimensions. Island Press, Washington

Maehr D, Noss R, Larkin J (2001) Large mammal restoration—ecological and sociological challenges in the 21st century. Island Press, Washington

Piatt JF, Nettleship DN (1987) Incidental catch of marine birds and mammals in fishing nets off Newfoundland, Canada. Mar Pollut Bull 18:344–349

Reaka-Kudla ML, Wilson DE, Wilson EO (eds) (1997) Biodiversity II: understanding and protecting our biological resources. Joseph Henry Press, Washington

Read AJ, Drinker P, Northridge S (2006) Bycatch of marine mammals in U.S. and global fisheries. Cons Biol 20:163–169

Ring I (2009) Toward a framework for integrating human dimensions in wildlife management. In: Manfredo MJ, Vaske JJ, Brown PJ, Decker DJ, Duke EA (eds) Wildlife and society—the science of human dimensions. Island Press, Washington, pp 90–103

Treves A, Wallace RB, Naughton-Treves L, Morales A (2006) Co-managing human-wildlife conflicts: a review. Hum Dimen Wildlife 11:383–396

Unsworth RE, Petersen TB (not dated) A manual for conducting natural resource damage assessment: the role of economics. Industrial Economics Inc., Cambridge, MA. http://www.fws.gov/policy/NRDAIndex.html. Accessed 13 Dec 2008

Westley F, Miller P (2003) Experiments in consilience—integrating social and scientific responses to save endangered species. Island Press, Washington

White RM, Fischer A, Marshall K, Travis JMJ, Webb TJ, di Falco S, Redpath SM, van der Wal R (2009) Developing an integrated conceptual framework to understand biodiversity conflicts. Land Use Policy 26(2):242–253

Wilson EO (ed) (1988) Biodiversity. National Academy of Sciences/Smithsonian Institution, Washington

Woodroffe R, Thirgood S, Rabinowitz A (2005) People and wildlife. Conflict or coexistence? Cambridge University Press, Cambridge

# Part I
# Lessons Learned from the Analysis of Model Conflicts

Grey seal (*Halochoerus grypus*) in Hel, Poland. *Photo:* Mateusz Włodarczyk

# Baltic Seal Reconciliation in Practice

## The Seal Conflict and its Mitigation in Sweden and Finland

**Karl Bruckmeier, Håkan Westerberg and Riku Varjopuro**

**Abstract** This chapter presents the results of case studies on human-wildlife conflicts conducted in Sweden and Finland. In both cases the conflict is between the conservation of the grey seal (*Halichoerus grypus*) and small-scale coastal fishing. The characteristics of the conflict between grey seal protection and fishery can be shown by way of a systematic comparison between the two countries and model regions as done here. Thus one can also better see what can be learned from the comparison of the cases. The main messages from both case studies in the Baltic Sea are formulated with regard to the significance of coastal fishery (as resource manager, not only resource user), with regard to single conflict mitigation measures, such as seal hunting and technical solutions to the conflict, and with regard to a combination of measures and overall approaches to conflict management. All these messages converge to the conclusion that much more can be learned from the management of the seal conflict than the resolution of that specific conflict: one by one the lessons learnt turn out to be steps of a more encompassing strategy of sustainable resource management in the coastal zone.

K. Bruckmeier (✉)
School of Global Studies, Human Ecology, University of Gothenburg,
Box 700 40530, Gothenburg, Sweden
e-mail: karl.bruckmeier@globalstudies.gu.se

H. Westerberg
Department of Research and Development, Swedish Board of Fisheries,
Box 423 40126, Gothenburg, Sweden
e-mail: hakan.westerberg@slu.se

R. Varjopuro
Environmental Policy Centre, Finnish Environment Institute (SYKE),
Box 140 00251, Helsinki, Finland
e-mail: riku.varjopuro@ymparisto.fi

R. A. Klenke et al. (eds.), *Human-Wildlife Conflicts in Europe*,
Environmental Science and Engineering, DOI: 10.1007/978-3-540-34789-7_3,

# 1 History and State of the Conflict

## *1.1 Emergence of the Grey Seal Problem in the Baltic Sea*

The number of grey seals in the Baltic declined during most of the twentieth century for several reasons (Härkönen and Hårding 2001). The Baltic grey seal is protected on the basis of the Convention on the Protection of the Marine Environment of the Baltic Sea (HELCOM) and the EU Habitats Directive. Due to the rapid growth of the seal population since the early 1980s the conflict with coastal fishermen has increased in Finnish and Swedish coastal waters because of damage to gear and loss of catch by seals. In the model regions, like in the whole Baltic Sea the number of seals is growing. In both countries the conflict has gained a lot of attention even though it is only one of several environment-related conflicts threatening coastal fishery. It can be seen as the last in a long chain of problems and conflicts. The coastal fishery is in decline because of its competitive disadvantages compared with the large-scale industrialized fisheries, but also because of the deteriorating quality of coastal waters and less availability of fish.

The problem with grey seals in the Baltic includes several aspects that successively have come into focus:

- *A problem of nature or species protection* Due to a combination of high hunting pressure and environmental toxins, such as DDT (Olsson et al. 1994; Hårding and Härkönen 1999; Nyman et al. 2003), the grey seal became endangered during the 1960—70s and was protected. However, the rapid recovery of the population since the 1980s has caused new conflicts and controversies about the continued necessity of protection.
- *A problem of resource use.* Seals, as a consequence of their growing number and changing behavior, compete with the coastal fishermen for the same resource.

International discussions and negotiations have taken place especially between HELCOM parties, resulting in a recommendation for protection of seals in 1988. In Finland, hunting became more and more restricted since 1975 until a total ban on hunting came into force in 1982, whereas the seal reserves were designated as late as in 2001. In Sweden the hunting of grey seal was banned in 1967 in Skagerrak and Kattegatt and from 1975 onwards in the Baltic, with the exception of protective hunting at fishing gear. Protective hunting was completely stopped in 1988. Parallel to the ban of seal hunting a number of seal reserves were established in the 1970s in Sweden.

The latest turn in protection has been that the ban of hunting begins to be modified—although still valid, the door is open for gradually increasing the number of seals hunted. (Protective) hunting of the grey seal has been allowed again since 1997 in Finland and 2001 in Sweden. The seal hunting can be understood in different ways and different interpretations prevail in the Swedish and Finnish model regions. It can be interpreted as still in accordance with the HELCOM rules by following an exemption introduced in 1995 under which

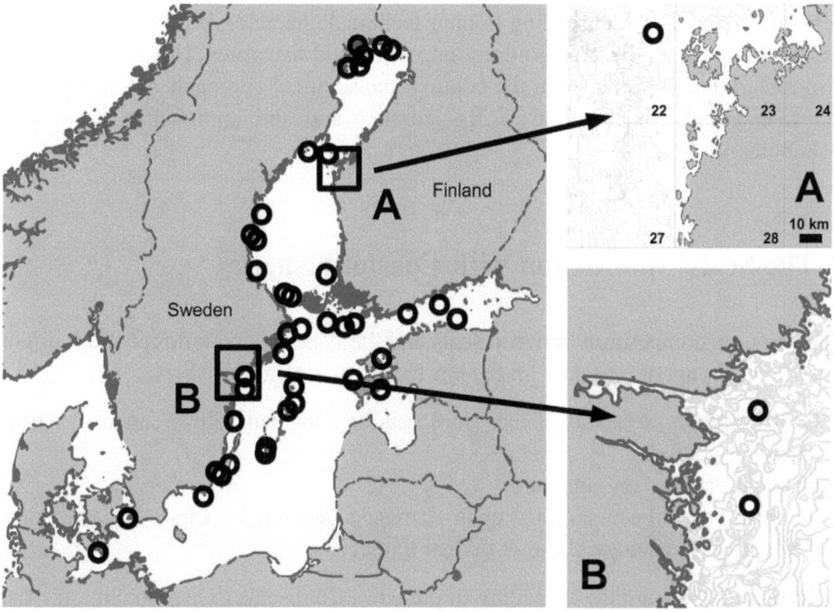


**Fig. 1** The Baltic sea model regions and the major haul-out sites of Baltic grey seals (marked with circles; data from the Biomad database, Museum of natural history, Sweden, and RKTL, Finland). The 10 m depth contours are shown in the model region inserts A and B. *Sources*: Modified version from Wahlberg et al. (2003)

limited hunting is allowed for specific purposes or it can be understood as a gradual return to a practice of free hunting as it has been until the 1970s.

## *1.2 The Case Studies: How to Reconcile an Increasing Conflict?*

For comparative conflict analysis two model regions were chosen, one in each country. The choice of the model regions is based on areas representing a characteristic example for the conflicts in each of the countries, and, for the Swedish region, also due to an increasing conflict with the southward expansion of the seal population. The Swedish region covers the archipelago of county Södermanland and Östergötland. In Finland the region called Kvarken is on the west coast in a narrow sea area between Sweden and Finland (see Fig. 1).

The case studies are presented together for several reasons among which the main one is, that the ecological features of the conflict are the same (see below). How the conflict is mitigated reflects the ecological peculiarities of the species-specific conflict and furthermore the socio-economic peculiarity of small-scale coastal fishery as the main economic sector affected. The mitigation measures generated so far have, according to their technical character, been similar for both countries, targeting the stakeholder group of coastal fishermen. However, there are also remarkable differences in the conflict management approaches with regard to the managerial coordination of the conflict and the approaches to involve stakeholders.

# 2 The Seals' Interaction with Coastal Fisheries

The resource competition between seals and the fishery is both direct and indirect. The direct competition takes place in a form of damage to fishery:

- Loss of catch—removal of fish from nets or damaged fish; escaped fish from damaged gear,
- physical damage on nets, and
- the extra work necessary to repair damaged gear or to empty the gear more frequently to decrease the risk of seal damages.

Indirect effects are the spreading of parasites from seals to fish, making the fish unfit for consumption, and a widespread loss of fishing grounds where the level of damage makes fishing impossible. Below we concentrate mostly on the direct damages, because that has been the main topic in debates and the only reason for mitigation activities in both countries.

## *2.1 Development of the Grey Seal Population*

The historic maximum of the grey seal population was around the turn of the nineteenth century, when the total number was 88,000–100,000 animals (Hårding and Härkönen 1999). A combination of bounties and more efficient hunting weapons caused a steady decrease during most of the twentieth century. In addition the burden of polychlorinated toxins in the Baltic decreased the condition and reproductive capacity of the seals from the 1950s and onwards (Olsson et al. 1994; Nyman et al. 2003), which led to a minimum population of probably just a few thousand animals in the mid 1970s. The Baltic grey seals are counted yearly from aircrafts and boats within Swedish, Finnish, and Estonian monitoring programs (Fig. 2). The major field efforts take place in March to count the number of newborn pups on the ice and during molt in late May, when the majority of grey seals are hauling out (Helander and Karlsson 2003). Since the early 2000s, monitoring has been coordinated between the Baltic countries to avoid double counts. The population has increased throughout the 1990s with an annual growth of 8.7 % (Helander and Karlsson 2003) except of the southern part of the Baltic Proper, where the growth is about 5.8 %. The development is illustrated in Fig. 3.

**Fig. 2** An aerial photograph of a grey seal islet, taken on a counting flight on 9th June 2005. *Photo:* Riku Lumiaro, SYKE

A major uncertainty with the monitoring program is to estimate how large the fraction of the population is, which is in the water and thus is uncounted during the census. By taking pictures of seals and analyzing a catalogue of identified individuals during a number of years, as was done in Sweden 1998–2002, an independent estimate can be made of the total population size. This shows that the counted number is at least 20 % below the actual population size. The seal combined count in 2004 was 17,640 seals, which means that the estimated amount of grey seals exceed 20,000 (Halkka et al. 2005).

The total number of seals is not the only relevant aspect in relation to interactions with coastal fishing, since the seals do not stay in one place during the year. The grey seal hauls out on rocky islands in the Swedish and Finnish archipelago. During the molting period in late May to early June the largest aggregations of grey seals are found on land and ice (Bonner 1981; Curry-Lindahl 1970). Compared to the extent of the vast archipelagos in the Baltic Sea, the present number of preferred haul-out sites (see Fig. 1) is surprisingly low (Sjöberg 1999). Satellite telemetry studies have shown that even though the movements of individual grey seals may extend throughout the entire Baltic Sea, most seals prefer to confine their movements to within 50 km of their favorite haul-out sites (Sjöberg and Ball 2000; Dietz et al. 2003). Grey seals of the Gulf of Bothnia prefer to forage in daytime and haul out at night (Sjöberg et al. 1999). Because seals move in such a large area their numbers in the model regions vary a lot.

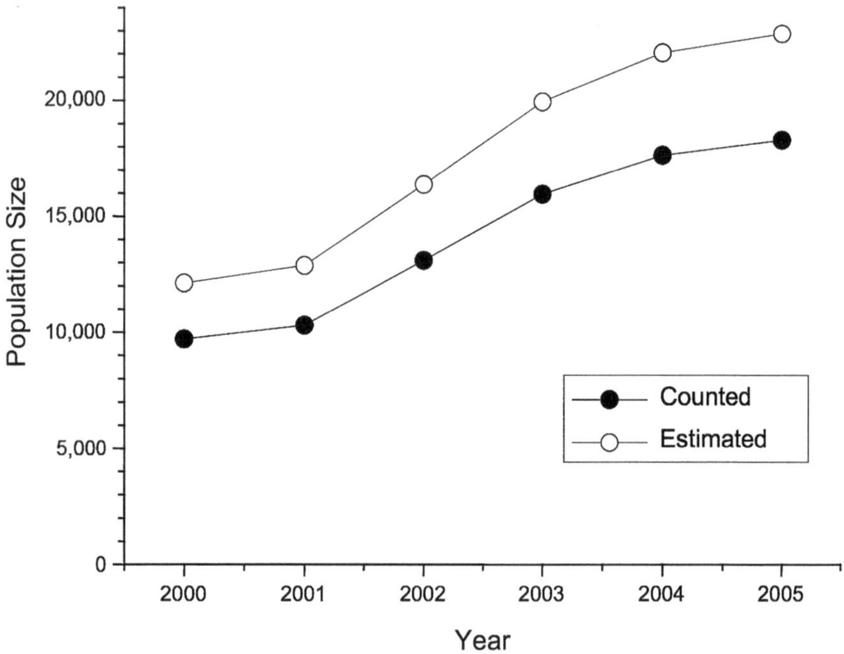


**Fig. 3** Development of the grey seal population between 2000 and 2005, the period when monitoring in the Baltic has been coordinated and using comparable methods. The estimate is based on a correction derived from the individual photographic identification project (Hiby et al. 2006)

## *2.2 Damage Assessment Relating to Diet of Seals*

In order to assess the loss of commercially important fish from predation by seals the diet composition and consumption rate has to be known. For the model regions in the Baltic grey seal case such data were lacking and a comprehensive study was conducted to quantify the present diet of the seals (for more details see Lundström et al. 2006). The digestive tract contents from 145 grey seals collected between 2001 and 2004 in the Baltic Sea were examined. By using additional hard-part structures other than otoliths, and species-specific size and numerical correction factors biases introduced by erosion of otoliths was compensated for. In the absence of numerical correction factors based on feeding experiments for some species, correction factors based on a relationship between otolith recovery rate and otolith width was used. A total of 24 prey taxa were identified but only a few species contributed substantially to the diet. The estimated diet composition was, independently of the prey number estimation method and diet composition estimation model used, dominated by herring (*Clupea harengus*), both by numbers and biomass. In addition to herring (*Clupea harengus*), common whitefish (*Coregonus lavaretus*) and sprat (*Sprattus sprattus*) were important prey, but cyprinids (*Cyprinidae*), eelpout (*Zoarces viviparus*), flounder (*Platichthys flesus*)

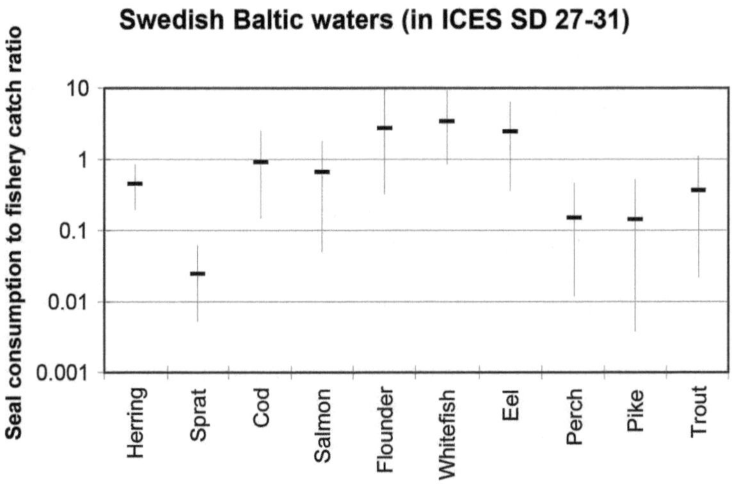


**Fig. 4** Exploitation of fish by grey seals relative to the total Swedish commercial and leisure fishery in Swedish Baltic Waters (data for the area north of latitude 56° 30′ N, corresponding to the statistical divisions SD 27-31 according to international council for the exploration of the sea)

and salmon (*Salmo salar*) also contributed significantly. The results indicate dietary differences between grey seals of different age as well as between seals from the northern (Gulf of Bothnia) and the southern (Baltic Proper) Baltic Sea.

With a sufficient sample size the mean species distribution and size composition of the diet can be established. Using the consumption rate—measured in physiological experiments or estimated from the feeding of captive animals—and the total population in the area of interest the amount of different fish species that is consumed can be calculated. As a first step in analyzing the degree of competition between human and seal use of the fish resource the total seal consumption can be compared to the landing in the fishery. Figure 4 shows the relative exploitation of different commercial species in the central and northern Swedish Baltic. The Swedish model region is a part of this, but the analysis was made for a larger area due to a lack of fishery and seal population data on a smaller scale. The fishery data is the sum of Swedish commercial and leisure fishery and the grey seal population is the part of the population counted on the Swedish side of the area.

In Fig. 4 a ratio larger than 1 means that the seal consumption is larger than the take by the fishery. This is the case for flounder (*Platichthys flesus*), eel (*Anguilla anguilla*), and Common whitefish (*Coregonus lavaretus*). For salmon (*Salmo salar*) and Atlantic cod (*Gadus morhua*) the seal consumption is close to the human removal. The core of the cod distribution is south of the area and the seal predation is likely marginal for the cod population. The fishing pressure on flounder is low and the high ratio does not necessarily mean that the seal predation is important for the Baltic flounder stock. But for salmon, herring, and in particular for whitefish both seal and fishing pressure are high and a true resource competition may be present. A detailed presentation of the diet studies is given by Lundström et al. (2006).

A simple comparison mass by mass between grey seal consumption and human fishery can be misleading, however. On the one hand a large consumption by seal may be targeted on small, non-commercial size classes. If there is a size dependent growth limitation of the prey, such predation may actually improve the production of fish of commercial size. On the other hand a relatively small quantity, which is consumed at a sensitive stage in the life cycle of the prey, can have significant effect on the fish stock. An example is the concentration of seal predation on salmon in the river estuaries during the spawning migration. Even if salmon constitutes a small part of the yearly consumption the effect on a threatened wild salmon stock can be disastrous. In the future an advanced approach to the analysis of diet data thus has to take an ecosystem approach.

### *2.3 Small-Scale Coastal Fisheries: a Threatened Community*

Our study concentrates on small-scale coastal fisheries. The Swedish and Finnish fisheries differ regarding the target species and, in spite of being "archipelago fisheries", they differ in their relative location between inner archipelago and outer archipelago.

## 3 The Coastal Fishery in the Finnish Model Region "Kvarken"

Back in the early 1960s coastal fishing in the Finnish model region called Kvarken was concentrated in inner archipelago areas, but environmental degradation of rivers and coastal waters lead to the decline of near shore fish stocks during that decade. Consequently, fishermen had to move to new fishing grounds, which were in the outer archipelago. They also had to switch the target species to herring, Common whitefish, and salmon that were less affected by environmental degradation. Technological development of nets and boats supported this change and by the 1980s, fishing in the outer archipelago and specializing on few species had become the dominant strategy (Tuomi-Nikula 1981; Österbottens fiskarförbund 1990). An unfortunate complication to this development was that the seals became abundant in 1990s on these new fishing grounds and the adopted fishing technology was vulnerable to seals.

In the 1960s the number of fishermen declined rapidly in the Kvarken region, but since that the decline has been more moderate as the geographical and technological adaptation became stabilized (Österbottens fiskarförbund 1990). In the 1990s the decline became stronger again as the economic situation worsened; in spite of increased catches the revenues decreased (Broman 1998). Today there are approximately 250 fishermen that get a substantial proportion of their income from fishing.

The value of the catch of coastal fishery in the region in 2002 was 1.3 million € (1650 tons) for the 18 different species caught. The Common whitefish was the most important species in economic terms (790,000 €).

The coastal fishery is nearly insignificant regarding GDP or employment of coastal areas, except on a small scale—in the archipelago villages: In 1999 fishermen formed 6.2 % of the "population in active age", that is the population at an age of between 16 and 64 of the archipelago villages of Kvarken (Registry of commercial fishermen; Leppänen and Rissanen 2000).

## 4 The Coastal Fishery in the Swedish Model Region

In the Swedish model region we similarly deal with a small-scale fishery but target species and gear differ between the Finnish and Swedish region (compare Table 1). Also the location of fishing differs being in the inner instead of the outer archipelago. The main Swedish target species in value are eel (*Anguilla anguilla*), European perch (*Perca fluviatilis*), and pikeperch (*Stizostedion lucioperca*). Locally whitefish is also important. A large part of the fishery is part-time with farming or seasonal occupations as a supplement. In this part-time fishing a long history of coastal fishery is still visible—it is not an exclusive source of livelihood for the coastal fishermen, but the coastal population in both former centuries and today had often combined several sources of income and production or resource use activities, although these may differ historically. The tradition in the Swedish Baltic coast was the combination of fishing and agriculture; when the agricultural population declined, fishing became combined with other income sources available at the coast; with fishers more and more depending on tourism related income. The characteristic fact nowadays still is the combination of several income generating activities in the households of coastal fishermen. The part-time fishery is conducted on private fishing rights and there is no official statistics of the amount and value of the catches. A rough estimate is that it is an order of magnitude larger than the catches by licensed fishermen or 2,000 tons if the leisure fishery is included (Anon 2003).

## 5 Common Development Trends in Baltic Coastal Communities

Judged on the comparison of the Finnish and Swedish model regions the specific differences in target species and location of fishery may account for local differences with regard to the pressure of the conflict for the single fishermen, but the following common trends and developments characterize the specific situation in coastal communities and of coastal fishermen, which are of importance for mitigating the seals conflict:

**Table 1** Summary information on the Swedish and Finnish model regions

| | The Vaasa Subregion, Finland | Baltic SE coast, Sweden |
|---|---|---|
| Counties | Part of county Ostrobothnia | Södermanland and Östergötland |
| Human population | 88,385[h] (2002) | 670,000[f] (2004) |
| Coastline | Scattered islands | Extensive archipelago |
| Region area, sea part | ~100,000 km$^2$ | ~60,000 km$^2$ |
| Length of coast | 223 km[j] | 120 km[f] |
| Period of ice coverage | November–April | January–March |
| Number of commercial fishermen | 280[p] (2002) | 46[a] (2004) |
| Target species | Whitefish, salmon, herring, and perch[k] | Herring, eel, whitefish, pikeperch, perch, cod, flounder[g, n] |
| Fishing gear | Gill-nets, fish traps, fyke nets[k] | Eel trap, gill net, trawling, trammel net[c, g] |
| Major fishing season | April–August[j] | February–December[c, g] |
| Coastal fishery annual catch | 1,650 tons (2002)[k] | 180 tons[n] (2004) |
| Value of catch in coastal fishery | 1.3 M € (2002)[k] | Not available |
| "Private water fishermen" and catch | Not available | >10,000 fishermen, > 1,000 tons[e] |
| Sport angling and catch | 20,000 fishermen, ~660 tons[b] (1992) | >100,000 fishermen, > 1,000 tons[e] (2000) |
| Seal species and population size | Grey (<500[l, o]) and ringed (<20[p]) seal | ~1,000 grey seals |
| Nature of seal conflict | Grey seal: Extensive[e]; ringed seal: growing | Grey seal: growing[d] |
| Seal Reserves | 1 | 2 |
| Seal hunting (2003) | Quota 110 grey seals and 95 were caught[q] | 25 grey seals allowed, 6 caught[m] |
| Compensation for seal damages | 416,000 € (2002)[r] | 154,000 € (2003)[i] |

References: [a] Bengtsson and Brantäng 2002; [b] Internet information; [c] Saulamo and Neuman 2002; [d] Westerberg et al. 1999; [e] Anon. 2000; [f] Internet information; [g] S&F logbook data, National Board of Fisheries Sweden; [h] Statistics Finland (http://www.stat.fi/index_en.html); [i] Swedish Environmental Protection Agency, Dnr 429-1072-03; [j] SYKE (measured in scale of 1:400,000); [k] RKTL Fisheries statistics (raw data); [l] Kvarken Council 2003; [m] Swedish Environmental Protection Agency, Dnr 412-2340-03; [n] EU Log book (National Board of Fisheries Sweden); [o] http://www.rktl.fi/riista/seuranta/harmaahylkeet_kesalla.html; [p] WWF Finland (pers. comm.); [q] www.riista.fi; [r] Ostrobothnia Centre for Employment and Economic Development, Fishery Unit; Sport angling is defined as fishing with hand-held gear, whereas "fishermen on private water" (in Swedish. *husbehovsfiskare*) fish with static gear, nets, and pots (Anon 2000)

- The dominant trend in the model regions, as in most Swedish and Finnish coastal regions, is: coastal areas are attractive places for living although they can often not provide for sufficient work places, so that large parts of the working population are commuting. This is reflected in the growing numbers of summer

houses and tourists in coastal areas, as well as in the growth of permanent habitants in coastal communities.
- The coastal fishermen and their families as the population group directly involved in the seal conflict at regional and local levels, represent a small part of the coastal population, with peculiar forms of work and employment.
- It is not their number or economic influence that makes the fishermen powerful but their specific social and economic roles for the maintenance of "living coastal communities", which is a prominent discourse in coastal areas in Finland (Peuhkuri 2004) and Sweden (Carlberg et al. 2005).

The coastal fishermen are also specific with regard to their demographic structure as a group, which is better visible in the Swedish model region: fishing is done exclusively by men and most of the fishermen are elderly; the number of coastal fishermen is rapidly decreasing all along the Swedish coast and there is a recruitment crisis. In the Finnish model region the situation is not that drastic, and there are more fishermen active, but the long-term trend is similar in both regions:

- In future coastal fishery will be different, probably with less commercial and more leisure and sport fishery; the role of fish farming (which is already more significant in Finnish than in Swedish fishery) may get more importance. All this may result in significant changes of the seal conflict. Presently being so clear a conflict between coastal fishermen and seals, it may still remain relevant for coastal communities, even though importance of the present forms of coastal fishing would diminish.

## *5.1 Policies and Institutional Framework Regarding Coastal Fisheries*

The Swedish model region is located in the part of the Swedish coast where private fishing rights exist. This right covers 300 m from the shoreline and bays inside straits narrower than 600 m. In the complex archipelago of the model region this covers a large fraction of the fishing ground, where detailed knowledge and official statistics of the fishery is limited. The legislation regarding fishing methods and quotas applies in the same way to private fishing rights as to fisheries elsewhere. The cod fishery is strongly restricted by quota as is the salmon fishery where in addition the fishing season is regulated. The other target species have no quota.

The Swedish fishery has received high attention on the political and research agendas during recent years. Coastal and inland small-scale fishery is meanwhile of high priority of the Swedish government, because it is considered more sustainable and of great importance for the employment in rural areas. A national policy to promote this fishery was adopted by the Parliament in 2004 (Anon. 2003). Whether and how far the policy succeeds in stabilizing coastal fishery is still not clear. However, there has been intensive research in recent years about Swedish fishery and coastal resource management, which included also the conflict between seals and fishery (SUCOZOMA research program about sustainable coastal zone management, see Carlberg et al. (2005); project “Seals and fishery” that has developed a large part of the technical solutions for the seals conflict (Westerberg et al. 2000; Lunneryd et al. 2003)

In Finland coastal waters up to 500 m from the 2 m depth curve are privately owned. The ownership is based on the system that land property by or near shorelines includes a right to certain water area. In archipelago areas like in the model region, the 500 m rule means that most of the coastal waters are in private ownership. In the model region there are 198,000 hectares of privately owned waters, which is 33 % of the whole water area. Private water areas are managed by statutory fishery associations (SFAs) in a sort of a co-management arrangement. Associations give or sell fishing permits on their areas, which determines where fishermen can fish.

Since 1982 the state introduced a new management system organized as fishery regions that aim at management of larger water bodies than the statutory fishery associations. Water areas of statutory fishery associations, water areas managed by individual owners, and state’s waters near the coast were combined into fishery regions that coordinate fisheries management, but can also order fishing regulations. There are three fishery regions in the model region. Water areas outside fishery regions are state owned, and fisheries authorities grant permits in these areas.

## 5.2 *Types and Severity of Economic Seal Damage*

Earlier in this chapter, we presented a damage assessment based on an ecological analysis of the seal diet. In the following, we aim at a damage assessment in economic terms. For this purpose, different estimates can be used. One of them relates to the relevant policy instruments such as damage compensation schemes and financial support of seal-safe gear and their actual public expenditures. In both countries compensation is given only to professional fishermen and only for direct damage by seals to the catch or the gear. The damage is calculated and reported differently in both countries (Fig. 5).

In the Finnish case a generally accepted monetary value of seal damage to fishery cannot be given under present knowledge, but there are at least three different estimates that could be used to describe the damage in the model region.

First, the actors in the model region estimated that losses in the whole Kvarken were in 2001 0.6—0.8 million € (Kvarken Council 2003). This estimate includes both damage to fishing gear and the catch taken from the gear.

Second, the Finnish Game and Fisheries Research Institute has studied damage that seals caused to catch in 2001. Results from the Gulf of Bothnia shows that damages in different locations vary from 2 to 42 % of the catch. These were the minimum estimates, because not all visits that seals make to fishing gears leave traces (Kreivi et al. 2002). However, none of the study sites of that research was in Kvarken.

**Fig. 5** Damaged cod on the long line. *Photo:* Sven-Gunnar Lunneryd

Third, the Finnish authorities compensated loss of catch that seals caused in 2000 and 2001. During the process regional fisheries authorities together with fisheries sector representatives estimated the proportion of fish that seals take from fishing gear (Box 1) shows the amounts of compensations paid in Kvarken). They estimated how much salmon, whitefish, and trout from different gear types in different locations[1] are taken by seals. The estimate was not based on scientific studies or monitoring of loss of catch and therefore it must be seen as an expert opinion. In addition, the process of estimating damage may have had a bias towards overestimation because the beneficiaries themselves estimated the damage. It is important also to notice that the estimated damage proportion, when it is

[1] So-called "statistical squares" that are used as basis for catch reporting in Finland. The squares are 50 × 50 km large squares following the ICES catch re-porting system. The model region extends to five of these squares (number 22, 23, 24, 27, and 28).

very high, is partly or totally hypothetical, since the fishermen do not fish anymore with vulnerable gear in the locations were seal damage is known to be high. Thus in these cases the estimate tells what the experts assume that seals might have damaged if fishing was conducted in that location.

**Box 1** Damage Compensation in the Finnish Case Study Area

The actual compensations were paid to professional fishermen, i.e. who got 30 % or more of their income from fishing. Only about one third of all commercial in Kvarken are "professional". Additional conditions were that the damage was at least 20 % of the fisherman's catch and fishermen had to cover 250 € of the damage per year as their own risk. The loss of catch per year was 426,000 € when the amount paid to fishermen and fishermen's reducible own risk is taken into account (Table 2).

**Table 2** Compensations to damages on catch in 2000 and 2001

| Municipality | Applied for compensations | Got compensations | Compensations, (€) | Range for individual fishermen (€) | Damages (compensation + own risk) |
|---|---|---|---|---|---|
| Mustasaari | 20 | 19 | 404,960 | 2,007–48,940 | 414,460 |
| Korsnäs | 3 | 3 | 51,540 | 5,664–36,103 | 53,040 |
| Maalahti | 9 | 9 | 133,049 | 3,723–31,380 | 137,549 |
| Maksamaa | 6 | 2 | 74,323 | 4,972–45,523 | 75,323 |
| Vaasa | 9 | 9 | 167,658 | 5,111–54,102 | 172,158 |
| Total | 47 | 42 | 831,532 | – | 852,532 |
| ~per annum | – | – | 415,766 | – | 426,266 |

*Source*: Ostrobothnia centre for employment and economic development

If the value of lost catch is calculated from the estimation made for the compensation by using the reported commercial catch in the respective locations in Kvarken and the average price paid for the three species in question in 2002, one gets as a value of lost catch 873,000 €. The value of landed commercial catch in respective locations and species was 856,000 €.

Loss of catch is only part of the direct damage that seals cause, because seals can break fishing gears, too. In the model region the professional fishermen can get damage to gear compensated by a state subsidized insurance company. In 2001 altogether 108 coastal fishermen had the insurance and the company compensated them damage worth of 118,000 € (Österbottens fiskeriförsäkringsförening 2002).

When the cost of technical mitigation efforts is conceived as another cost of damage compensation then a subsidy to technical measures is one of the main costs in Finland. The subsidy became available in 2004 when professional fishermen were allowed to apply for it from the regional fisheries authorities. They will be subsidized only once and the subsidy is 70 % of the value of the first two sets of gear and 50 % for the rest (MoAF 2005). In the model region 20 fishermen applied the subsidy for

**Fig. 6** Seal safe pontoon trap for salmons with doubled net walls. *Photo:* Pekka Salmi

50 trap-nets. In the region the total costs were estimated to be 458,000 € of which the state covered 304,000 €.

The fishermen in the model region in Sweden estimate the economic losses that seals cause in the compensation applications that they send to nature conservation authorities. Their estimate summed up to 182,000 € in 2004. However, the authorities compensated half of that amount to the fishermen.

In addition to the compensations, the seal conflict causes other direct costs to the Swedish society in form of economic support to technical measures to avoid seal damage (so-called "seal safe trap-net", Fig. 6), but also in form of information, education, and monitoring (see Table 3).

We made an independent analysis of the damages based on a compilation of fishing effort from the EU logbook to the Swedish Board of Fisheries combined with general data on the loss of fish and gear damage detailed by voluntary journals kept by fishermen in the area (Lunneryd et al. 2005). To the resulting figures we added estimates of the hidden, i.e. invisible seal damage (Königson et al. 2005; Sundqvist 2005). In this way the direct damage of catch in the county of Sörmland and Östergötland was estimated to be 79,000 € in 2004. To this figure the damage to gear and indirect cost to minimize seal damage (e.g., by more frequent visits of the gear) should be added. This is estimated to be approximately 50 % of the direct damage (Westerberg et al. 2000). Altogether this calculation gives an estimate of 120,000 € in the counties together, which can be compared to the 90,000 € actually compensated (Table 4).

**Table 3** Estimated seal damage costs (compensation payments and costs for society) in the Swedish model region (Östergötland and Södermanland counties)

| Cost type | Amount |
|---|---|
| Compensation payments (2004) (approx. 50 % of applied amount) | 90,925 € |
| "Actual" damage (2004) (estimated by fishermen from applications) | 181,850 € |
| Support for seal safe gear (2003) | 45,860 € |
| Education, information and monitoring (2003) | 10,893 € |

*Source*: Own compilation based on information from regional administrations

**Table 4** Compensation for seal damage in the Swedish model region during 2000–2004. The compensation is intended for seal damage incurred during the previous year and includes support for acquiring seal-safe gear. *Source* Own compilation based on data from regional administration, Södermanland 2/12-2004; regional administration, Östergötland 2005; and Naturvårdsverket (see Bruckmeier and Höj Larsen 2004)

| Year | Södermanland | | | Östergötland | | | Total, whole Sweden (€) |
|---|---|---|---|---|---|---|---|
| | Cost (€) | Applicants | Granted applications | Cost (€) | Applicants | Granted applications | |
| 2000 | 57,600 | 14 | 12 | 9,000 | – | 8 | 1,720,000 |
| 2001 | 63,000 | 13 | 12 | 29,000 | – | 8 | 2,320,000 |
| 2002 | 55,400 | 11 | 9 | 37,000 | – | 10 | 2,970,000 |
| 2003 | 50,500 | 9 | 9 | 48,900 | – | 5 | 2,920,000 |
| 2004 | 50,500 | 9 | 9 | 40,200 | – | 4 | 2,780,000 |

## 6 Context and Approaches to Conflict Mitigation in Sweden and Finland

The main differences for the mitigation of the seal conflict in Sweden and Finland derive more from the institutional contexts of regulation and the socio-economic differences in coastal fishery than from the ecological components of the conflict, which are very similar. Since the social factors influencing the conflict account mainly for the peculiarities found in the mitigation approaches, they need to be described in more detail.

For both countries there are restraints on possible mitigation measures due to the agreements made in HELCOM. Primarily this makes population regulation by hunting in principle out of the question as long as HELCOM countries in the southern regions of the Baltic oppose such action. This is likely to remain the case for the foreseeable future. A decision by HELCOM Nature protection and biodiversity group was made in 2006, which opens for a continuation of the limited hunting already practiced in Finland and Sweden, but the population increase must not be threatened and the target population size shall be an, as yet undecided, percentage of the population level at the carrying capacity of the Baltic.

A model for population viability under different management regimes was developed within the FRAP project (Alberti, Wahlberg, Westerberg and Frank.

**Table 5** Policy instruments and mitigation measures in Sweden

| Type of policy instrument | Mitigation measure | Implementation | Purpose |
|---|---|---|---|
| Command and control | Protective hunting | National administration | Limit damage by seals |
| Economic | Compensation payments for seal damage | Regional administration | Compensate losses caused by seals |
| | Financial support for seal-safe gear | Regional administration | Limit damage by seals |
| Civil society | Stakeholder forum | Voluntary cooperation | Mitigate conflict between interest groups |

*Source*: own compilation

unpublished). This kind of model is an essential tool to ensure that a hunting regime allows for a continued increase of the population. An important lesson from the model is that the by-catch rate seems to be the limiting factor for the growth rate of grey seals in the Baltic.

In the following, we will present the policy mixes for conflict mitigation in each of the countries, and describe the various stakeholder groups and their positions in the conflict. Finally, we propose an evaluation of the different measures. For this purpose, we look at the measures and situate them in the social, administrative, and ecological context of the conflicts studied. Social context means the consideration of benefits mostly to the society by mitigating the losses of fishermen. Costs are considered mainly as a burden to administration and in comparison to the volume of the fishery sector. Costs are also compared to other mitigation measures in each country, but not between them.

## *6.1 Policy Instruments and Seal Management*

In Sweden the main approach to resolve the seal conflict is the grey seal management plan that has been implemented since 2001. It is based on the compromises achieved to mitigate the conflict for a specified time period and includes all the main instruments and measures analyzed here. The following policy instruments (see Table 5), with the exception of the stakeholder forum, are found in the management plan (for further analysis see Bruckmeier and Höj Larsen 2004; 2005).

The Swedish Grey Seal Management Plan (see Box 2), the core document for present conflict mitigation, has been formulated under the guidance of national level institutions and actors, although, according to its nature and geographical applicability, it is primarily a regional management system.

**Box 2** The Swedish Grey Seal Management Plan (2001–2005)

Overall principles:

The grey seal population should be protected from threats such as extensive hunting, human induced contamination and by-catches in fishing gear.
The grey seal population should be able to recover further from threats listed above. Hence the management plan aims at strengthening already existing populations and supports the reestablishment of populations in areas less populated by seal.
According to the management plan the estimated population consists of at least 10,000 seals growing at a rate of 6.5 % a year (Naturvårdsverket 2001, p. 69). Hence, the grey seal population is now stable enough to endure protective hunting to minimize the conflict between coastal fishery and seals.
The protective hunting is limited to 180 grey seals a year. It is pointed out that the hunted population is reduced most efficiently if the animals targeted are adult females.
Development of seal-proof techniques is still preferred as the long-term conflict reducing instrument in the management plan. The game-damage legislation ("Viltskadeförordning", SFS 2001:724, §11) states that protective hunting should be considered first and foremost as damage-reducing action. Given the protected status of the grey seal and the limits put by the HELCOM hunting exemption this is a necessary priority.

The management plan points to the future possibility of extending the seal hunting so that the seal can be a resource that can be used in a sustainable way. There is no indication what population level would grant such an extended hunting. The management plan comprises the following measures related to the mitigation of the conflict: population monitoring of seals (including inventory, photo-monitoring, counting and calculation of population size, growth, pathology), fishery technology and statistics (including different activities regarding gear development and seal behavior, see p. 64 of the plan), seal protected areas, HELCOM-cooperation, hunting, information.

As hunting of seals is the most controversial issue in the management plan, it was through this measure that the different institutions positioned most clearly within the conflict. Still different views and assessments continue to exist in the institutions and organizations involved. The great seal management plan is only a compromise but it is accepted by all institutions and stakeholders and it integrates the main instruments and measures necessary to solve the problems (Fig. 7).

In Finland the approach has been less coordinated than in Sweden. Several instruments have been introduced since the late 1990s (Table 6), but only in 2004 the Ministry of Agriculture and Forestry initiated a working group to

**Fig. 7** Grey seal in the south Baltic. *Photo:* Sven-Gunnar Lunneryd

**Table 6** Policy instruments and mitigation measures in Finland

| Type of policy instrument | Mitigation measure | Implementation | Purpose |
|---|---|---|---|
| Command and control | Protective hunting | Ministry of Agriculture and Forestry, delegated to regional game managers | Reduce seal-induced damage |
| | Compensations | Ministry of Agriculture and Forestry, delegated to regional fisheries administration | Eases economic losses |
| Economic | Fishery insurance | Regional insurance company | Eases economic losses |
| | Financial support for diffusion of innovative technical measures 2004 | Ministry of Agriculture and Forestry, delegated to regional fisheries administration | Encourage investments in technical measures |
| Civil society | Stakeholder forum | Voluntary collaboration of regional actors | Better relationship between stakeholders |
| | Training of seal hunters | Regional game managers | Hunting becomes more effective and ethical |

prepare a national management plan for the Baltic Sea seals. The plan was published in 2007.

The Department of Fish and Game of the Ministry of Agriculture and Forestry has introduced most of the mitigation measures, namely protective hunting, compensation of lost catch and subsidy to new trap-net models. The Ministry decides annually the hunting quota and its distribution among regions as well as hunting seasons. Protective hunting was started in 1997 with a small quota (30 seals), but the quota has been increased since then: for the hunting year 2007/2008 it was 685. The compensation for loss of catch covered the damage that occurred in 2000 and 2001. Only temporary compensation, in this case two years, is allowed according to EU state aid regulations (Similä et al. 2006). New compensation system was launched in 2008. In 2004 the new types of trap-nets that can prevent seals from causing losses became eligible for subsidies from EU's FIFG funds (Financial Instrument for Fishery Guidance) by the decision of the Ministry. These subsidies were paid in 2005 and the subsidy system has been extended under the European Fisheries Funds' national program for 2007–2013.

A state-subsidized fishing gear insurance compensates the loss of nets by seals. It has existed since the 1930s and today the main cause of gear damage compensated in coastal fishing is seal-induced damage (Österbottens fiskeriförsäkringsförening 2002).

Implementation of all four above-mentioned state-governed mitigation measures is delegated to regional authorities. Seal hunting licensing and monitoring is the responsibility of regional game management authorities to which hunters can apply for licenses. Hunters have to have a general hunting license and for seals they also need a special license for each seal. Hunters pay fees for both of these licenses. The fishery unit of regional centre for employment and economic development administered the assessment and payment of compensation and is now managing the subsidies for technical solutions to fishermen. The insurance system operates also in regional units like normal insurance companies with the exception that state subsidizes the insurance compensations.

In addition to the above four nation-wide policy instruments, there are two policy instruments in Kvarken of the type "civil society" which are based on voluntary action: a stakeholder forum and training of hunters (Table 6). The stakeholder forum was based on a project "Grey Seal in Kvarken" that the Kvarken Council launched in 2001 (Kvarken Council 2003). The project brought together the regional authorities from fishery, hunting, and nature conservation sectors and regional fishermen's and hunters' organizations. The project aimed at reaching a common understanding of the seal's role in the region. The same project also conducted training of hunters in order to make protective hunting more efficient and ethical. The training was in practice given by the regional game managers.

The Ministry of the Environment does not have a big role regarding the mitigation measures, because the grey seal is classified as a game animal in Finland, which means that its protection and management is on the responsibility of the highest game management authority, which is the Ministry of Agriculture and Forestry. The Ministry of the Environment is, however, consulted while annual decision about hunting quota is made.

Finnish Game and Fisheries Research Institute—a government research institute under game and fisheries administration—is also one of the official institutes actively involved with mitigation measures. The institute has the responsibility of

**Table 7** Comparison of instruments in Sweden

| Measure | Costs | Benefits | Recommendations |
|---|---|---|---|
| Protective hunting | Low costs for the society, all costs taken on by the hunter<br>High costs in terms of time and effort to shoot and land the seal | Benefits the society by mitigating conflict (positive effect on fishermen), no economic incentive for individual hunter | Improve incentive to hunt (directly or indirectly)<br>Investigate effectiveness of protective hunting<br>Investigate regulations for protective hunting/inform on the reasons for them |
| Compensation payments | High (and increasing) costs for the society<br>50 % of the reported losses are not compensated; these costs are taken on by the fisherman | Benefits the society by mitigating the conflict in a short-term perspective (positive effects on fishermen), thus even supporting the coastal fishery and the individual fisherman | Shift more towards financing of seal safe gear provided that increasing resources are allocated toward local development and improvement of seal safe gear |
| Financial support for seal safe gear | Moderate costs for the society compared to compensation payments, but high compared to low catches<br>Moderate costs for fishermen if they are efficient. High costs during local adaptation and development | Benefits the society by mitigating the conflict on a long-term perspective<br>Benefits the fishermen by finding new methods and gear<br>Benefits the seal by reducing mortal by-catches | See recommendations for compensation payments above |
| Civil society action | Low costs for society<br>Moderate costs for stakeholders | Could benefit the society as well as the coastal communities on a long-term perspective through networks and local interest groups that can participate in future planning and management | Further strengthen the local communities in finding methods to plan, manage, and evaluate local development and conflict solution |

*Source*: own compilation

monitoring the seal populations and they give the scientific advice for hunting decisions. They are also giving scientific advice to the Ministry in all fisheries related decisions. For instance, their expertise in developing fishing gears that could help to minimize the losses that seals cause was used when subsidies for these gears were planned (Table 7).

## 7 The Stakeholders

The Swedish conflict is characterized by unsymmetrical stakeholder participation since not all conflicting interests are present at local level. The interests are not as differentiated as in the Finnish conflict between interests of fishermen, hunters, fishery organizations, and environmental interests although similar stakeholders exist—there are five main groups of stakeholders in both countries.

In Sweden the following groups could be identified:

*Local fishermen.* The small-scale coastal fishermen fishing in the archipelago are the main stakeholders in the conflict. The conflict is concentrated on this small group of professional licensed fishermen, for which the damages by seals cause major economic problems. The interests and perception of the conflict by these fishermen differ remarkably from the explanations given by the scientists involved. However, their local knowledge and their interpretation of the behavior of seals are up to now not accepted by most of the other stakeholders.

*Regional administration—Fishery and game unit.* Fishery and game is often managed at the same unit at the regional administration. Since the fishery unit is familiar with the problems the fishermen face they have often tried to negotiate a high number of seals in the quota for the county when talking to the Environmental Protection Board. This was also the case for the counties in the model region. The fishery units do not think of the protective game management as a means of reducing the seal population, neither do they necessarily believe that it will have any impact on seal behavior; it is rather seen as a means for the individual fisherman to act when frustrated by seal damage.

*Regional administration—Nature protection unit.* The nature protection units have generally accepted the restricted protective hunting of seals. This is often done with reference to the Environmental Protection Agency and the management plan issued by the agency. The protection units do not perceive the protective seal hunting as a threat to the seal population; neither do they think that it will result in behavioral changes of the seals. The thoughts expressed by some fishermen and parts of the fishery units to issue a more general hunting on seals is frowned upon. This follows the perception of the Environmental Protection Agency.

*Swedish Environmental Protection Agency.* The Environmental Protection Agency is responsible for the management plan, in which the possibility to hunt seals is foreseen. This was done to relieve the conflict tension on the fishery side without reducing the grey seal population. The complaints by fishermen and seal hunters as to restrictions concerning hunting on haul-outs or from boats etc. is not accepted by the Agency as it seems not to be in line with an understanding of protective hunting as an exceptional and limited form of hunting. The hunting should be carried out in a protective manner only, thereby restricted to the vicinity of the gear.

*National Board of Fishery.* The National Board of Fishery was advisory partner to the Environmental Protection Agency when the management plan was developed. The Board regards protective hunting as a strategy for the individual fisherman to protect his gear and catch—still the hunting is not meant to affect the population size at this point.

*Scientists involved in research on seal, seal safe gear* etc. Scientists in Sweden have had a central role in formulation of the national management plan. The perception of seal hunting differs among scientists involved in seal related research that had a decisive role in the formulation and implementation of the grey seal management plan. Some see the protective hunting of seals as a mitigation measure with great psychological impact that relieves the conflict by lending the individual fisherman the opportunity to defend his property. However, hunting should be limited to the marginal number of seals described in the action plan since the population is still not able to endure a more general hunting pressure. Again others think that the seal should be treated as a resource and not as a pest that must be shot protectively.

On the basis of the interviews that explored different actors' perspectives and activities in relation to the model conflict in Finland five main stakeholder groups could be identified. They all are on local or regional levels.

The first group is *fishermen*. These are the local coastal fishermen. For fishermen it is clear that the seals are the biggest problem for the industry. They feel strongly that their views are not listened to in the formulation of seal policies.

The second group is *hunters*. The regional game manager's views are so similar to the views of hunters so that the game manager is, despite being a regional level actor, included in this group. Hunters did agree that the seal problem is a difficult one for the fishermen. In fact many of the fishermen have hunted seals and they talked a lot about hunting, too, in their interviews. In the hunters' interviews hunting had, of course, an important role. Hunters were especially happy about the hunting tradition's revival in Kvarken and they very strongly feel that hunting should aim at utilization of seals as a resource. They would like to see hunting less restricted.

The third group is *fishery organizations*. It includes fishermen's organizations as well as fisheries authorities mainly operating at the regional level. Their views were rather similar to each other and, in relation to fishermen, there were certain differences. They recognized the seal problem as the most difficult problem that the industry is facing. However, compared to fishermen, these interviewees had different views about some of the mitigation measures, especially about technical measures that the fishermen did not find useful at the time of the interviews.

*Environmentalists* is the fourth group. This group consists of the environmentalists and regional environmental authorities. Similarly to the Swedish case, the environmentalists are mainly regional level organization activists. The stakeholder group "environmentalists" agreed that seals are a big problem for the fishermen. It was also mentioned that in some occasions it seems that fishermen, when talking about seal damages in the media, are exaggerating the damages to gain sympathy. Of all the interest groups, "environmentalists" had the most cautious view about seal hunting, although none of them opposed it completely.

The last stakeholder group in Kvarken that could be identified is *tourism entrepreneurs*. This group has not been involved in activities in the region or nationally. From the tourism entrepreneurs' point of view many of the issues that other stakeholders relate to the conflict do not seem to be relevant. Their interest in this issue comes from their close cooperation with fishermen in organizing "seal-safaris" and potential harms of seal hunting to their business.

## 8 Evaluating and Improving the Policy Mixes

In this chapter we concentrate on the actual policy measures found in each country. We look at the measures and situate them in the social, administrative, and ecological context of the conflicts studied. Social context means the consideration of benefits mostly to the society by mitigating the losses of fishermen. Costs are considered mainly as a burden to administration and in comparison to the volume of the fishery sector. Costs are also compared to other mitigation measures in each country, but not between them. We also outline how the approaches in mitigating the seal conflict in both countries could be improved in short and long terms.

In the assessment of the Swedish policy mix, the following limitations regarding effectiveness, cost-effectiveness, dynamic efficiency, and participation of stakeholders can be pointed out:

- it is a rather conventional policy mix in which two traditional classes of instruments are used, "command and control" instruments and economic instruments, and both of these need to be financed from public budgets and implemented by governmental agencies. The instrument mix has unfolded to dynamic efficiency mainly with one measure (seal safe gear), but not developed far in the direction of participation of regional and local stakeholders;
- although some instruments can be improved with regard to their cost-effectiveness, especially protective hunting and seal safe gear, the improvement of cost-effectiveness of the whole instrument set is rather difficult and limited;
- the policy mix does (up to now) not stress the instruments that stimulate more active involvement of stakeholders (civil society action and participatory decision-making), although consultation of stakeholders has been done in formulation of the grey seal management plan.

Beyond these limitations given within the Swedish management plan, also more specific factors in the model region should be taken into account to allow for a context-bound assessment of the policy mix:

- the low number of professional fishermen that can benefit from the measures,
- the weak economic status of these fishermen (their fishery is hardly sufficient as the only income source; to survive the fishermen's families are dependent on further income sources), and
- the weak presence and activity of nature protection stakeholders at the regional and local levels.

A lot of mitigating activities in Finland have been tried and are practiced today (see Table 8). Even though there are many relevant instruments available, the problem is still severe. Since the damage remains severe it is evident that none of the practiced policy instruments alone is very effective in reducing the damages that seals cause and also that together they do not form an effective combination of policy instruments at the moment.

The mitigation approach suffered for a long time from temporality of the measures and weak coordination. Only the protective hunting and the insurance system are measures that have been practiced for a long time. Coordination should now improve as the national management plan has been launched.

However, the situation is a dynamic one. Many of the instruments are such that their effects may materialize in future. For instance, seal hunting has been practiced in a relatively high volume only very recently. Hunting is argued to bring effects when seals learn to be afraid of humans and move away from fishing areas, but so far such effects have not been reported. Similarly, technical measures that may also change the behavior of seals have not been used very commonly so far.

A positive thing about these two measures is that they both may bring long-term preventing effects, on which the emphasis should be placed if one wishes to bring sustainable solutions to the problem. It is important to notice also that all of the instruments have positive effects and are accepted by the stakeholders.

When we talk about conflict mitigation we should also pay attention to relationships between stakeholders. One of the mitigation measures in the model region is deliberately addressing this issue. The Kvarken Council's grey seal project has done valuable work in this respect and succeeded to bring the views of stakeholders closer to each other. The project is continuing, which should help to find common ways of working together also in the future.

## 9 Recommendations for Sweden

In the short run a combination of financial and technical measures is satisfying for conflict mitigation in Sweden. An improvement through a better combination of instruments can be expected if new measures are introduced.

In the long run technical measures to improve fishing methods and gear and to influence the behavior of seals are to have lasting effects for mitigating the conflict (and these measures can also be developed to become more cost-effective). Improvement of the instruments through a more active involvement of regional and local stakeholders is limited—there is already intensive involvement of the few stakeholders in different phases of the management plan and conflict mitigation. Additionally, gains won through technical measures are lost when the seal population increases rapidly and the conflict expands to new areas and fisheries. In the long run population regulation has to be considered in the Swedish seal management.

**Table 8** Comparison of instruments in Finland

| Mitigation measure | Cost | Benefit | Recommendation |
|---|---|---|---|
| Protective hunting | Low costs to administration (financed by the license fees). The monitoring of seal population is part of the overall biological monitoring of seals<br>Seal hunting as such is expensive, but it is conducted voluntarily as a hobby | May reduce the damage that seals cause to fishing in the long run<br>Controlled hunting does not jeopardize the favorable conservation status of the grey seal<br>Seals are seen now as a resource, not only as a pest<br>Welcomed in some coastal areas as revitalizing the hunting tradition. Some economic benefits to hunters | Should be continued, but its effectiveness as a mitigation measure should be studied<br>Some restrictions like the license's validity only in one game management district could be reassessed |
| Compensations 2000–2001 | High immediate costs to society in relation to revenues from fishery | Only temporary benefits | In the long run the emphasis should be placed on preventive measures |
| Fishery insurance | Medium high costs to society in form of subsidy to the fishing gear insurance companies | Important: allows fishers to buy and repair broken nets<br>Risk-sharing between fishermen and society | The same as with the compensation<br>Should include an incentive to reduce damages by matching funds of fishermen |
| Funding spread of technical measures | Moderate costs to society since it is only a one-time subsidy for gears that will be used a few years | Benefits the society by mitigating the conflict on a long-term perspective<br>Benefits the fishermen by finding new methods and gear<br>Reduced by-catches of seals | Subsidies should be available also for other technical measures than trap-nets if such will be developed |
| Stakeholder forum | Low costs to society, since it is a voluntary action<br>Moderate costs to participants to the collaboration | High benefits at the regional level, better collaboration among actors. Low benefits to the whole society | To expand the participation |

(continued)

**Table 8** (continued)

| Mitigation measure | Cost | Benefit | Recommendation |
|---|---|---|---|
| Training of seal hunters | Low costs to hunters and the game manager | Benefits if the training has changed the hunting practices and effectiveness. Unclear results | Training should be continued in the future, especially if hunting quota will increase |

The combination of nature protection measures with the support of coastal fishery through different economic and technical measures will be a dominant pattern of conflict mitigation in the foreseeable future. It can be improved stepwise, through experience, evaluation, and development of new technical measures.

Stakeholders view the instruments with regard to their mitigating effects, and this creates a base for compromise and consensus between them: *nearly all stakeholders accept the mitigation measures as necessary, even when they do not meet their direct interests and when knowledge about their effectiveness and cost-effectiveness is limited.* This means, that the stakeholders involved in conflict management have accepted that mainly coastal fishermen need to be supported or compensated in this conflict. Through the framework policy of HELCOM for protection of seals and ban of hunting the nature protection interests seem to be satisfied.

*Research and development as specific features of the Swedish solution through a national management plan*: The Swedish seal conflict shows its innovative quality with regard to the combination of policy instruments for the purpose of conflict mitigation in the strong role of research and development, important for conflict mitigation and for supporting coastal fishery. This priority (including the way of seeking technical solutions for fishery gear) can be understood as learning from prior conflicts or seeking promising solutions. Technical measures such as development of seal-proof gear are effective as indirect mechanisms to conflict mitigation. Also this may be understood as an innovation through learning from prior experience: direct conflict management via negotiations and compromises may not always be easy or best. Creating technical solutions for fishery is not only a means to satisfy the interests of fishermen but also a stimulus of development and innovation in fishery.

## 10 Recommendations for Finland

The main recommendation regarding immediate actions in Finland is a better coordination between policy instruments. There are a lot of instruments available, but they could be enhanced by more effective use especially in a way that brings synergies between them. The policy instruments can be matched better to support each other; especially a connection between economic and technical measures can be reached. For instance, the bonus mechanism in the insurance system can encourage these investments in the future. If fishermen can decrease damages the insurance payment is lower. The national management plan is likely to bring improvements regarding the coordination of measures. The policy should consider strengthening preventive measures to bring sustainable long-term solutions to the problem, for instance by fishermen adopting technical preventive measures and by changing fishing methods. However, support for expensive testing and new technical solutions are needed. Technical solutions can be supported with economic measures and training of fishermen. Compensation for losses caused by seals seems to be necessary in the short-term to help coastal fishery to survive until long-term preventing measures or adaptive strategies are available. A problem with compensations is, of course, that they do not support prevention of damages, unless the compensation scheme includes such a mechanism. If a new compensation scheme will be established it should be closely related to technical measures.

The protective hunting of seals is an important measure that is thought to decrease damage in the long run. It should be continued, since it does not seem to threaten seal population viability in the present hunting volume. The hunting management could be improved by closer studies on the effects of hunting. Another issue important for good management of seal populations is to monitor all mortality of seals: hunting, natural mortality, and by-catch.

In Finland the conflict has evolved in a similar way as in Sweden: the first years' active debate about the seal damage and possible measures towards calmer atmosphere, *in which all measures are more or less accepted.* There are a lot of different actors on local, regional, and national levels involved with the conflict and, what is important, the actors represent different sectors (fisheries, environmental, and hunting). This clearly is a positive aspect of the Finnish case: involvement of many kinds of actors has brought resources in mitigating activities and kept the scope of activities large and, even so the seal problem has been conceived as a "fishermen's problem", the nature conservation interest has limited the selected actions. Another important aspect in the Finnish model conflict, especially for future development, is that most of the practiced mitigation measures are accepted by the stakeholders. Today the conflict is not clearly polarized between certain stakeholder groups, but rather a problem of handling negative side-effects of a successful seal conservation policy.

For future conflict mitigation, if the policy does not go through important changes, the present reasonably good relationships between the stakeholders will continue. More immediate risks for the relationship between stakeholders come from outside the seal-fishery interaction, especially the economic drivers of coastal fishing. The economic situation of the coastal fishing is an important aspect for the seal-fishery conflict. The seal population's growth will continue in the near future and, therefore, it seems unlikely that the basic conflict between seals and fishing can be solved. Consequently, fishery has to adapt to a new environment in which seals are abundant, but here the economic terms become crucial by setting limits on what fishermen can afford to do. Economic support and incentives to testing and developing adaptation strategies seems necessary.

## 11 What can be Learned from Comparing the Finnish and Swedish Cases?

Major lessons refer first, to coastal fishery in a broader sense and second, to the mitigation of the conflict between coastal fishery and seal protection. With regard to coastal fishery as such, one main lesson to be learned is the importance of the conflict for sustainable fisheries management and living coastal communities:

Small-scale coastal fishery is a weak and economically declining sub-sector of commercial fishery in both countries and the prospects for such locally adapted and target-specific fishery are not good. In Sweden efforts through national policy have been made to safeguard the future for small-scale coastal fishery. Coastal fishery is not of importance at the levels of national economy or because of its contribution to GDP (it is nearly insignificant in terms of regional or national income statistics), but it is a supporting factor for maintaining living coastal communities also in future, when the coast is more and more changing from a space of production into a space of consumption of natural resources through non-productive use by urban populations for leisure and tourism purposes. Small-scale coastal fishery has an important function not only as resource user, but also—through its traditions and experiences in managing local fish species—for maintaining and managing in sustainable ways the coastal fish species. This role of coastal fishery has to be relearned in science and policy and the society. It is not that exceptional a lesson to learn, but resembles in many respects similar experiences that have stimulated the debates about agriculture in modern society under the recent debates and policies for multifunctional agriculture (Young et al. 2005): After many years of seeing agriculture and fishery only as producers of food for growing populations, it is rediscovered that both resource use activities by tradition and in history had much more functions to fulfill—beyond production the tasks of managing and maintaining the local resource base so that it can be used over long time, taking into account as well the necessities of ecosystem management as that of social sustainability.

With regard to the mitigation of the seal conflict, lessons of experience can be distinguished referring to single mitigation measures or to the combination of measures and the overall approach. The importance of seal hunting as a single means to mitigate the conflict is revealed through the conflict as a measure of "strategically formulated inexactness": Hunting of seals as it is presently done in Sweden and Finland can be interpreted differently—as an exceptional means that is severely limited to stay within the HELCOM protection of seals (protective hunting) or as a gradual return to unlimited hunting as it has been practiced formerly, and these both aspects can be learned from the two case studies:

The form of *restricted protective hunting in Sweden* cannot be seen as a means of reducing the seal population, neither does it necessarily have an impact on seal behavior; it can rather be understood as a means for the individual fisherman to act when frustrated by seal damage. Still, seal hunting could be made more effective, for example through training courses.

From the *Finnish practice of seal hunting* (where the seal is classified as a game animal) one can learn more how hunting needs to be understood and practiced when it is seen as a return to formerly practiced hunting: Then the seal should be seen as a resource for human consumption, not as a competitor of coastal fishery or as a pest. General seal hunting can become possible only when the seal is no longer a protected and endangered species and when seal products are again introduced into human consumption as it had been before modern industrial society, when skin, oil, and meat of seals were still used and markets for seal products existed. Finding a place for seal hunting fitting a modern industrial society is a long process that also requires rethinking the hunting itself. This process has been started in Kvarken and the future will prove or disprove its vitality.

The *importance of technical solutions* as alternatives to killing seals is shown best with the development of seal-safe gear (independent of what economic instruments are used to support the use of such gear). Seal-safe gear, although including expensive gear such as the push-up trap, and still not offering solutions sufficient for all types of coastal fishing, is often not only a means to solve the conflict, but a technological innovations that strengthens coastal fishery economically. In the ideal case an indirect and technical measure of conflict mitigation, such as seal safe gear, combines the advantages of social sustainability (to maintain coastal fishery as a component of living coastal communities), economic sustainability (improving the yield and economic situation of the fishermen), and environmental sustainability (by effective target-specific fishing that allows to maintain a viable seal population as well as the maintenance of the target species). Still, there is not yet a consequent policy of active experimenting, learning, and innovating of resource management as it would be required not only for the limited purpose of conflict mitigation but for the overarching goal to maintain the natural resource base through adaptive management as it is demanded by ecologists (Westley 2002).

Two lessons regarding the combination of mitigation measures or the overall approaches to conflict management are important according to the Swedish and Finnish case studies:

*Choice of top-down or bottom-up approaches*: The Swedish and Finnish cases differ and do not clearly allow for complementary lessons. Although there was not a clear bottom-up approach with focus on local conflict management initiated by the stakeholders in Finland, the decentralized approach included more bottom-up components and resulted in the only innovation so far in both countries with regard to civil society action: the emergence of a stakeholder forum where the regional stakeholders actively sought for solutions that have not been provided before through governmental policies.

The Swedish centralized and top-down approach to conflict management allowed for the early formulation of an overall approach in a national grey seal management plan with considerable progress in conflict mitigation. However, it kept the solution under control of some powerful actors and did not distribute the costs and benefits for solutions more equally between the stakeholders. In contrast, the absence of coordinated central governmental activity in the first phase of the Finnish conflict mitigation has stimulated both civil society action and further new instruments. But the introduction of now also the Finnish national seal management plan seems to say clearly that there is no way to bypass coordinated solutions in which governmental agencies take a leading role. However, it still needs to be decided whether national management is done through a standardized approach or through a lose coordination of regionally differentiated approaches, of which the management plan gives a clear signal: it divides Finnish sea areas into three "seal management regions". National governments should formulate framework plans with considerable variation and differentiation for regional and local mitigation—as has been found out through ecological research (Walters and Holling 1990; Jentoft and McCay 1996; Ostrom 1999; Westley 2002).

*The course of conflict management:* Is there a characteristic process model to be derived from the history of the relative advanced conflict management in the Finnish and Swedish seal conflicts? An element of joint learning is the building of trust between the stakeholders with different interests In both the Swedish and Finnish cases a rather controversial debate and dissent about the conflict and its solution in the first years of the conflict has been followed by more trust, consensus, and partial or temporary solutions that emerged as soon as not just single measures have been opted for, but more balanced solutions through combinations of different kinds of measures. This has already happened with formulation of national management plans. Through this process many of the initial controversies in the conflict have been weakened and more willingness to develop and learn from the functioning of several policy instruments and mitigation measures exists.

**Acknowledgments** The authors are grateful to fishers, hunters, nature conservationists, civil servants and other individuals in the Kvarken region. Without their willingness to take part in interviews and a focus group writing this chapter would not have been possible. This work was financed by the EU 5th Framework Program (5th FP) Project "FRAP" (Development of a procedural Framework for Action Plans to Reconcile conflicts between large vertebrate conservation and the use of biological resources: fisheries and fish-eating vertebrates as a model case), contract number EVK 2-CT-2002-00142-FRAP.

## References

Anon (2003) Kust- och insjöfiske samt vattenbruk. Regeringens proposition 2003/04:51. http://www.regeringen.se/content/1/c6/01/50/40/fbc5a1ec.pdf. Accessed 28 Nov 2008

Bonner WN (1981) Grey seal *Halichoerus grypus* Frabricius, 1791. In: Ridgway SH, Harrison RJ (eds) Handbook of marine mammals, vol 2., SealsAcademic Press, London, pp 111–144

Broman H (1998) Grön Bro over Kvarken—en utredning om skärgårdsnäringarnas utvecklingsmöjligheter. Kvarkenrådets publikationer, 7 Kvarken Council, Vaasa, Finland

Bruckmeier K, Höj Larsen C (2005) FRAP-Project, work package 9: development of cost-effective policy instruments: Human Ecology, Göteborg University, Sweden

Bruckmeier K and Höj Larsen C (2004) FRAP-Project, work package 9: best practice report Sweden. Human Ecology, Göteborg University, Sweden, p 32

Carlberg A, Bruckmeier K, Elmgren R, Frizell B, Hill C, Neuman E, Sterner H (2005) Kustmiljöns framtid – Erfarenheter från forskningsprogrammet SUCOZOMA. Göteborg: Länsstyrelsen Västra Götalands Län, Rapport, Sweden, p 10

Curry-Lindahl K (1970) Breeding biology of the Baltic grey seal (*Halichoerus grypus*). Zool Gart NF 38:16–29

Dietz R, Teilmann J, Henriksen OD (2003) Movements of seals from Rødsand seal sanctuary monitored by satellite telemetry. National Environmental Research Institute Technical Report No. 429

Halkka A, Helle E, Helander B, Jüssi I, Jüssi M, Karlsson O, Soikkeli M, Stenman O, Verevkin M (2005) Numbers of grey seals counted in censuses in the Baltic Sea, 2000–2004. In: Helle E, Stenman O, Wikman M (eds) Symposium on biology and management of seals in the Baltic area. Kala- ja riistaraportteja, 346. Finnish game and fisheries research institute, Helsinki, pp 16–17

Helander B, Karlsson O (2003) Gråsäl. In: Bottniska Viken 2002. Årsrapport från den marina miljöövervakningen, Umeå Marina Forskningscentrum, Sweden, pp 24–25

Hiby L, Lundberg T, Karlsson O, Watkins J, Jüssi M, Jüssi J, Helander B (2006) NAMMCO Sci. Publ 6:163–175

Hårding K and Härkönen T (1999) Development of the baltic grey seal (Halichoerus grypus) and ringed seal (Phoca hispida) populations during the 20th century. Ambio 28:619–627

Härkönen T, Hårding K (2001) Sälbestånden i Östersjön. Vad kan vi lära oss av historien för att lösa dagens problem? Kungl. Skogs- och Jordbruksakademiens Tidskrift 140:9–36

Jentoft S and McCay B (1996) From the bottom up: participatory issues in fisheries management: issues in institutional design. Soc Natl Resour 9(3):237–250

Kreivi P, Siira A, Ikonen E, Suuronen P, Helle E, Riikonen R, Lehtonen E (2002) Hylkeen aiheuttamat saalistappiot ja pyydysvahingot lohirysäkalastuksessa vuonna 2001. Kalatutkimuksia 185. Riista- ja kalatalouden tutkimuslaitos, Helsinki

Kvarken Council (2003) Kustbefolkningen och gråsälen i Kvarken. Handlingsplan för bevarandet och utnyttjandet av gråsälbeståndet i Kvarken. Kvarken Council, Vaasa

Königson S, Fjälling A, Lunneryd S-G (2005) Impact of grey seals in the herring net fishery along the Swedish Baltic coast. ICES CM 2005/X:12, ICES Annual science conference, Aberdeen

Leppänen M, Rissanen E (2000) Replotsbron socioekonomiska effekter. Uppföljning 2000. Vaasan yliopiston tutkimuslaitos, Vaasa

Lundström K, Hjerne O, Alexandersson A, Karlsson O (2006) Estimation of grey seal (*Halichoerus grypus*) diet composition in the Baltic Sea. NAMMCO Sci Publ 6:177–196

Lunneryd S-G, Fjälling A, Westerberg H (2003) A large-mesh salmon trap: a way of mitigating seal impact on a coastal fishery. ICES J Mar Sc 60:1194–1199

Lunneryd S-G, Hemingsson M, Tärnlund S, Fjälling A (2005) A voluntary logbook scheme as a method of monitoring the by-catch of marine mammals in Swedish coastal fisheries. ICES CM 2005/X:04, ICES Annual science conference, Aberdeen

MoAF (2005) Valikoivien ja hylkeenkestävien rysien tukemiseen varattiin 1.6 miljoonaa euroa. A press release 25/1/2005. Ministry of agriculture and forestry, Helsinki

Naturvårdsverket (2001) The Swedish grey seal management plan (2001–2005). Stockholm

Nyman M, Bergknut M, Fant ML, Raunio H, Jestoi M, Bengs C, Murk A, Koistinen J, Bäckman C, Pekonen O, Tysklind M, Hirvi T, Helle E (2003) Contaminant exposure and effects in Baltic ringed and grey seals as assessed by biomarkers. Mar Environ Res 55:73–99

Olsson M, Karlsson B, Ahnland E (1994) Diseases and environmental contaminants in seals from the Baltic and the Swedish west coast. Sci Total Environ 154:217–227

Österbottens fiskarförbund (1990) Österbottens fiskarförbund 1930–1990. Österbottens fiskarförbund, Vaasa

Österbottens fiskeriförsäkringsförening (2002) Verksamhetsberättelse 2001. Österbottens fiskeriförsäkringsförening, Vaasa

Ostrom E (1999) Coping with tragedies of the commons. Annu Rev Polit Sci 2:493–535

Peuhkuri T (2004) Tiedon roolit ympäristökiistassa. Saaristomeren rehevöityminen ja kalankasva-tus julkisen keskustelun ja päätöksenteon kohteena. Ann Univ Turku, Sarja C, Osa 220. University of Turku, Turku

Saulamo K, Neuman E (2002) Local management of Baltic fish stocks–significance of migrations. Finfo 2002: 9 Fiskeriverket, Sweden

Similä J, Thum R, Varjopuro R, Ring I (2006) Protected species in conflict with fisheries: the interplay between European and national regulation. J Eur Environ Plan Law 3:432–445

Sjöberg M (1999) Behaviour and movements of the Baltic grey seal. Ph. D. thesis, Swedish University of agricultural sciences, Umeå

Sjöberg M, Ball JP (2000) Grey seal, *Halichoerus grypus*, habitat selection around haulout sites in the Baltic Sea: bathymetry or central-place foraging? Can J Zool 78:1661–1667

Sjöberg M, McConnell B, Fedak M (1999) Halout patterns of grey seals *Halichoerus grypus* in the Baltic Sea. Wildl Biol 5:37–45

Sundqvist F (2005) An Assessment of the true damages caused by grey seals, *Halicoerus grypus*, in the Swedish Baltic net fishery after Atlantic cod, *Gadus morhua*. Masters thesis in marine biology. Department of biology, Lund University, Swedish national board of fisheries. https://www.fiskeriverket.se/download/18.2fd63c72114a6399bf680008l9/B16-damage+codnet.pdf. Accessed 11 June 2011

Tuomi-Nikula O (1981) Kalastus Pohjanmaan joissa 1800—ja 1900-luvulla. Kokolan vesipiiri, Kokkola

Wahlberg M, Karlsson O, Lunneryd S-G, Varjopuro R, Virtanen S, Westerberg H (2003) FRAP-project, Workpackage 2: preliminary list of landscape factors and visiting rates: Baltic grey seals. Fiskeriverket, Göteborg

Walters C, Holling CS (1990) Large-scale management experiments and learning by doing. Ecology 7:2060–2068

Westerberg H, Fjälling A, Martinsson A (2000) Sälskador i det svenska fisket. Fiskeriverket Rapport 3. Fisheriverket, Göteborg

Westerberg H, Fjälling A, Wahlberg M (1999) Evaluation of an acoustic seal scarer at salmon trap nets in the Baltic. Conference on Baltic Seals, 18–21 November, Pärnu

Westley F (2002) The devil in the dynamics: adaptive management on the front lines. In: Gunderson LH, Holling CS (eds) Panarchy—Understanding transformations in human and natural systems. Island Press, Washington, pp 333–360

Young J, Watt A, Nowicki P, Alard D, Clitherow J, Henle K, Johnson R, Laczko E, McCracken D, Matouch S, Niemela J, Richards C (2005) Towards sustainable land use: identifying and managing conflicts between human activities and biodiversity conservation in Europe. Biodivers Conserv 14:1641–1661

Eurasian otter (*Lutra lutra*). *Photo:* André Künzelmann

# Reconciliation of the Conflict Between Otters and Fish Farmers

## Lessons Learned from Sado Estuary in Portugal

**Margarida Santos-Reis, Rui Santos, Paula Antunes, Teresa Sales-Luís, Jorge Gomes, Dália Freitas and Luisa Madruga**

**Abstract** In Portugal coastal fish farming is a growing economic activity receiving financial incentives from the EU Community Structural Policy as a response to the decrease of sea fish stocks. At the same time Portugal is known to hold one of the most viable otter populations in Europe, setting the ground for a conflict scenario between fish farmers and conservationists. The Reconciliation Action Plan presented here, and illustrated by the study—case of Sado estuary, is structured in three phases, outlining respectively the ecological and socio-economic contexts of the conflict, the following assessments, and the resulting recommendations applicable to this conflict for successful reconciliation. Using ecological and

M. Santos-Reis (✉) · T. Sales-Luís · D. Freitas
Centro de Biologia Ambiental, Departamento de Biologia Animal, Faculdade de Ciências, Universidade de Lisboa, Campo Grande, Bloco C2–5 Piso, 1749-016 Lisboa, Portugal
e-mail: mmreis@fc.ul.pt

T. Sales-Luís
e-mail: tsl@fc.ul.pt

D. Freitas
e-mail: difreitas@fc.ul.pt

R. Santos · P. Antunes · J. Gomes · L. Madruga
Centro de Investigação em Ambiente e Sustentabilidade, Faculdade de Ciências e Tecnologia, Universidade Nova de Lisboa, Campus da Caparica, 2829-516 Caparica, Portugal
e-mail: rfs@fct.unl.pt

P. Antunes
e-mail: mpa@fct.unl.pt

J. Gomes
e-mail: jvg@fct.unl.pt

L. Madruga
e-mail: lmcm@fct.unl.pt

R. A. Klenke et al. (eds.), *Human-Wildlife Conflicts in Europe*,
Environmental Science and Engineering, DOI: 10.1007/978-3-540-34789-7_4,

socio-economic parameters, the study concerned 14 estuarine fish farms. Results indicate high visiting rates in most fish farms although only in few, species stocked were the most consumed prey. The conflict was quite consensual among fish farmers, although large gaps between effective and perceived predation were identified. No specific instruments exist in Portugal to address the conflict, but some not specifically targeted can have an effect, although with limitations to effectively contribute to its mitigation. Formulating solutions is the following step, using a participatory approach to the development and evaluation of mitigation/compensation strategies, capable of providing an effective reconciliation of the conflict.

## 1 Background and Aims

The otter (*Lutra lutra*) is one of the few Eurasian predators who evolved the ability to actively swim and forage in water. However, this same capacity often brings otters in direct competition with humans, since they eat mainly fish, a resource that simultaneously is the object of human exploitation, namely as food (e.g., aquaculture) or recreation (e.g., angling).

For some the presence of otters is an indicator of environmental preservation and a value of biological diversity, but for others it is a pest that should be controlled or even eradicated. Legally protected by international and national laws, these different view points have led to a long lasting conflict of interests that has resulted in illegal persecution, with unknown effects to otter populations.

In Portugal fish farming is a growing economic activity that has been receiving financial incentives from the EU Community Structural Policy as a response to the decrease of sea fish stocks (Source DGPA). The vast majority of the operative fish farms are located in estuarine areas, which in turn are included in the national network of protected areas due to their high conservation value (http://www.icn.pt). This has resulted in restrictive attitudes to fish farming related activities, leading to a scenario of conflict of interests with nature conservation.

The Sado river estuary (see **Box 1**) accounts for almost half of the existing marine fish farms in the country (INE 2002) and is included in a nature reserve that also holds otters (Trindade et al. 1998), although their numbers are not known.

The Reconciliation Action Plan (RAP) presented here gives guidelines and recommendations to ensure a long-term conservation of otters in the Sado estuary and their coexistence with human activities. It is intended to be a tool for wildlife managers to cope with conflicts which occur due to otter predation on fish. Structured in three subsequent phases, it outlines relevant approaches, results, and recommendations applicable to this conflict that should be taken into account for successful reconciliation.

First, the basic ecological and socio-economic contexts of the conflict are summarized, the conflict potential is discussed, and the needs of further assessments are identified (1-screening phase). Following, the scientific approaches undertaken to analyze the ecological and socio-economic dimensions of the conflict, and resulting conclusions, are described (2-assessing and analyzing phase). Finally a series of guidelines and recommendations to reconcile the conflict at different levels of implementation (species management and protection, site safeguard and management, ecological research and monitoring, policy instruments, and communication and publicity) are presented (3-resolution and implementation).

**Box 1** Study Area

The Sado estuary, in the central west part of the country, is included in two planning regions-Lisboa e Vale do Tejo (LVT) and Alentejo, and part of an economically depressed region with above average unemployment rates and below average income. It is mainly a rural area with a strong environmental protection status, although very close to urban and industrialized sites. Besides fish farming, an important area of the estuary is used for salt production and intensive agriculture, especially rice plantation. Other important economic activities in the area are forestry and fishing, along with manufacturing and services (INE 2002). There are also several industrial plants, such as chemical, electricity, pulp and paper, and ship building and repair. Faced with the threats of agricultural intensification as well as industrial and tourism development projects, which may cause significant impacts related to habitat disruption and pollution, the designation as protected area aimed at the preservation of the ecological values, such as a resident population of Atlantic bottlenose dolphins (*Tursiops truncatus*), greater flamingos (*Phoenicopterus ruber*), water birds [e.g., black-winged stilt (*Himantopus himantopus*), Eurasian curlew (*Numenius arquata*), dunlin (*Calidris alpina*), grey plover (*Pluvialis squatarola*)] and birds of prey [e.g., marsh harrier (*Circus aeroginosus*)], along with the Eurasian otter (ICN 2001). Due to the protection status of the estuary, some constraints on land use and management practices are enforced in order to preserve habitats and species. Such an area has a strong potential for the occurrence of conflicts between nature conservation and socio-economic development.

# 2 Ecological and Socio-Economic Contexts of the Conflict

## *2.1 Ecological Context*

### 2.1.1 Otter Distribution and Abundance

Portugal is considered to have one of the most viable otter populations in Europe (Foster-Turley et al. 1990). Otters may be found in a wide range of aquatic environments, from very small streams and large rivers to dams and, also, to coastal and estuarine environments (Santos-Reis et al. 1995). In 1995 a nation-wide otter survey was conducted, using a 10 × 10 km UTM grid system, and the resulting map showed a continuous distribution with very few absences in the coastal area (Trindade et al. 1998).

In 2002, the otter distribution was reevaluated in the river Sado basin at three different resolutions of scale (10 × 10, 5 × 5 and 2.5 × 2.5 km). The results confirmed the wide presence of otter. At the finer resolution absences were evenly distributed along the basin, with exception of the estuarine area, where they are concentrated on the west side, near the coast. Here absences are mostly related to the absence of freshwater (Sales-Luís et al. 2012).

Otter densities in Portugal are unknown but suspected, at least in some areas, to be high, as evidenced in 2000 in the area to be flooded by Alqueva dam (river Guadiana, SE Alentejo), where, in the frame of a Catalonian reintroduction program (Spain), 19 otters were captured with a trapping effort of about 2,500 trap/nights (Santos-Reis et al. 2003). Sizes of territories are also largely unknown. Available data only refer to otters that live along the SW steep coastline, using both marine and freshwater resources, whose territories range from 4 to 15 km (Beja 1989, 1992).

### 2.1.2 Otter Diet and Damage Assessment

Most of the available information refers to local studies on food habitats and is mainly spread across technical national reports (Santos-Reis et al. 1995). Some of these studies analyzed otter diet in estuarine and marine environments (Beja 1991; Chambel 1997; Gomes 1998; Pedrosa 2000) but only one focused on otter predation in fish farms in the Mira estuary (Trigo 1994). The reported impact on commercial species was considered low and limited in time, but visiting rates were not assessed and the correlation with fish numbers inside ponds was not investigated. Concerning wild prey availability, the situation is slightly different in the Sado estuary, where an inventory of the fish community already existed (Cabral 1999).

To overcome the general lack of information on otter predation in fish farms, in 1990 a questionnaire was sent by the Portuguese Institute for Nature Conservation to 208 fish farmers, aiming to determine the impact of predation on their businesses and the financial losses (Trindade 1991). Otters were the most commonly referred predator, causing damages all year round, and some fish farmers perceived the species as a pest. Complaints referred to predation and stress imposed upon the fish, and the losses reported were high. However, producers could not quantify damage levels and nearby productions reported very contrasting damages. In the frame of the FRAP project (Henle et al. 2013), in 2003, phone inquiries were performed at the national level to both marine and trout farms and visits conducted to selected fish farms to survey for otter signs (Santos-Reis et al. 2007). Of the 57 (55 % of a total of 103 known fish farms in Portugal) marine fish farms approached, eight (14 %) were not operating, 28 (49 %) reported absence of otter signs, and 21 (37 %) reported regular otter visits.

The assessment of damage, on the basis of eaten fish and possible secondary losses (e.g., wounded fish, stress-disturbed fish), is quite difficult and sometimes not even accurately perceptible to the fish farmer on harvesting occasions. Losses by diseases and production hazards, as well as non-otter predation, may also have a strong influence on damages reported. Many piscivorous birds are common near Sado marine fish farms, such as little egret (*Egretta garzeta*), grey heron (*Ardea cinerea*), and great cormorant, (*Phalocrocorax carbo sinensis*). Cormorants occur in the Sado estuary only for a brief period during wintering time but in fairly high numbers, whereas egret and heron presence is more frequent and regular throughout the year (Carss 2003).

Specific characteristics of fish farms that may promote the risk of otter predation are unknown but the type of production, the reared species, and landscape and habitat factors, such as freshwater and wild prey availability, are surely involved.

### 2.1.3 Mitigation Measures

To avoid otter predation, mitigation methods such as ordinary and electric fencing are available and fairly effective when correctly applied. The most common method in Portugal is pond-fencing, using several types and heights of fences, including fishnets. Fish farmers consider the method to be expensive but prefer installing a fence to having the otters causing damage. The effectiveness of fencing is perceived as high, although mesh size and depth of the net into the ground may significantly influence the result. Smaller meshes and burrowing deeper the net increase the costs. The cost-effectiveness of the electric fence is perceived as very high, although regular maintenance is necessary. Some other methods, including sound or visual stimuli, are used to discourage predators (e.g., recorded alarm or distress calls, gas cannons, scarecrows, guard dogs, and human surveillance). These deterrent methods reduce but do not eliminate predation, since predators become easily familiar and ignore them. Lethal methods such as traps, shooting, or even poisoning, although illegal in Portugal, are sometimes used, but again their long-term effectiveness is questionable.

## *2.2 Socio-Economic Context*

### 2.2.1 Legal Framework for Otter Protection and Economic Relevance of Aquaculture

Apart from being a nature reserve, the Sado estuary is also classified as a special bird protection area (EEC Birds Directive) and a Ramsar Site. Moreover, the area is included in the national list of NATURA 2000 sites, partially due to the presence of otters.

The otter is protected by the national legislation, transposing the Habitats Directive, being included in two different categories:

1. as a species of community interest, whose conservation demands the designation of a Special Conservation Area;
2. as a species of Community interest that requires strict protection, for which capture, killing, and appropriation, no matter the method used, as well as deliberate disturbance of animals, are prohibited.

As an economic activity, aquaculture plays a modest role at the national level. However, coastal aquaculture, producing mainly marine fish species, is considered important in the Sado estuary, with a significant socio-economic impact at the local level [(6 % of aquaculture farms in Portugal are located in the "Lisboa e Vale do Tejo" (LVT) region; considering only marine fish farms, i.e., excluding bottom culture and cages, this region accounts for 56 % of the 58 units existing in the country)]. Fish farms operate mainly in semi-intensive regime, often recovering old salt production tanks, and produce four marine species—gilthead sea bream (*Sparus aurata*), European sea bass (*Dicentrarchus labrax*), Senegal sole (*Solea senegalensis*) and common sole (*Solea solea*). The main product is the gilthead sea bream, which is bought and stocked at very early stages of their life cycle as is the case for the sea bass (INE 2002), whereas soles are opportunistically captured as juveniles in the estuary and further raised inside fish ponds (Fig. 1).

### 2.2.2 Relevant Stakeholders

The issue of predation in the Sado Estuary fish farms may be responsible for a social conflict notoriously bipolarized around two key actor groups. On one hand fish farmers, mostly owners of small or medium-size production facilities, aware of the existence of otter predation and privately using whatever means they could to keep otters away, including illegal killing. On the other hand, technical staff and managers of the Sado estuary nature reserve, the authority responsible for the management of the protected area, also aware of the predation as well as of the weakness of the monitoring system for control of illegal killing, but forbidding the use of any mitigation measures to prevent predation within the borders of the reserve.

Local and central administration, industries, nature tourism operators and environmental organizations, that do not have a direct stake in the issue, were also identified as relevant actors and potential partners in solutions for the conflict.

### 2.2.3 Potential for a Conflict and Needs of Assessment

In Portugal, over the past years, there is evidence that otters are a potential threat for the fish farming activity (Trigo 1994). However, the available information is not sufficient to support the quantification of damages. The Portuguese Nature Conservation Institute survey from 1990 (Trindade 1991) has reported important

**Fig. 1** Fish species stocked at Sado river estuarine fish farms, Portugal. *Photos:* Dália Freitas

losses at a national level, but individual fish farmers in the Sado Estuary cannot quantify the damages in their farms and there are no statistical data on fish losses, at the national, regional, or local level.

The lack of quantitative data does not reduce the conflict. Fish farmers have the perception, sometimes overestimated, that there is considerable damaging predation and the nature conservationists, namely from the nature reserve, perceive that aggressive unauthorized measures, namely illegal killing, are sometimes taken against these predators; hence, several aspects have to be taken into account to identify the real scope of the conflict.

The individual perception of the predation and consequent damage were assessed for a sample of relevant fish farmers and compared with data about the predation and damage in the corresponding fish farm, resulting from fieldwork developed by the project's ecological team. It was also important to assess whether the social conflict among the different actors was only related to this specific issue or if other issues were also relevant.

Another aspect identified as relevant in the assessment phase, was to assess the knowledge that the actors had about the different types of technical mitigation measures available to control predation, as well as their effectiveness and impacts. Finally, the assessment of the role of past and present policy instruments with direct or indirect influence on the conflict, as well as the evaluation of their performance, were relevant to guide the development of new policies to address the conflict.

In summary, the incentives to growth that fish farming has been receiving in the past, namely at the European level, in order to reduce open sea fish stock pressure, associated with the operational and spatial restrictions resulting from the protected area status, have been leading to a growing conflict over several issues. Otter predation on fish farms is only one of them. It was therefore essential to promote an in-depth evaluation of the ecological and socio-economic dimensions of the conflict, namely to clarify potential misunderstandings among the different actors, to identify the real magnitude of the problems and to support the development and implementation of instruments and measures to promote the reconciliation between aquaculture and nature conservation in the Sado estuary.

# 3 Ecological and Socio-Economic Assessment of the Conflict

## *3.1 Ecological Assessment*

In order to improve the understanding on the predation of otters in the Sado estuary fish farms the focus was on otter visiting rates to farms, the consumption of reared fish by otters, and the landscape factors promoting those damages (Fig. 2).The study followed two lines of research: a coarse approach (landscape factors affecting otter visiting rates) and a detailed approach (otter diet and loss of

**Fig. 2** Collection of otter scats in the Sado river estuary. *Photo:* Teresa Sales Luís

commercial fish), using existing data and field research (Freitas et al. 2007). Fourteen semi-intensive fish farms were selected for the study (Figs. 3, 4).

### 3.1.1 Otter Visiting Rates

From July 2003 to June 2004, otter visiting rates (*VR* = number of positive visits/total number of visits) were assessed with weekly surveys for otter signs (e.g., spraints and footprints) along the fish farm perimeters. In each visit all spraints were collected for diet analyses and footprints were destroyed (Fig. 3). Spraints found were classified as fresh (from the previous night), dry (two or three nights old), or old (up to one week old).

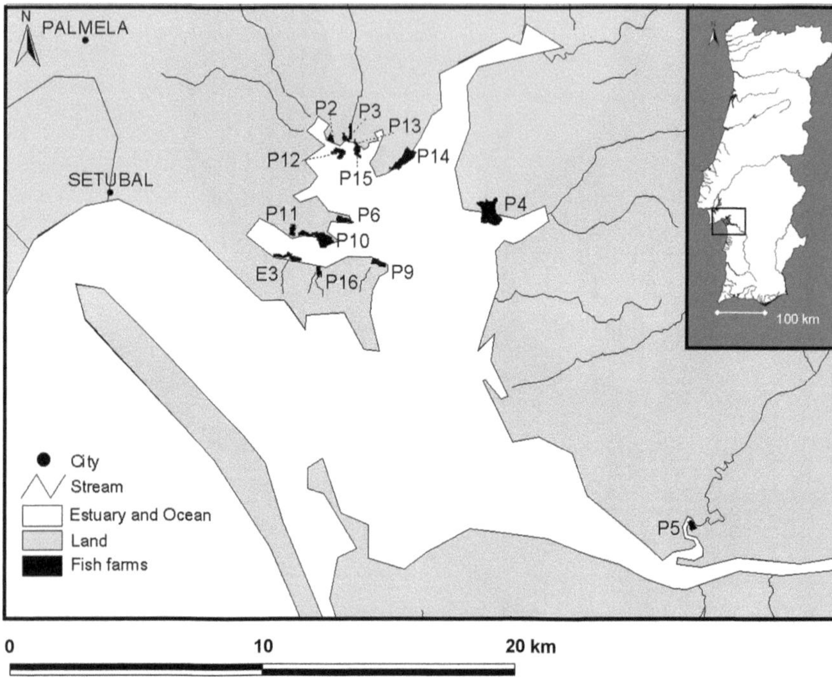


**Fig. 3** Surveyed fish farms in the Sado river estuary. Fish farms are identified with codes for purposes of comparison with the socio-economic analysis

Estimators were developed [a maximum likelihood function estimate (*VRMLE*)—Gruber et al. 2008] to maximize weekly information collected by transforming it into daily visiting rates.

Otter weekly visiting rates were very high (76 % on average; $VR = 0.76$), with half of the fish farms being visited by otters in more than 80 % of the occasions ($VR = 0.8$), although varying between 20 and 100 % (Fig. 5). However, estimated daily visiting rates (*VRMLE*) best represent the real variability observed between fish farms (0.03–0.68). In seasonal terms mean visiting rates were also high, varying from 0.69 (spring) to 0.82 (winter).

### 3.1.2 Otter Numbers

In the 2004/05 winter, additional surveys were conducted to evaluate the number of otters visiting each fish farm. For an average of 8 days, less than 18 h old spraints were collected for molecular analyses, the most recent non-invasive technique for individual identification (Dallas et al. 2003). This technique, although having a low proportion of spraints yielding preserved DNA and

**Fig. 4** Marine fish farm in Sado river estuary, Portugal. *Photo:* Teresa Sales Luís

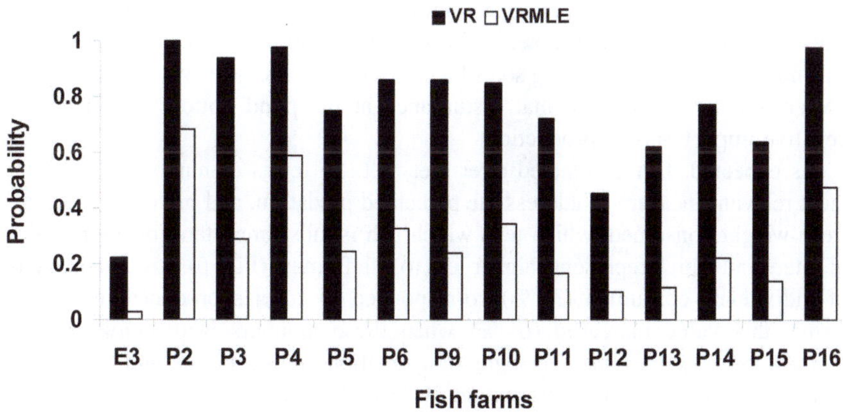


**Fig. 5** Probability of otter visiting rates by fish farm: VR—weekly visiting rates and VRMLE—estimated daily visiting rates

requiring a large effort in field sampling and DNA typing, has shown to yield unbiased estimates of population composition (Dallas et al. 2003).

The molecular spraint analysis, using 5 microsatellites and a success rate of 25 %, revealed that the number of otters visiting each of the studied fish farms varied from 1–7 individuals. Overall a minimum number of 15 individuals were identified in a 100 $km^2$ area, indicating a high density otter population (Sales-Luís et al. 2009).

### 3.1.3 Otter Diet

To assess otter diet, and consequently damages due to predation, the standard approach of scat analysis was followed using a minimum sample of 30 spraints per season and fish farm. Fish remains were identified to the species level and the diet composition expressed in terms of numeric frequencies (*P.O.* = percentage of occurrence = number of a particular prey item in all spraints/sum of all prey items * 100) and ingested biomass (*P.B.* = percentage of biomass = biomass of a particular species in all spraints/sum of biomass of all species * 100). To estimate size and mass of fish consumed by otters, dimensions of non-digested bony structures were used and regression equations were developed (Freitas et al. unpublished data).

Besides quantifying the spatial and seasonal variation in diet, the abundance of commercial species and of alternative prey was assessed. Prey diversity and size range in the estuary were assessed in winter and spring, using a beam trawl, and abundance data of commercial fish inside ponds were provided by fish farmers; size range inside ponds was evaluated in winter and spring, using a small seine net, called "chinchorro". This procedure allowed to evaluate whether consumed commercial fish species (gilthead sea bream and sea bass) had similar sizes to those existing in the ponds, in an attempt to understand where predation occurred, that is, if fish had been consumed inside or outside the fish farm. This fishing gear was not adequate for capturing soles but no other specific gear was allowed by the owners on the assumption that disturbance at the pond bottom would have a negative impact on fish production.

As expected, fish dominated otter diet, both in terms of numbers (*P.O.*), the most relevant measure to address the perceived predation, and biomass (*P.B.*), i.e. fresh weight consumed, which tells which fish is most important for the nutrition of otters and which represents higher loss to fish farmers (Fig. 6). Around one third of individuals consumed (31 %) corresponded to species produced in the fish farms; this value increased (61 %) when those numbers were converted into biomass by a species specific regression function. Wild, marine and freshwater, fish species (all together representing 36 % of biomass consumed), crustaceans, and amphibians were other prey regularly eaten by the otter.

The most important commercial species preyed upon was gilthead sea bream, being only surpassed by the soles in terms of percentage of biomass (Fig. 7).

The presence of sea bass in otter diet was very low, both in numbers and biomass, being only occasionally preyed upon in autumn and winter. Soles were regularly eaten all year round, whereas the consumption of gilthead sea bream varied seasonally (higher in autumn/winter than in spring/summer), but differences were observed for both species at the spatial (fish farm) level (Fig. 7), with the

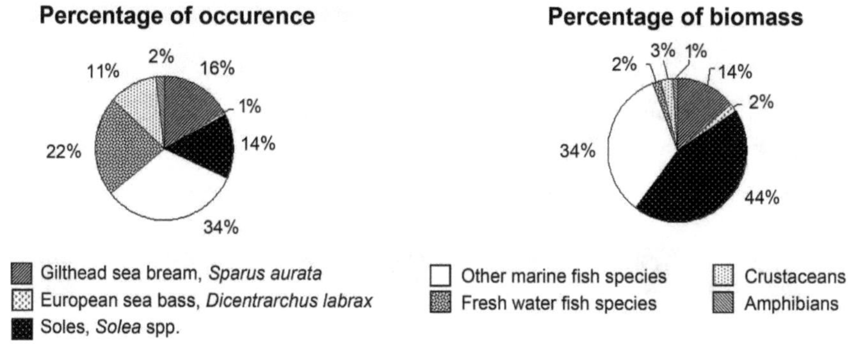


**Fig. 6** Composition of the diet of otters in Sado estuary expressed as percentage of occurrence and percentage of biomass

consumption of stocked fish varying from negligible to significant. Sizes of consumed fish were relatively small (gilthead sea bream: 50–150 mm; sea bass: 150–200 mm; soles: 100–250 mm); although larger sizes were available, they were not eaten.

The fish species assemblage occurring in the estuary outside fish farms was not diverse (14 species) and was clearly dominated by Senegal sea bream (*Diplodus bellottii*) and golden grey mullet (*Liza aurata*). On the other hand, sea bass was not detected during sampling procedures, Senegal sole was much less abundant than the dominant species, and gilthead sea bream was one of the least abundant species.

Although soles in the Sado estuary were previously considered relatively common, especially *Solea solea* (Cabral 1999, 2000), during this study the observed density was low, but this may result from the species' habits and sampling bias (e.g., benthonic habits, nocturnal activity, and type of nets used). However, their consumption was very high despite stocks inside fish farms being low, when compared to the bream or even to the bass, and this suggests that otters may exhibit a prey selective behavior searching for soles in the wild environment. Gilthead sea bream and sea bass, however, proved to be uncommon in the estuary, as it was described previously (Amorim 1982; Lopes da Cunha 1994). Therefore, it can be assumed that otters are capturing these species inside the ponds.

Gilthead sea bream consumption was more intensive in autumn and winter, when the temperatures are low, leaving fish more vulnerable to predation. Moreover, in these seasons, prey availability in the estuary, in the small freshwater streams, and in the rice fields is lower (own unpublished data).

In summary, fish farms are an important food source for the otter in the Sado estuary, mainly in what concerns the gilthead sea bream. However, it is still uncertain whether the most important prey in terms of biomass (soles) is caught in the wild or inside fish ponds. Surrounding freshwater bodies can provide alternative prey [Louisiana crayfish (*Procambarus clarkii*) and black bullhead (*Ameiurus melas*)] to the otter and, in some cases, even greatly reduce the predation pressure over farming stocks.

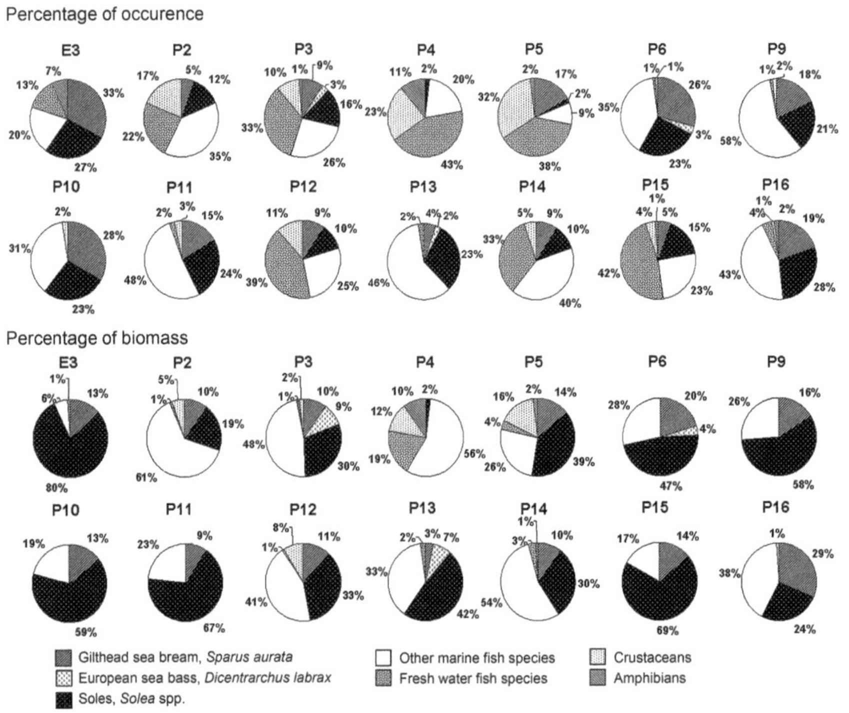


**Fig. 7** Spatial composition of the diet of otters in Sado estuary expressed as percentage of occurrence and percentage of biomass

# 4 Key Landscape Factors for Damage

Bank vegetation, bank structure, hinterland use, alternative prey availability, and existence of preventive measures (e.g., fences, electric fences, dogs) were recorded as descriptive factors for landscape analyses. Several metric variables (distances, areas in buffers) were also assessed using a Geographical Information System (GIS) data base.

Association between landscape descriptors and potential surrogates for otter damage (Visiting Rates, Percentage of Occurrence, and Percentage of Biomass) were tested using correlation analysis. Multiple linear regression (backwards stepwise) was used to identify key landscape factors for damage.

No clear relation was found between visiting rates (*VR*, *VRMLE*) and the consumption of the stocked species (neither *P.O.* nor *P.B.*) ($r < 0.25$; $p > 0.37$; $\alpha = 13$). Because percentage of biomass best reflects the real damage inflicted by otters, it was used as a surrogate for damage. A possible explanation to why visiting rates do not correlate with consumption of reared species is that otters may regularly visit fish farms near their home streams, not only to predate, but also to scent mark; these fish farms might be important territory sites, such as passing ways to other important resource areas. This explanation is supported by the visiting rates being negatively influenced by the distance to the nearest streams and preventive measures and positively related to the refuge cover area in a 1 km buffer.

The highest associations ($r > 0.6$; $p < 0.03$; $\alpha = 13$) between landscape descriptors and the best surrogate for otter damage (*P.B.*) were: distance to running freshwaters [ditches (−); streams (+)]; distance to refuge cover areas (+); alternative prey (−); natural bank structure (+). Nevertheless, the best model for key landscape features only includes the positive influence of the distance to the nearest stream (*DNS*) and the distance to refuge cover area (*DRCA*) (term 1):

$$P.B. = 13.7\ \mathrm{DNS} + 68.1\ \mathrm{DRCA} \quad (1)$$

with $R^2 = 0.84$; $F_{11;3} = 28.2$; $\alpha = 0.0001$.

## 5 Damage Assessment

The impact of otter predation was assessed considering the overall production of this fish farming area (347.5 ha) and taking into account the local otter density (15 ind./100 km$^2$), the average stocking level for the three produced species (gilthead sea bream: 20,153 ind./ha; sea bass: 11,147 ind./ha; soles: 676 ind./ha) as reported by fish farmers, the average daily visiting rates (0.28), and the daily food consumption of an adult Iberian otter using captive individuals (0.75 kg/day—J. Ruiz-Olmo pers. com.). In addition, otter damage on stocked fish was assessed per fish farm using the respective number of otters (1–7 individuals) visiting the fish farm, the stocking level (gilthead sea bream: 12,045–46,428 ind./ha; sea bass: 1,852–19,358 ind./ha); soles: 100–1,090 ind./ha), and the percentage of consumed biomass (0.4–69.0 %), combined with the daily visiting rate (0.10–0.68).

Otter damage on the fish farming area represents 0.60 % of the overall production and the economic loss per fish farm varied from 0.01–0.30 % for gilthead sea bream, from 0.03–0.11 % for sea bass, and from 1.04–28.20 % for soles.

Damage is higher in fish farms further away from water lines of some importance (streams and brooks) that usually offer better refuge and alternative prey, especially if the fish farm is close to ditches that may act like "roads" for otters, but which alone have a lower carrying capacity.

Although damage assessment in economic terms should consider the balance between costs and benefits, data considering just fish weight loss suggest that the total damage on commercial fish, considering the overall area, is meaningful, especially because fish farmers expect an overall mortality (predation, diseases, etc.) of up to 10 % of stocked fish. At the fish farm level, otter predation varied significantly between fish farms, but it was low in all cases, especially for gilthead sea bream and sea bass. Very likely the damage on bream stock is not higher because soles are available almost all year round, since they are winter and spring spawners (Cabral 2000), and easily captured both inside the ponds and in the estuary. Despite greatest damage being observed in soles stocks, at a first glance, it must be kept in mind that operation costs are low for the fish farmer, since the majority do not buy juveniles of these species from hatcheries but capture them opportunistically in the estuary in the course of other fishing operations. However, damage is perceived as locally high due to the high market value of soles.

### *5.1 Socio-Economic Assessment*

The socio-economic dimension of the conflict was assessed considering the following aspects: (1) study of the institutional and legal framework of relevant sectors; (2) social impact assessment, with a discourse analysis of stakeholders interviews; (3) socio-economic analysis of the study region to assess the role and importance of each sector, mainly those related with the conflict; and, finally, (4) analysis and evaluation of the instruments and policies that potentially address the conflict.

## 6 Institutional and Legal Framework

The institutional and legal framework analysis focused mainly on the identification of relevant directives, laws, and policies, as well as decision-making mechanisms, discussing their advantages, problems, and limitations. General policy goals and relevant regulations for a broad variety of issues, such as species protection, protected areas, fisheries (and agriculture) policy, land-use planning, coastal management, hunting, regional development programs, and state aid, were analyzed. Some aspects analyzed turned out to be relevant to understand the conflict and to influence the development of policy instruments to mitigate it.

The centralized decision making process in Portugal associated with the lack of power and resources of regional authorities, the absence of integrated regional policies, and the legal and institutional framework of protected areas can hinder the implementation of effective solutions, even if there would be consensus between regional authorities and stakeholders. For example, the Sado nature reserve does not have a management plan in place, due to many factors including the scarcity of human and financial resources, but also as a result of the centralized policy approach adopted in the past.

The lack of this type of instrument creates relevant problems to manage the conciliation of sustainable economic activities with the conservation objectives of the Sado reserve. The rules of the game for the economic actors are not well defined, namely, for example, in what concerns zoning of areas for aquaculture development, the density of fishes allowed in each aquaculture or the identification of actions and measures which can be adopted to mitigate the impact of otters. This lack of guidelines creates problems to find the most efficient economic decisions and favors the adoption of solutions based mostly on individual and short-term objectives. For the reserve it is also difficult to justify the imposed constraints to the economic activities as well as to define, implement, and enforce effective conservation measures and instruments.

## 7 Stakeholder Perspectives Assessment

A participatory social impact assessment (SIA) was developed for the Sado case study. This approach allows for articulation of impacts from a local perspective, reflecting a unique understanding of an individual's community (Becker et al. 2003). Like an environmental impact assessment, a SIA should be anticipatory and carried out before any action or research has taken place (Barrow 2002; Kranz and Rauschmayer 2013). The social impact assessment performed in the Sado region was based on open format inductive interviews performed with 31 stakeholders [according to the SIA methodology proposed by IAICGP (1994)].

SIA provided the basis to perform a discourse analysis that could help answer the sociological related questions raised during the screening process. [1] Discourse analysis can help in illuminating the different interests being articulated and struggles between interest groups thus providing a starting point for a negotiation process (MacDonald 2003). It also emphasizes, in a perspective that Wood and Kroger (2000) call "critical discourse analysis", how text relates to context, allowing the identification of ideas and messages that are not literarily written.

This analysis was followed by an exercise that aimed to aggregate stakeholders with similar positions around a set of themes, providing not only a picture of the attitudes and feelings of the several actors towards the important issues, but also a ground for a future consensus building process. Actors were divided into groups (fish farmers, other economic agents, governmental bodies, such as local and regional authorities or the nature reserve administrations, environmental NGOs and scientists), and their opinions (agree/disagree) were positioned towards several themes (a total of thirteen) related to the conflict gathered during the discourse analysis.

In general, there is no consensus regarding the threat due to otter predation, but most fish farmers have the feeling that otters predate significantly on their fish farms. The conflict between otters and aquaculture in the Sado estuary has a data/facts type component, since there is a strong discrepancy between what the fish farmers perceive and what the ecological team assessed as damage (Table 1). However, it is possible to find also a conflict of interests and values, where the conservation of the reserve clashes with the expansion of fish farms.

It was observed that some actors' positions were closer to actors that do not belong to their base groups (e.g., fish farmers and environmentalists in Fig. 8),

[1] The NUD*IST text analysis software package was used to perform the discourse analysis.

**Table 1** Total damage by fish farm (in % of final mass harvested) and the conflict perception of otter impact found through FRAP in each fish farm (P2-P16) (disagreement situations are highlighted in the shaded rectangles)

| | Fish farms | | | | | | | | | | | |
|---|---|---|---|---|---|---|---|---|---|---|---|---|
| | P02 | P03 | P05 | P06 | P09 | P10 | P11 | P12 | P13 | P14 | P15 | P16 |
| Total damage | 5.33 | 0.08 | 28.25 | 0.25 | 1.06 | 4.23 | 13.87 | 1.63 | 6.28 | 1.41 | 3.07 | 14.96 |
| Conflict perception | None | None | High | Low | High | Low | Low | High | High | Low | High | High |

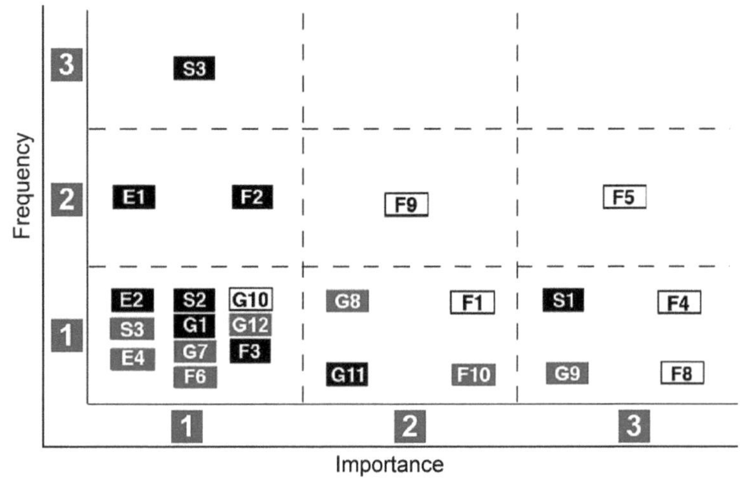


**Fig. 8** Positioning of stakeholders towards the theme "Otters are a threat to fish farms". White boxes represent stakeholders who agree with the theme, black boxes represent those who disagree, and grey boxes represent stakeholders who do not have a clear opinion or do not know whether the theme is true or not. The codes are used in order to keep the anonymity of the stakeholders and are related to the base group the stakeholder comes from (F—Fish farmers, S—Scientists, E—Environmentalists, G—Governmental representatives)

which may point to the conclusion that opinions are divided in the whole. This is mainly due to different ideas and disagreement regarding management methods and solutions to solve the conflict in the different dimensions.

# 8 Socio-Economic Relevance and Impacts

The socio-economic analysis broadly covered the past development, current situation, and future perspectives of the economic sectors, with a special focus on the conflicting sectors (fisheries and biodiversity conservation). The main objectives included the evaluation of the economical and social importance of fisheries in a local and national level, as well as the economic and social impacts of a potential

**Fig. 9** Pond fishing in a fish farm in Sado river estuary, Portugal. *Photo:* Hugo Matos

conflict between aquaculture and biodiversity conservation (Santos et al. 2013). The data collection strategy for the socio-economic analysis of the study region was primarily based on available socio-economic statistical data complemented with literature reviews and field interviews. A time horizon of 10 years to the past was adopted and since censuses in Portugal are done every 10 years, the analysis focused mainly on the period 1991–2001. The analyzed descriptors cover aspects, such as the general economic structure, the fisheries sector, population data, income situation, labor market characteristics, and protected areas.

Covering 1,581 km$^2$, the Setúbal Peninsula comprises the northern shore of the Sado River Estuary, an area of great interest for the study of the conflict due to the importance of the aquaculture activity and to the presence of otter.

The Setúbal Peninsula is covered primarily by agriculture, forest, and semi-natural areas, which represent 84 % of the area. However, urban parishes (where 95 % of the population lives) account for 57 % of the total area. The area is highly industrialized, tourism is developing, and agriculture undergoes intensification. This creates problems and conflicts with nature conservation. Pollution from urban centers and industry located in the study area is identified by fish-farmers as an important threat to their activity and for protected species.

Aquaculture ponds are situated in the Sado estuary, dominated by a rural landscape, in spite of being close to urban and industrial areas (Fig. 9). This activity presents peculiar features in the studied area. Coastal aquaculture is dominant, and it is carried out in ponds built according to the features of old salt marshes and producing marine fish species in extensive and semi-intensive regimes due to the classification of the Sado estuary as a nature reserve. The study area has about half of its area under environmental protection. Considering the proportion of protected areas, nature conservation is very important in the study area: 14 % of the Setúbal Peninsula is protected by national legislation (6 % is the nature reserve of the Sado River Estuary).

The Setúbal Península shows some economic depressed indicators, with above average unemployment rates and below national average income. In this context small-scale economic activities can have a significant local importance. Aquaculture has a low relevance for fisheries at the national level, since that marine fishing is dominating in Portugal, but has a great growth potential in the study area and it is relevant at a local level. However, inside the reserve the establishment of new fish farms and the enlargement of existing ones, among other activities, are subject to approval by an Installing Commission chaired by the Institute of Nature Conservation.

In this context, the need for an integrated management of all these diverse and sometimes conflicting land uses—particularly fishing, aquaculture, agriculture, salt production, industry, navigation, urbanization, and tourism and leisure—is acknowledged.

## 9 Policy Analysis

A policy analysis was performed, aiming to identify and collect relevant information to describe the past use and present implementation of instruments to address the conflict between economic activities and protected species, as well as to develop a preliminary analysis of their effectiveness. Three types of instruments were analyzed: "command and control", "economic", and "action directed to civil society".

This analysis unveiled that no specific instruments exist in Portugal to address the conflict between otter conservation and fish farming and therefore there are no cases that can be used to derive "Best Practices". Specific instruments for otter protection have not been implemented in Portugal due to its suspected abundance (Trindade et al. 1998). Although the otter has a strictly protected status due to the Habitats Directive and was considered potentially threatened in the past (SNPRCN 1990), there is today evidence of the existence of a large population of otters throughout Portugal. The status of otters in the Red Book of Vertebrates of Portugal has been recently reassessed and is currently included in the "least concern" category, in the frame of the new IUCN (International Union for the Conservation of Nature) criteria (Cabral et al. 2005, http://www.icn.pt).

However, some instruments not specifically targeted to address the conflict may have an effect on it, although with several limitations to effectively contribute to its mitigation (e.g., aquaculture licensing; subsidies, namely EU financial support, to aquaculture development; legislation on species protection and on creation and management of protected areas). A relevant conclusion is, therefore, that some of the existing instruments can be improved in order to better address the conflict. For example, licensing and EU financial support to aquaculture could give an incentive to a sustainable co-existence with biodiversity, and the protected area management could develop goals and guidelines to conciliate aquaculture development with the conservation objectives.

The policy analysis also made clear that it is important to develop new instruments with participation of all relevant stakeholders in order to incorporate different visions and knowledge, as well as to involve them in a joint effort to mitigate the conflict. The development of voluntary instruments, such as a sustainable production label for fish products from aquaculture in the Sado Estuary, was considered as a promising challenge.

## *9.1 Uncertainties and Conflict Resolution*

Ecological data suggests that soles are the main prey in terms of biomass for otters in the Sado estuary. So far, it was not possible to assess whether they are caught mainly outside or inside the ponds. However, as these captures do not represent a significant part of the fish farm production, this loss is not threatening the economic viability of the activity. It is nevertheless not so certain that the predation on the other commercially valuable species might, or will, be threatening the economic sustainability at some point.

The policy analysis concluded that there are no instruments in place to ensure the attainment of the objectives of maintaining viable otter populations and sustainable economic activities. Stakeholders consider that it is important to find new solutions and to increase the dialogue between the different actors, but agree also on the existence of a significant uncertainty about the capacity to put new instruments in place. It was also found that serious misunderstandings exist between key actors, not limited to this specific issue (e.g., aquaculture licensing, allowed densities for production, allowed mitigation measures), that might also impeach the achievement of sustainable and strong solutions.

These two findings led to the decision to promote the development of a participatory process for the proposal and joint development of new instruments, to which all stakeholders were invited to join. Participation strategies have been advocated as a way to tackle the complexity that environmental policy-making implies (Santos et al. 2006). Such a process has also the potential to bring a greater understanding and an increase in trust among the major actors, while providing for sounder solutions to the conflict.

# 10 Guidelines and Recommendations for Conflict Reconciliation

Lessons learned show that the main obstacle to conflict resolution is a lack of dialogue between the two main stakeholders (fish farmers and nature reserve officials).

Scientific results proved that the impact of otters is not important at the study area level but varies from one fish farm to the other and may be of significance at the fish farm level. This suggests that different solutions should be recommended for each fish farm, such as to promote the use of mitigation measures in those farms in which damages are high, no matter how they are perceived, and to invest in educational activities on those in which damages are not significant but perceived as such.

Although this may be seen as the more effective approach to reconcile the conflict in the ecological perspective, it may, however, be unacceptable in social terms. Fish farming in the Sado estuary is highly clustered and fish farming owners, although not working cooperatively, do interact with each other and even with additional explanations will not easily accept different individual solutions.

The prohibition of using any mitigation measures by the local authorities has been the main cause of the lack of dialogue among the main stakeholders. A change in the attitude of the Sado nature reserve can prevent the conflict escalation and improve the relation between the interest parties, opening ground to the development of other short- to long-term tools towards the conflict reconciliation, as further described.

## *10.1 Species Management and Protection*

Although the otter has been considered a species of Community interest that demanded protection (see above), no action plan was ever designed at the national level, nor management plans were implemented in critical areas, such as the Sado estuary, where socio-economic activities could be conflicting with the species safeguard. It is therefore recommended in the short-term to prepare a management plan for the Sado Estuary nature reserve. This instrument is specially important now considering that the otter conservation status was downgraded at a national level to the category of Least Concern (Cabral et al. 2005) and that, as a consequence of this project, fish farmers have additional knowledge about the issue and are waiting for the definition of clear rules to develop their activity and for proposals to mitigate the problem and conciliate their interests with the conservation objectives.

Fish farmers should be allowed to use electric fences around production ponds, to prevent predation even when real damages are not high, since this will not have a major effect on the fairly well distributed and apparently dense otter population inhabiting the area, which is therefore not at risk. This will prevent the use of lethal methods, such as traps, shooting, or even poisoning, that, although illegal in Portugal, are sometimes used.

Moreover, it is also recommended to evaluate the cost-effectiveness and feasibility of the use of compensation instruments, mainly those directed to compensate for the provision of environmental services. An example is the construction of deviation ponds, as suggested by the nature reserve, which also may provide services not strictly otter-related (e.g., bird observatories), since these ponds also aim to attract other fish-eating species (e.g., particularly cormorants) that may be causing more damages than the otter itself. It should also be considered that some instruments are only cost-effective in a long-term perspective. So, the time horizon adopted to evaluate the different measures (e.g., electric fencing as a predation deterrent) and instruments (e.g., site-specific predation risk maps) is a key variable.

One of the identified limiting factors of the otter distribution in the estuary was freshwater availability, since very few sources, i.e. streams or ditches were found in the study area and many of them were highly polluted as a consequence of the urban and industrial wastes released to the water lines. For the otter safeguard it is essential the maintenance of small drainage lines and ponds, and preservation guidelines should be prepared and enforced by the local authorities if otters are the key issue. Alternatively, limiting fresh water around ponds could make them less attractive to otters, and reduce visiting rates, but the collateral consequences of such a measure, such as the reduced availability for agriculture purposes, should be first investigated.

## *10.2 Site Safeguard and Management*

Estuaries are highly productive areas that, as a consequence, hold high levels of biodiversity, including otters. For this reason, its conservation role was recognized, benefiting nowadays from the classification as nature reserves. On the other hand, these areas have attracted, in the past, many important economic activities generating significant income and employment (e.g., as salt producers). Currently, they have a high potential as fish farming areas that may produce fish and reduce the pressure over the depleted sea stocks. Safeguard and management of these estuaries depend therefore on the ability to find solutions to conciliate both aptitudes (nature conservation and sustainable economic development). The way forward is to promote best practice guidelines that will raise awareness to the interest and need of biodiversity conservation and prevent operations that would have an adverse impact upon the otter population and other important wildlife values.

Fish farming, and specifically semi-intensive fish farming, may be reconcilable with conservation if some precautionary management measures are taken in terms of site safeguard and of precautionary development of the farming activities. Fish farms operating in extensive regime are getting less common, but the semi-natural ponds, if well integrated in the landscape, could be a pleasant place for recreational fishing and also an area where a high degree of biodiversity is maintained.

To preserve the high levels of biodiversity, fish farming activities must not reduce the carrying capacity of the estuary by competing excessively for space, dramatically altering the refuge areas or significantly reducing feeding opportunities, either directly (e.g., fencing all ponds irrespectively of the predation risk) or indirectly (e.g., disturbance). Moreover, persecution and illegal killing are not acceptable in terms of the prevailing legislation, and, in general, are not effective and desirable practices to control wildlife populations.

On the other hand, to allow the sustainable development of fish farming activities, a, so far inexistent, management plan must be developed to: (1) define the maximum area that can be devoted to exploitation; (2) provide the rules of establishment including the need of accurate stocking quantities and compensation measures; (3) give guidelines for pond construction and maintenance; and (4) inform about allowed mitigation measures to avoid stock predation.

Examples of possible guidelines refer to the need of quality control of the ponds' effluents and of ways to maintain pond margins as naturalized as possible but simultaneously avoiding excessive predation (e.g., by electric fencing). The initiative, taken by the nature reserve authority, of obliging the largest fish farms to leave a pond for bird use, therefore kept with low depth and renewing water, is of course a compensation measure with clear benefits in terms of bird biodiversity. However, fish farmers must be informed of its usefulness, namely that they may also have something to gain with it (e.g., bird watching may be a complementary income source that could be promoted and organized).

Small commercial fish species are an easy prey, not only for otters but also for many fish-eating birds (beyond egrets there are also kingfisher, gulls, terns and, seasonally, cormorants), since they tend to occupy the shallow areas in ponds, becoming more vulnerable to predation. Therefore, in the juvenile phase, fish need special attention from fish farmers that should keep a high water level in the pond and solutions should be looked for to prevent birds from settling at the margins. Location of ponds inside the fish farm is also another factor to take into account, since it influences the risk of predation. Peripheral ponds are visited more frequently by predators, so keeping juveniles in the interior ponds can contribute to reduce predation.

Predation is just one of the factors that may have a negative impact on fish farming activity. There are other factors that were not investigated in the project. For a full understanding of the role of predation induced damages relatively to other possibly interacting factors, fish farmers should keep updated records of losses due to the evasion of commercial species, disease, and temperature instability, since they may influence the total damage. Moreover, fish farmers should regularly control pond water quality (e.g., pH, oxygen level) and fish health condition, besides counting the number of fish-eating birds (like *Ardea cinerea*, and *Phalacrocorax carbo*) and of carnivorous fish species, such as eels (*Anguilla anguilla*) and sea bass.

## *10.3 Ecological Research and Monitoring*

Our project represented a step ahead in the understanding of the role of otter predation in the human-wildlife conflict existing in the Sado estuary. However, time and scope limitations do not allow the full comprehension of the nature and scale of the conflict. Otters are one of the elements of the conflict, but more research and monitoring is needed to promote sustainable solutions.

Research is still needed to obtain data on otter population dynamics, as such data do not exist in Portugal, nor in the estuary, or in the remaining freshwater basins. In the short term, research should be concentrated on obtaining more reliable data on otter density in the Sado basin as a starting point for future monitoring. It is also important to understand to which degree fish farming activities in the estuary are acting as attraction points to otters and increasing the carrying capacity of the Sado river basin as a whole; this means an urgent need of evaluating the spacing patterns of otters using radio-tracking techniques. The long-term effects of several potential changes occurring in the estuary (e.g., increase in the number of fish farms) as well as in the remaining areas of the Sado river basin (e.g., decrease in rainfall, longer and hotter summers, increase in water pollution due, mainly, to agriculture) on otter population dynamics are still unknown and should therefore be further investigated.

Besides otters, several fish-eating birds are attracted to fish farms and also cause damage to fish production. To fully assess the role of predation on fish losses, including the share of otter predation, a multi-species assessment should be faced as a priority in terms of research.

The physical relation between fish farming ponds and the estuarine environment and the fact that the species stocked in the ponds are also present in the estuary were major constraints in the otter damage assessment. Moreover, inside the ponds it is also possible to find non-commercial benthonic species like European eels and several species of gobies (*Gobius sp.*, *Pomatoschistus sp.*). Different abundances and size-ranges inside and outside the fish farming ponds may be used as good indicators to ascertain the origin of the predated individuals. However, this information could not be obtained for all species involved, especially for the individuals inside ponds. To estimate the damage caused by otters on the stocks of commercial fish, which was an important objective, it was considered that all individuals of those species eaten by otters were caught inside ponds, a conservative scenario from the point of damage assessment that most probably is not realistic.

Future studies need also to concentrate on the fish community in the estuary to assess the availability, and not just the abundance, of alternative prey to fish-eating predators, as well as their size range. Special attention should be given to shallow zones, i.e. to the inner estuarine areas, where predation is more efficient and some fish species, such as soles, are more abundant (Cabral 2000). It is relevant to obtain accurate information in the fish farms about stocking composition, densities, and size ranges. A major challenge is to convince the fish farmers of the usefulness of such information.

The finding that otters are regularly visiting fish farms and that they apparently are not going there just for feeding purposes (visiting rates do not correlate with diet results) is a puzzling result of the project. A combination of radio-tracking and video surveillance techniques could be a useful tool to study otter behavior inside fish farms and indirect damage, such as fish disturbance. More research is also needed to improve the exclusion of predators by non-lethal methods.

The design of a predation risk map of Sado estuary, if possible involving not only the otter but other predators as well, would be informative to fish farmers and the nature reserve administration, helping to decide about the adequate mitigation measures or compensation schemes (either *ex post* or *ex ante*—Schwerdtner and Gruber 2007) or to give guidelines during site selection for the installation of new fish farming facilities.

Monitoring programs of fish-eating predators will allow the analysis of the population trends in a long-term and the establishment of scenarios according to changes induced in the landscape.

### *10.4 Policy Instruments*

To identify the potential for the application of new policy instruments for the Sado estuary case, a participatory conflict reconciliation process was initiated as part of the FRAP project. This has resulted in the development of a collaborative strategy, involving fish-farmers, the nature reserve administration, and the municipality.

## 11 Platforms for Participation

The participatory conflict reconciliation process initiated has combined the use of formal participation techniques—consultation workshops—with an informal approach of information gathering and building of trust, based on individual meetings. This process was important to understand the relation between the ecological and socio-economic dimensions of the conflict. It also allowed the inclusion of the actors' perceptions, perspectives, and knowledge in the policy design.

It was interesting to notice, throughout the process that was implemented in the Sado estuary, how the perceptions and views of the different stakeholders have changed during the interaction. For instance, fish farmers have changed their perception of the relations between their activity and nature conservation and have acknowledged the importance of adopting a collaborative approach.

Another striking fact is that the participatory process is still ongoing, even though the FRAP project has finished. The actors acknowledge that there is still a need for a participation platform to support ongoing efforts for conflict mitigation. This platform, that has started in a rather *ad-hoc* way through the FRAP research, may (hopefully) evolve into an institutionalized participation forum to continuously address the main issues related to biodiversity conservation and fish-farming in the Sado estuary.

## 12 Increase of Aquaculture Products' Value and Quality

During the participatory process, fish farmers in the Sado estuary have recognized that they are in a difficult economic situation and that their survival depends on the capacity to differentiate their products, gaining competitive advantages in the market. In this context of economic difficulties, it is more difficult for the fish farmers to understand the need to preserve otters, which they feel as a threat to their economic activity and as causing economic losses. Having this in mind, as well as the need to conciliate this economic activity with conservation, there is a good opportunity for the creation of a scheme aimed at promoting the ecological and economic sustainability of fish farming in the Sado estuary.

One of the policy instruments that has been appointed as more promising during the course of the participatory process was the creation of a "sustainable production" fish labeling scheme, to be implemented on a voluntary basis and managed by the fish-farmers. To join the scheme, and to be able to display the sustainable production logo on their products, fish farms have to comply with a set of environmental and product quality assurance criteria.

Currently, the work in this instrument is ongoing with the collaboration of the Setúbal municipality that is providing support in the definition of the legal format for the fish farmers association and is actively involved in the promotion of a solution to the development of a fish packaging unit, which is essential for the implementation of a certification scheme to assure product traceability.

The successful implementation of this instrument will require a cooperation approach among the fish farmers, and between them and the nature reserve administration, what can be considered as an important condition to guarantee a sustainable economic sector and the conservation of biodiversity in the Sado area.

## 13 Communication and Publicity

Otters, like dolphins, have a popular public image in the Sado estuary. Environmental education can contribute to change the attitude towards otters. If fish farmers consider otter protection as a setback in fish production and management, the mere aim of protection will be put at risk. Ideally fish farmers should be able to value the existence of the species even if mitigation measures have to be applied to prevent it from causing damages. This can be achieved when certification schemes associated to environmental issues are implemented, as mentioned above, or by increasing the potential for the development of new activities as a result of the presence of a high biodiversity (e.g., eco-tourism).

Above all, and especially when there are no ways to add value to economic activities from species conservation, the imposition of restrictive measures to fish farm management should always be very clear, scientifically supported, knowledge wise, and firmly and explicitly stated. The imposed rules must not be questionable or varying in time and space without any reasoning.

The dissemination of scientific results to the local community should be supported by the authorities closest to fish farmers. It would be a good opportunity for the Sado nature reserve administration to prove that they are not only worried with the predators' survival but also interested in controlling the damage caused by predation and therefore that they are pledged in the resolution of the conflict.

Otters are not the only biodiversity element menaced in estuaries. Illegal fishing in the estuaries, among many other factors, is an additional cause of fish population decline (Costa and Cabral 1999), affecting especially juveniles. The Sado estuary is not an exception and despite being closely controlled, this activity is a common practice. So, a different approach should be attempted, namely promoting information campaigns about the importance of conserving the areas of juvenile fish' concentration, since an increased fish abundance will not only benefit otters, but also the fishermen that use the estuary as a way of subsistence or just for recreation.Tracks from the Eurasian otter (*Lutra lutra*) in a thin layer of snow on a frozen pond. *Photo:* Lukáš Poledník

**Acknowledgments** The authors of the Sado Estuary fish farms chapter are grateful to all the fish farmers: Mr. David Dias Bexiga (Bonita), Mr. Américo Ângelo (Anjo), Mr. Joaquim Modesto (Deus-Assim-O-Quis), Dr. Artur Pires e Eng. Ana Raposo (Pescomex), Dr. Leonídio Alexandre (Bocage e Apóstolos e Monte Cabras), Mr. Silvino Neves (Neves e Longa), Mr. António Neves (Alfinetes), Mr. Elias Mouta (Palma), Mr. António Pereira (Musgos), Mr. António Reis Santos; Mr. José Garrelhas (Turino); Mr.Mário Vinhas/Eng. Saraiva/Mr. Fernando Camilo (Zambujal), Mr. Reinaldo Mendonça/ Mr. António Leal (Sapalsado - Soc. Aquícola do Sado), Dr. Henrique Simas (Esteiros do Sado), Eng. Manuel Serra and Mr. Valdemar (fisherman). We would also like to thank the Sado Estuary Nature Reserve staff and to all the colleagues that helped with valuable information and advice: Dr. Ana Neves, Dr. Mário Boieiro and Mr. Carlos, Dr. Henrique Cabral, Dr. Filipe Ribeiro and Dr. Rui Rebelo. This work was financed by the EU 5th Framework Program (5th FP) Project "FRAP" (Development of a procedural Framework for Action Plans to Reconcile conflicts between large vertebrate conservation and the use of biological resources: fisheries and fisheating vertebrates as a model case), contract number EVK 2-CT-2002-00142-FRAP.

## References

Amorim A (1982) Contribuição ao estudo da ictiofauna pelágica do rio Sado. Graduation thesis. Faculdade de Ciências da Universidade de Lisboa, Lisboa, Portugal

Barrow CJ (2002) Evaluating the social impacts of environmental change and the environmental impacts of social change: an introductory review of social impact assessment. Environ Stud 59:185–195

Becker DR, Harris CC, McLaughlin WJ, Nielsen EA (2003) A participatory approach to social impact assessment: the interactive community. Environ Impact Assess Rev 23:367–382

Beja PR (1989) Coastal otters in southern Portugal. IUCN/SSC Otter Spec Group Bull 4:2–7

Beja PR (1991) Diet of otters (*Lutra lutra* L.) in closely associated freshwater, brackish and marine habitats in southwest Portugal. J Zool 225:141–152

Beja PR (1992) Effects of freshwater availability on the summer distribution of otters *Lutra lutra* in the southwest coast of Portugal. Ecography 15:273–278

Cabral HN (1999) Ictiofauna do estuário do Sado. Relatórios Científicos e Técnicos do Instituto de Investigação das Pescas e do Mar 47:1–30

Cabral HN (2000) Distribution and abundance patterns of flatfishes in the Sado estuary, Portugal. Estuaries 23:351–358

Cabral MJ, Almeida J, Almeida PR, Delinger T, Ferrand De Almeida N, Oliveira ME, Queiroz AI, Palmeirim JM, Rogado L, Santos-Reis M (eds) (2005) Livro Vermelho dos Vertebrados de Portugal. Instituto de Conservação da Natureza, Lisboa, Portugal

Carss DN (2003) Reducing the conflict between cormorants and fisheries on a pan-European scale (REDCAFE), vol 1: Pan-European Overview. Report of a concerted action of European Union. Centre for Ecology and Hydrology Banchory, Scotland

Chambel I (1997) Situação actual da lontra (*Lutra lutra*) na Reserva Natural do Sapal de Castro Marim e Vila Real de Santo António. Technical Report, Instituto da Conservação da Natureza, Programa Life, Lisboa, Portugal

Costa MJ, Cabral HN (1999) Changes in the Tagus nursery function for commercial fish species: some perspectives for management. Aquat Ecol 33:287–292

Dallas JF, Coxon KE, Sykes T, Chanin PRF, Marshall F, Carss DN, Bacon PJ, Piertney SB, Racey PA (2003) Similar estimates of population genetic composition and sex ratio derived from carcasses and faeces of Eurasian otter *Lutra lutra*. Mol Ecol 12:275–282

Foster-Turley P, Macdonald S, Mason C (1990) Otters: an action plan for their conservation. IUCN Species Survival Commission, Gland Switzerland

Freitas D, Gomes J, Sales-Luís T, Madruga L, Marques C, Baptista G, Rosalino LM, Antunes P, Santos R, Santos-Reis M (2007) Otters and fish farms in the Sado estuary: ecological and socio-economic basis of a conflict. Hydrobiologia 587:51–62

Gomes N (1998) Distribuição e ecologia trófica da lontra (*Lutra lutra*) na Ria de Aveiro. MSc thesis. Faculdade de Ciências da Universidade do Porto, Porto Portugal

Gruber B, Reinking B, Calabrese JM, Kranz A, Poledníková K, Poledník L, Klenke R, Valentin A, Henle K (2008) A new method for estimating visitation rates of cryptic animals via repeated surveys of indirect signs. J Appl Ecol 45:728–735

Henle K, Ring I, Klenke RA, Kranz A, Jepsen N, Rauschmayer F (2013) Introduction. In: Klenke R, Ring I, Kranz A, Jepsen N, Rauschmayer F, Henle K (eds) Human wildlife conflicts in europe—fisheries and fish-eating vertebrates as a model case. Springer, Berlin. doi:10.1007/978-3-540-34789-7_2

IAICGP (Impact Assessment Interorganizational Committee on Guidelines and Principles) (1994) Guidelines and principles for social impact assessement. Environ Impact Assess 12:107–152

ICN (2001) Reserva Natural do Estuário do Sado. Instituto da Conservação da Natureza, Lisboa

INE (2002) Statistics on Fisheries 2001. Instituto Nacional de Estatística, Lisboa, Portugal

Kranz A, Rauschmayer F (2013) Module 1—screening of the conflict. In: Klenke RA, Ring I, Kranz A, Jepsen N, Rauschmayer F, Henle K (eds) Human wildlife conflicts in Europe—fisheries and fish-eating vertebrates as a model case. Springer, Berlin. doi:10.1007/978-3-540-34789-7_2

Lopes da Cunha P (1994) Estrutura e dinâmica da ictiofauna do estuário do Sado. PhD thesis, Universidade de Lisboa, Lisboa, Portugal

MacDonald C (2003) The value of discourse analysis as a methodological tool for understanding a land reform program. Pol Sci 36:151–173

Pedrosa M (2000) Aspectos da ecologia da lontra (*Lutra lutra*) em ambientes aquáticos de Vilamoura (sul de Portugal). Graduation thesis. Faculdade de Ciências da Universidade de Lisboa, Lisboa, Portugal

Sales-Luís T, Freitas D, Santos-Reis M (2009) Key landscape factors for Eurasian otter *Lutra lutra* visiting rates and fish loss in estuarine fish farms. Eur J Wildl Res 55:345–355

Sales-Luis T, Bissonette JA, Santos-Reis M (2012) Conservation of Mediterranean otters: the influence of map scale resolution. Biodivers Conserv 21:2061–2073

Santos R, Antunes P, Baptista G, Mateus P, Madruga L (2006) Stakeholder participation in the design of environmental policy mixes. Ecol Econ 60(1):100–110

Santos R, Antunes P, Ring I (2013) Module 5—regional economics and policy analysis. In: Klenke RA, Ring I, Kranz A, Jepsen N, Rauschmayer F, Henle K (eds) Human wildlife conflicts in Europe-fisheries and fish-eating vertebrates as a model case. Springer, Berlin. doi:10.1007/978-3-540-34789-7_2

Santos-Reis M, Ferreira JP, Pedroso N, Baltazar C, Matos H, Pereira I, Grilo C, Sales-Luís T, Santos MJ, Cândido AT, Sousa I, Rodrigues M (2003) Projectos de Monitorização de Mamíferos. Monitorização de Carnívoros. Relatório Final. 2ª Fase de Monitorização. (Programa de Minimização para o Património Natural). Centro de Biologia Ambiental (FCUL) e Centro de Estudos da Avifauna Ibérica (CEAI), Lisboa, Portugal

Santos-Reis M, Trindade A, Beja PR (1995) Situation et état des recherches sur la loutre au Portugal. Cahiers d'Ethologie 15 (2–3–4):181–194

Santos-Reis M, Freitas D, Sales-Luís T (2007) Man-otter conflict in Portuguese fisheries: introducing an ecological approach. In: Conroy JWH, Yoxon G, Gutleb AC, Ruiz-Olmo J (Eds.). In: Proceedings of the European otter conference "The return of the Otter in Europe—where and how?" J Int Otter Survival Fund, No 2. Broadford, Isle of Skye, Scotland

Schwerdtner K, Gruber B (2007) A conceptual framework for damage compensation schemes. Biol Conserv 134(3):354–360

SNPRCN (1990) Livro Vermelho dos Vertebrados Terrestres de Portugal, vol I: Anfíbios, Répteis, Aves e Mamíferos. Ministério do Ambiente e Recursos Naturais, Lisboa, Portugal

Trigo MI (1994) Predação por lontra (*Lutra lutra* Linnaeus, 1758) em pisciculturas do estuário do Mira. Graduation thesis, Faculdade de Ciências da Universidade de Lisboa, Lisboa, Portugal

Trindade A (1991) Fish farming and otters in Portugal. IUCN/SSC Otter Spec Group Bull 6:7–9

Trindade A, Farinha N, Florêncio E (1998) A distribuição da lontra ( *Lutra lutra*) em Portugal: situação em 1995. Instituto da Conservação da Natureza/Divisão de Espécies Protegidas/ Programa Life, Lisboa, Portugal

Wood L, Kroger R (2000) Doing discourse analysis: methods for studying action in talk and text. Sage Publications, Inc. Thousand Oaks, California, USA

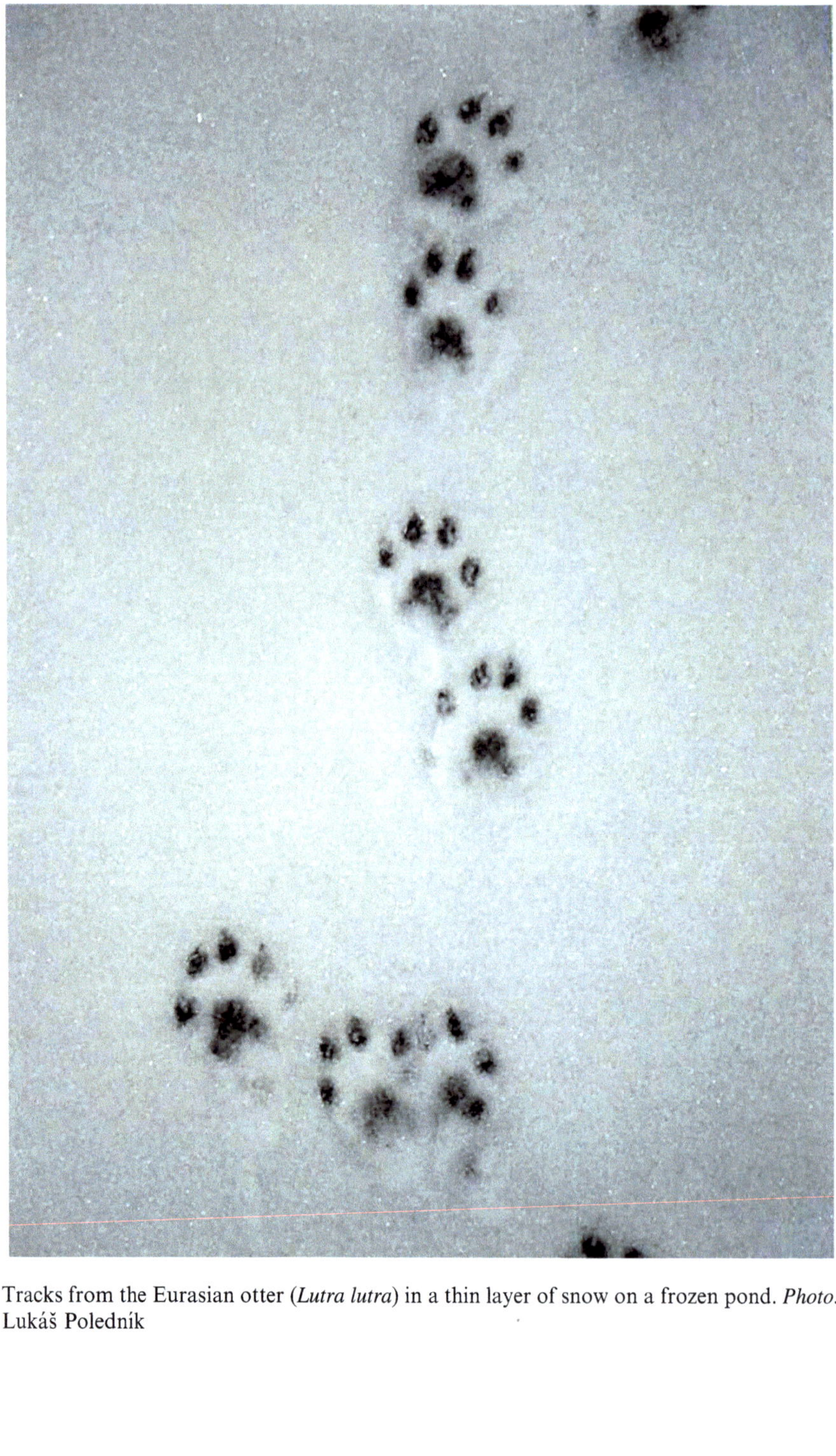

Tracks from the Eurasian otter (*Lutra lutra*) in a thin layer of snow on a frozen pond. *Photo:* Lukáš Poledník

# Otters Causing Conflicts

## The Fish Farming Case of the Czech Republic

**Kateřina Poledníková, Andreas Kranz, Lukáš Poledník and Jaroslav Myšiak**

**Abstract** In the Czech Republic fish farming is common throughout the country with a stronghold of carp production in southern Bohemia. The conflict arising from otter predation on commercial fish rapidly increased due to the political and social changes since 1989. Nature protection authorities undertook a set of measurements to appease the conflict. It included compensation schemes for losses, public relations etc., however, the conflict has continued. Within the FRAP project, social and ecological research was carried out and provided further recommendations how to mitigate the conflict. These are a continuous assessment of the compensation scheme, a simplification, differentiation and decentralization of the compensation payments, joint data collection, and setting up an organization of small pond farmers, a better involvement of stakeholders in conflict mitigation and further ecological research to better understand the biological background of the conflict.

K. Poledníková (✉) · A. Kranz · L. Poledník
Institute of Wildlife Biology and Game Management,
University of Natural Resources and Life Sciences Vienna,
Gregor Mendel Street 33, 1180 Vienna, Austria
e-mail: katerina.polednikova@alkawildlife.eu

A. Kranz
e-mail: andreas.kranz@aon.at

L. Poledník
e-mail: lukas.polednik@alkawildlife.eu

J. Myšiak
Department of Economics, UFZ—Helmholtz Centre for Environmental Research,
Permoserstraße 15, 04318 Leipzig, Germany
e-mail: jaroslav.mysiak@cmcc.it

R. A. Klenke et al. (eds.), *Human-Wildlife Conflicts in Europe*,
Environmental Science and Engineering, DOI: 10.1007/978-3-540-34789-7_5,

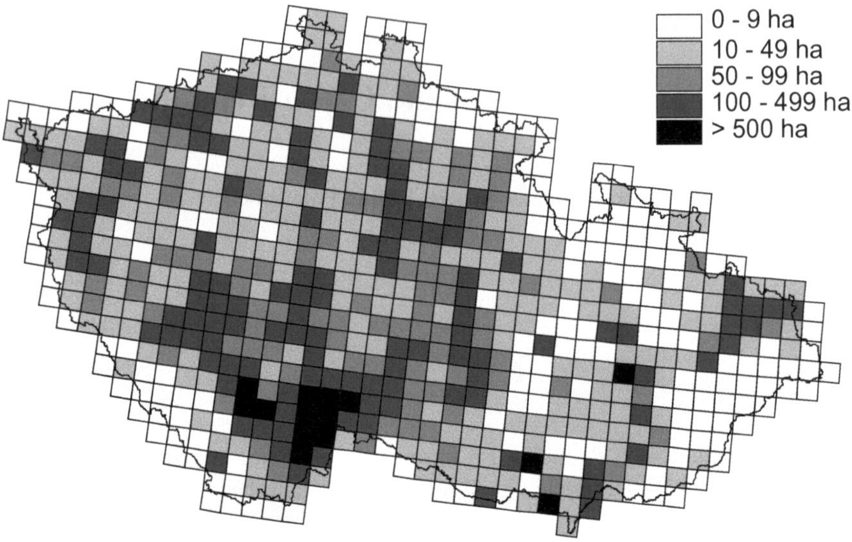


**Fig. 1** Amount of fish ponds in the Czech Republic expressed in surface area (hectares). Size of grid: 11 × 12 km

# 1 Introduction

Conflicts arising from otter (*Lutra lutra*) predation on commercial fish are nowadays a common phenomenon in many Central European countries (Kranz 2000; Kloskowski 2005; Kranz et al. 2007). In this publication we focus on the situation in the Czech Republic, where fish farming is particularly common in the central southern part of the country (Fig. 1). The case of the Czech Republic allows insights into a conflict with special reference to the political and social changes in the past 20 years. This includes the transformation from communism to a market based economy and the EU accession in 2004.

The country is famous for producing common carp (*Cyprinus carpio*) in ponds with extensive management (Box 1). These ponds are part of the cultural landscape and at the same time are of outstanding significance for wetland biodiversity of the country, where large-scale transformations of natural wetlands into agricultural and forestry land have occurred in particular in the course of the twentieth century.

Traditional land use, particularly fish farming, creates habitats and supports biodiversity, but its further development may have negative effects on biodiversity because conservation of some species living and thriving in these habitats may cause economic burdens for the land user.

**Box 1** Fish Farming in the Czech Republic

Czech Republic has an old tradition of growing carp in ponds that goes back to the thirteenth century. The maximum expansion of fish ponds, with about 1,800 km$^2$ of water surface, occurred during the sixteenth century. Currently, there are more than 50,000 ponds with a total area of about 520 km$^2$. Carp is farmed for human consumption within the country and abroad. Most families eat carp as traditional Christmas meal, similar to turkey or geese in other countries, and therefore the carp also has a cultural and emotional dimension for people.

Ponds are artificial water bodies (up to 2 m deep), which may vary considerably in size, ranging from less than 0.1 ha to about 700 ha and are usually scattered across the landscape according to prevailing natural water supplies, such as streams and rivers. In many cases they are organized in cascades of ponds, which form clusters of water bodies. Old ponds established decades and centuries ago usually have natural banks, providing cover and dens for otters, as well as habitat for otter prey other than commercial fish (Fig. 2). During the last three decades many new ponds were built and old ones were reactivated. They also provide suitable habitat for otters and this development may have increased the carrying capacity of otters considerably.

Ponds are usually stocked and harvested in spring and autumn. During the winter, some ponds (23 %) remain empty, while both juvenile and one year old carp and carp not yet sold are kept in special ponds for over wintering. After winter they are redistributed to other ponds for growing. The productivity of ponds varies between 300 and about 600 kg/10,000 m$^2$.

Differences in productivity reflect the two contrasting carp farming areas of the country: the lowlands along the River Lužnice in the south of the country (Třeboň Basin Biosphere Reserve, the region of South Bohemia) and the highlands found northeast of Třeboň (Czech-Moravian Highlands, the Vysočina Region). In both areas otters are present and cause conflict, though with different accents. In the highlands, carp farming is rather suboptimal due to climatic reasons and losses of fish stock are more common (Kranz 2000). Geomorphological aspects are responsible for the prevalence of small sized ponds in the highlands.

Apart from the traditional carp farming, other fish may be reared in these ponds, namely tench (*Tinca tinca*), pike (*Esox lucius*), and pikeperch (*Stizostedion lucioperca*).

**Fig. 2** Small pond with diverse vegetation structure at the bank side in South Bohemia (Czech Republic). *Photo:* Lukáš Poledník

There is a long history of fish farmer-otter interactions in the Czech Republic. Otters[1] were regarded as a pest species at fish ponds since the early thirteenth century. With otter hunting methods becoming more sophisticated and with incentives fostering a large-scale reduction of otters (Hell 1980), the species became rare in most of the country by the nineteenth century. Habitat alterations and water pollution accelerated the drastic decline of the Czechoslovakian population of the Eurasian otter in the twentieth century (Kučera 1980). Otters became extinct in large parts of their previous distribution area and rare in traditional core areas, such as the large-scale pond farming areas in southern Bohemia (Baruš and Zejda 1981). Damage caused by otters ceased and otters became a species of conservation interest. In 1947 the otter became fully protected throughout Bohemia and Moravia (Hell 1980).

The otter conflict in the Czech Republic started by the end of the twentieth century, when damages caused by otters increased as a consequence of the

[1] The Eurasian otter is a medium-sized carnivore in the family Mustelidae with a high degree of adaptations for a semi-aquatic life (Kruuk 1995). In context of conflict resolution the following aspects of its biology are crucial: (1) otters are piscivore (= depend on fish as food), (2) they are mainly nocturnal, thus direct observations are only seldom possible and excrements give the main cues (3) they may breed all year round and thus any kind of regulation is highly controversial.

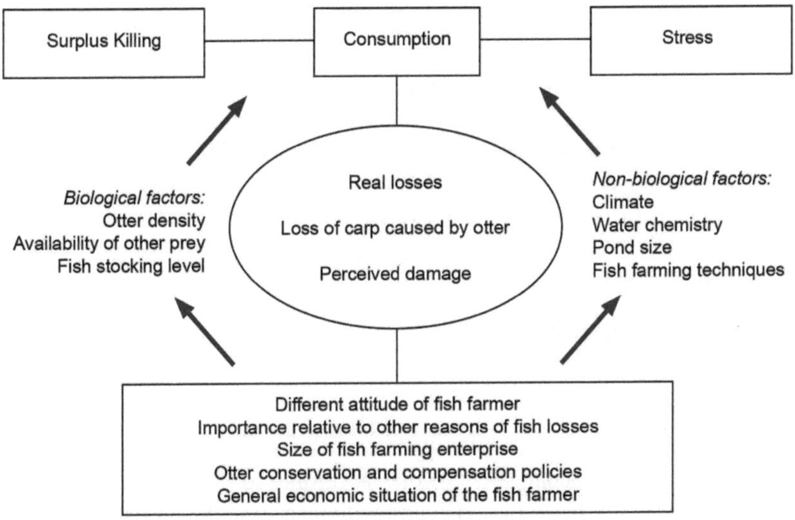


**Fig. 3** Factors responsible for otter damage in carp ponds (modified after Kranz 2000)

recovery of this species (Toman 1992; Kučerová et al. 2001). On one side stands a clear conservation commitment reflected in several laws and a generally high interest in species conservation; on the other side stands property rights to use the land and to reduce losses. The conflict got a special drive due to the change in ownership structure of fish farming enterprises from public to private (Toman 1998), and escalated.

In consequence, the number of otters killed illegally in South Bohemia was estimated to exceed 100 individuals annually (Kranz et al. 1998). For conflict resolution, it was recommended that studies on the following topics should serve as fundamentals for political decisions: (1) clarification of ecological questions, in particular causes of mortality of carp and quantification of otter numbers, and (2) a human dimensions study.

Instead of investing in these open questions, a compensation law was put in force in 2000. Various stakeholders are—not surprisingly—not satisfied with this law and its implementation (Moravcová 2002; Culková 2004). Within this context, the FRAP project focused on the following questions: damage quantification, indirect losses, otter abundance, habitat factors influencing the extent of damage, spatial distribution of damage, and testing known as well as new ecological mitigation devices. With respect to social context the legal and institutional framework were analyzed to understand better legal possibilities and limitations, in particular the compensation system and its implementation. Also an analysis of stakeholder perception was done. To a lesser extent the economic role of the fish farming sector was studied, and the same holds true for policy analysis (past and present implementation of instruments used in the conflict) and participatory decision strategies. Concerns of river anglers were excluded due to project resource limitations.

This dual approach (ecology and human dimensions) takes into account that damage is only possible to occur in context with humans. A prerequisite for damage is the existence of a resource competed for by wildlife and humans (Bath 2005). Factors responsible for otter predation and its perception as damage in carp ponds are schematically described in Fig. 3.

# 2 Human Dimensions

## *2.1 Legal Context*

The otter is a severely endangered species according to Decree 395/1992, which implements the central piece of legislation with respect to the protection of nature and the landscape, the Czech Act No. 114/1992. The Act transposes the provisions of the Bern Convention and the EU Habitats Directive into national legislation. It prohibits catching, killing, and disturbing the listed species, as well as damaging and disturbing their habitats. The otter is also listed as a protected species with an all-year-round closed hunting season under the national hunting law (Act No. 449/2001 and Decree 245/2002). Derogation to the strict protection status is possible in principle, but only if there is no satisfactory alternative and the derogation is not detrimental to the maintenance of the populations of the species concerned at a favorable conservation status in their natural range[2]. The derogation is issued by the Czech Ministry of Environment (MoE) in the case of critically and severely endangered species, and by the regional administration in case of endangered species. In practice, no derogation or exception has been granted so far, although similar provisions have been hypothesized[3]. Damage caused by protected species, including the otter, is compensated according to Act No. 115/2000 (Box 2).

Municipalities and regions are territorial self-administrative units with specific state-delegated functions in the field of nature protection. When executing them, such as in the case of damage compensations caused by protected species, they are bound both by law and by central government decisions and guidelines. The Municipalities Act and the Act on District Offices regulate the relationship between local governments and territorial public administration. The regions correspond to the NUTS[4] 3 level while the municipalities build the NUTS 5 level.

---

[2] The derogations and conditions under which they apply are listed in Article 16 of the Habitats Directive (Council Directive 92/43/EEC of 21 May 1992 on the conservation of natural habitats and of wild fauna and flora).

[3] Personal communication with Hlaváč V. and Roche, M., Toman, A.

[4] The Nomenclature of Territorial Units for Statistics (NUTS) of the EU.

## 2.2 Economic Relevance

The Czech Republic belongs to countries with the highest carp production among the members of the Federation of European Aquaculture Producers (FEAP) and it is the second largest carp producer among the EU member states. With a production of about 17,000 tons the Czech Republic makes up about one fourth of the EU carp production (Brožová 2005). The overall fish production in ponds amounts to about 19,000 tons of market-size fish per year, with no signs of increase or decrease during the last 15 years. About 96 % of market fish are produced in fish ponds. The average production per hectare of pond area is about 450 kg (Brožová 2005). Common carp dominates the production (about 87 %), followed by salmonids (about 5 %), herbivore fish (about 3.3 %) and tench (*Tinca tinca*) (2 %). About 43–44 % of the produced fish is exported, mainly within the European Union.

The Region of South Bohemia (10,000 $km^2$) and the Vysočina Region (7,000 $km^2$) comprise almost one half of the fish farming area of the country, and it is mainly here that the conflict flourishes. Both regions are rural areas, South Bohemia with 52 inhabitants/$km^2$ and Vysočina with 90 inhabitants/$km^2$ and with a gross domestic product per capita 5 % below the average of the whole country. The population slightly increased in the region of South Bohemia (0.9 %) and slightly decreased in the Vysočina Region (0.4 %) from 1995 to 2002. Unemployment is 5.2 and 5.3 %, respectively, significantly below the figure for the whole country (9.7 %). In the region of South Bohemia, there are 131 companies officially registered as fish producers and many more hobby and part-time farmers. From the 131 companies, less than ten have more than ten employees and just one has over 100 employees (Fig. 4).

## 2.3 Stakeholders

Fish farmers, anglers, environmentalists, administrative authorities, scientific experts, and the laymen community are the main actors in the conflict when narrowed down to the disputes concerning the damage instigated by otters. The groups are far from being homogeneous; some members of the same group hold more radical views on how the damage should be prevented or compensated.

Professional fish farmers who make their living on fish (mostly carp) are congregated in large and medium-sized farms widespread in the lowland part of the region. Small size farms owned by hobby or part-time farmers mostly consists of a few small ponds scattered around a given water course. Many of these smaller ponds are situated in the upland areas with suboptimal conditions (see Box 1) and less suitable for commercial pond farming. Although the Czech Fish Farmers Association is open to all farmers it assembles predominately large and medium-sized farms from all over the country.

**Fig. 4** Fishermen in southern Moravia harvesting carp (*Cyprinus carpio*) after pond draining in autumn. *Photo:* Lukáš Poledník

The Czech Fishing Union (CFU) and Moravian Fishing Union (MFU) represent together more than 300,000 anglers. The unions are organized in local and regional groups, coordinated by a central board. Angling is practiced in running waters and selected ponds. Apart from angling many local groups posses or rent fish ponds for fish production (mainly carp). Produced fish are then stocked into the angling waters or sold at local market.

In terms of administration, the Ministry of Environment (MoE) has a central position. It has the responsibility to declare decrees that refine laws and has responsibility for derogation to otters. Local and regional authorities are involved in the conflict by implementing the compensation law (Box 2).

The Agency for Nature Conservation and Landscape Protection of the Czech Republic (ANCLP) as a deliberative and executive organ of MoE defines the protection status of species, prepares management programs, expert reports etc. Moreover, in 1988 ANCLP established a Station of Fauna Protection (SFP) dedicated mainly to captive breeding of otters, rehabilitation of orphan or injured individuals, research and public relations. ANCLP, with experts from NGOs and research institutes, prepared the Management Program for the Eurasian otter in the Czech Republic for the period 2009–2018 (Poledník et al. 2005, 2007). One of the goals of the document is to establish a consultative board of representatives from all stakeholder groups involved in the conflict concerning otters. The Management Program was submitted for approval by the Ministry of Environment by the end of 2005.

Besides nature protection authorities, the Ministry for Agriculture (MoA) is a key player in the conflict. Fish farming and aquaculture belong to agriculture and fish farmers benefit from subsidies managed by the MoA.

Among the non-governmental organizations engaged in otter protection and public awareness, the Czech Otter Foundation (COF) is most distinctive. Founded in 1993, it has to date organized workshops and exhibitions, published books and educational materials for schools, and carried out research and monitoring of otter populations. The foundation provides expert reports required by the compensation law 115/2000 in South Bohemia and beyond.

Finally, various institutions of secondary and higher education, as well as research institutions, are worth mentioning: the Fishery Vocational Schools in Třeboň, Palacký University in Olomouc, Masaryk University in Brno, University of South Bohemia in České Budějovice, Charles University in Prague, University of Applied Life Sciences in Vienna (Austria), and Research Institute of Fish Culture and Hydrobiology in Vodňany.

### *2.4 Damage Compensation*

Currently, only one instrument to compensate otter damage or to assure or stimulate tolerance of otters exists in the Czech Republic (Box 2). Although the paid compensations are recorded, we are not aware of any assessment of the scheme, neither in terms of the number of farmers who adhered to the scheme, nor with respect to the paid compensations and the certified damage. Therefore, we analyzed the expert reports obligatory for the damage claims. Most of the reports are provided by COF and ANCLP.

Not all damage claims have been recognized as legitimate by the regional authority. Most claims refer to the South Bohemia region. Only a small number (up to 200) of potentially eligible beneficiaries already applied for the compensations.

The amount of money paid for otter compensations in 2005, 2006, 2007 was around 200,000 € per year. In the region of South Bohemia, the damage paid per pond averaged 74 € (minimum 15 €, maximum 962 €), which corresponds to 63 otter visits per 180 days (minimum 13, maximum 420).

The biggest part of claims was lodged by the largest group of hobby farmers (60 % of damage claims), followed by professional fishery companies (20 %), fishing unions (10 %), and by other companies (10 %), in which fish farming is not the main activity (e.g., hunting association, farmers).

**Box 2** The Current Otter Damage Compensation Scheme

The compensation scheme (Act No. 115/2000) for damages caused by protected species, including the otter, has been introduced in 2000 and subsequently changed in 2001, 2002, and 2006. The law covers damages caused on fish stock in ponds and water courses. It determines the right for compensation, thus there are no upper limits or ceiling as for the extent of compensated damage. The compensation takes into account the intensity of otter presence in the pond or water course. The claimants have to report the damage within 48 h to the competent local authority, which inspects the fish pond and confirms the presence of otters. An expert is required to assess the extent of damage; the expert report is an obligatory part of the claim. There is no special provision as who can count as an expert for the purposes of the law, generally the expert reports done by the Czech Otter Foundation and the Agency for Nature Conservation and Landscape Protection of the Czech Republic are accepted by the authorities. The damage claims must be submitted to the responsible regional authority up to 10 days after the claimant got to know the damage or at latest up to six months after the damage occurred.

The methodology applied to assess the extent of damage was developed by the Czech Otter Foundation Fund and the Agency for Nature Conservation and Landscape Protection of the Czech Republic (Roche and Toman 2003). It distinguishes between a detailed and a simplified assessment. The detailed assessment is based on regular monitoring of water quality, climate factors, fish diseases, and presence of other fish predators, but has not yet been applied due to the high costs associated. The simplified technique is a very rough expert guess. The amount of otter field signs, size and stocking of the pond, expected commercial fish consumption by otter, and market price of farmed fish are taken into account. The following equation is used to calculate the amount of compensation: $Z = c \cdot p \cdot n \cdot d$ (Z: compensation; $c$: average price of fish stocked; $p$: coefficient of diet composition (it is assumed that an otter consumes between 0.5 and 0.75 kg of commercial fish per day); $n$: number of otters using the pond; $d$: number of days of otter presence). The parameters $n$ and $d$ are based on otters signs (tracks, spraints, food remains) found at the particular fish pond or, in the case of fish pond networks, on an estimation of otter numbers found by snow tracking. In case of a single pond or a small complex of ponds, the pond area is considered. The damage is assumed higher in small ponds; therefore, the estimated damage is increased by 20 % in case of ponds smaller than 2 ha and decreased by 20 – 50 % in case of ponds larger than 5 ha. Damage assessment covers only fish actually eaten by the otter, not secondary damages caused by injuring or stressing the fish in winter.

## 2.5 Stakeholders' Perceptions of the Conflict

Various analyses have been carried out in the past to assess the extent and identify the main drivers of the conflict. According to Kranz (2000), fish farmers in uplands perceived the otter as the greatest threat, whereas in lowland regions other fish predators, especially great cormorants (*Phalacrocorax carbo sinensis*), were more feared. With respect to the type of damage, secondary losses caused by stress due to otter predation were ranked highest (50 % of answers), followed by surplus killing (27 %); direct consumption ranked last. Regarding the measures to prevent and compensate for large-scale losses, the reduction of otter population and introduction of a damage compensation scheme[5] were recommended most often (Roche 2003). More than 60 % of the interviewed persons believed that the damage caused by otter has steadily increased during the last years and that the damage is between 5 and 30 % of the produced fish (Roche 2003). These perceptions varied considerably across the surveyed administrative districts and the groups of respondents (anglers and fish farmers). The same survey revealed that the current damage compensation payments in South Bohemia are well known, but about 40 % of the people interviewed did not believe that this scheme provides a solution to the conflict. The highest discontent with the damage compensation scheme was reported among the owners of small fish ponds. In Spurný et al. (2003) the acceptance of piscivorous protected species among the anglers has been investigated, revealing a rather high acceptance of the otter (55 %), while the discontentment with the protection of cormorant is reported as being high (75 %). Additionally, Novotná (1998) reports that many respondents confused the otter with mink (*Neovison vison*), which may suggest an overestimation of the damage caused by otters.

We conducted 20 semi-structured interviews involving about 35–40 people[6] with representatives of the main groups presented in the previous section. The interviews, each about one and a half hours long, were recorded, transcribed, and analyzed qualitatively. Here we include a summary of the main features of the discourses, especially ones that are relevant for the policy recommendations discussed later in this document. The interviews were complemented by an extensive review of scientific and media articles.

The actors hold divergent beliefs about the population size and extent of damage occurring, blame various drivers for the increase of damage, and propose different policies to prevent or reduce the potential harm. However, they agree that the ponds give rise to a distinctive landscape worth preserving and conserving thanks to the tradition of pond farming. Yet fish farmers and environmentalists refer to the unique cultural and ecological value of the region for different reasons and the interpretation of what is a sustainable pond and landscape management is at the core of the conflict.

---

[5] The survey had been carried out before the compensation law 114/2000 was introduced.

[6] On average two persons took part in a single interview.

The fish farmers stress the importance of traditional practices of pond management for maintenance of the typical landscape and stress that a professional pond management is based in first place on economic profitability and long term sustainability of the farms. The emphasis on professionalism preserve the status of fishermen and separate out the hobby farmers and anglers, who are commonly believed little knowledgeable and often foreign to the region. The place identity and local knowledge is frequently pleaded to contest scientific knowledge and the authority of experts. In principle, fishermen see themselves as environmentalists, but ones, which make explicit a trade-off between economical viability of the fish farming practice and ecosystem preservation.

On the other hand, the environmentalists, who encompass mainly officers of state agencies, experts, and the NGOs engaged in protection of otters, stress the value of ponds and their littoral zones as unique ecosystems endangered by intensive farming practices. According to them, the high ecological value of ponds is inversely proportional to the density of the farmed fish, and higher proportion of non-commercial species in the ponds is indispensable both for the overall species diversity and as a measure to reduce damage on commercial fish population. The place identity is also connected to the value of the landscape (and especially the Biosphere Reserve situated in the region) as a refuge for many endangered species.

Another driver of the conflict is the private ownership of the ponds—most of the formerly state-owned ponds were privatized—and the different viewpoints regarding the exercise of property rights. The fishermen sustain that the ponds are functional facilities built for the purpose of fish production. Their high value for conservation is a welcome by-product, but the productive capacity of the pond—the purpose for which the pond has been built—ought not to be compromised. The limitation imposed by environmental laws and directives is challenged as inappropriate (because of violations of the property rights) and not reasonable for the maintenance of the ecosystem (because it is not balanced). The prevailing perception among fish farmers is that restrictions, if imposed by the state, should be accompanied by compensations for foregone benefits. Furthermore, the way environmental constraints have been imposed creates additional tension, in particular the selection of NATURA 2000 sites.

The environmentalists consider the ponds and the ecological values/services provided by them as a common (or community) good and assert the rights and obligation to treat them correspondingly. The compensation is seen with suspicion and many claim that the compensation payments are counterproductive, as they spoil the attitude towards nature protection and reduce the acceptance of wildlife. In their views the losses due to predation by otter are part of the production process.

The damage assessment is complicated by various practical and ethical issues. Both fishermen and anglers assert that all damage attributed to otter ought to be compensated. In this matter, the fishermen and anglers see compensation of secondary damage as legitimate, while the environmentalists do not agree with it.

For fishermen, otters may become a threat to pond farming (and subsequently to the preservation of the pond landscape). The otter is seen as a "killer" able to wipe out a pond within a very short time-period. The killing is not a subject of survival;

the otter is believed to kill for leisure. Less radical fishermen accept the presence of otters as part of nature and as a native species, but still call for protection that would include regulation of the population. This is strictly opposed by environmentalists for two reasons. First, the population is perceived as not strong enough to sustain the illegal killing and species management. Second, female otters can be pregnant and have offspring practically anytime throughout the year.

Larger fish farms are better satisfied with the current compensation schemes than small and hobby farmers. This is believed to be due to the fact that large farms are better able to cope with the transactional costs of the compensation claim and can also exercise considerable power (lobbying) either by themselves or through their representative bodies. Small farmers, on the other hand, feel poorly represented and thus powerless, insufficiently informed and little convinced of the merit of the compensation scheme.

The fishermen and environmentalists engaged in practical management emphasized positive experience from their collaboration. The mistrust and negative attitude were substantially higher towards hobby environmentalists, who were often not further specified. Some fishermen considered them as an instrument of foreign interest groups, aiming to make the Czech farms less competitive. Uncertainty was omnipresent in the discourses. This is partly because the region has experienced fundamental political and economic changes in the last two decades (the transition to market based economy, splitting up of the former federal state into Czech Republic and Slovak Republic, and the EU accession), which were not free of concerns and anxiety. These changes are seen as a threat to the farm practices, partly because the farmers fear loss of control (and subsequently additional restrictions and bureaucratic burden), and partly because of deep mistrust. Suspicions involve the "hidden" motivation of the rules, with which the state had to comply during the EU accession. Once again, the mistrust against the changes was interwoven with the place identity and the imperative to preserve traditional practices.

# 3 Ecology

## *3.1 Distribution and Otter Densities*

The distribution of otter population and its changes in the Czech Republic in the last decades are well know from several national surveys (Toman 1992; Kučerová et al. 2001; Poledník et al. 2007; see Box 3) based on the modified standardized IUCN Otter Specialist Group method. However, the reliability of various published estimates of size of the population (e.g., Kučerová et al. 2001; Brožová 2005) is questionable. Here we provide an estimate of the population size based on a statistical approach (Poledník 2005).

Otter densities were identified by snow tracking of seven 10 × 10 km squares in different parts of the Czech Republic comprising contrasting habitats. Otter numbers

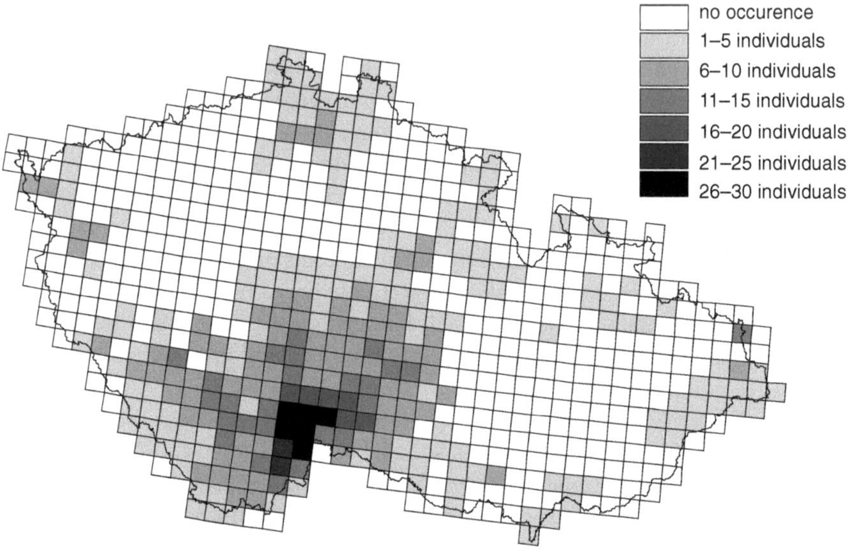


**Fig. 5** Otter density in the Czech Republic in 2000 according to the national survey in 2000 (Kučerová et al. 2001), known densities based on snow tracking and the relationship between otter numbers and pond

varied highly among snow-tracked squares: from 1 to 28 adults. Subsequently, the densities were plotted against various parameters describing aquatic habitats. Total length of pond banks explained most of the variation of adult otter numbers (backward multiple regression: $R^2 = 0.9742$). Based on the relationship between otter density and length of pond bank, the total population size of otters in the Czech Republic was estimated to be between 1,600 and 2,200 adult individuals at the end of the twentieth century (Fig. 5). The correlation may be used in context with otter numbers and damage quantification at a given area as well. The existing correlation also indicates that otter numbers appear primarily food limited and the carrying capacity has been reached in many areas of the Czech Republic.

**Box 3** Otter Distribution, Abundance and Diet in the Czech Republic

The population of otters was fragmented in the last decades (Toman 1992; Kučerová et al. 2001), with the largest metapopulation occurring in the southwest of the country and expanding its range into Austria and Bavaria (Kranz 1995). This metapopulation also coincides with the main fish farming area. Other otter metapopulations occur in the very east, along the border to Slovakia and Poland, and in the north on the border to Germany. In addition, otters were reintroduced in the 1990s to one mountain range in northern Moravia (Hlaváč et al. 1998). During the last 15 years otters have expanded their range considerably (Poledník et al. 2007).

Population estimates in terms of numbers, based upon comprehensible data, were not available. Therefore, some authors offered figures that should be considered rather as expert guesses: e.g., Kučera (1980) indicated about 330–350 individuals living in the whole country at the end of the 1990s, whereas Kučerová et al. (2001) suggested 800–1,100 individuals. Reliable figures were only available for single plots of 10 × 10 km squares. According to Kučerová and Roche (1999) and Kranz et al. (2002), the average number of otters in a single square amounts to approx. 3 individuals outside fish pond areas and between 15 and up to more than 30 otters in pond areas.
The diet composition of otters in the Czech Republic was intensively studied in the last decade, covering a wide range of otter habitats from mountain streams through lowland rivers to fish ponds (e.g., Knollseisen 1995; Roche 2001; Poledník et al. 2004). Otters showed an opportunistic feeding behavior. The dominant fish species and size in the diet reflected abundance and availability at a given site. Since young fish are more numerous, most of the fish predated upon by otters were smaller than 15 cm. Damage caused by otter predation on fish ponds was considered highly variable from negligible to reaching up to half of the fish stock of a pond (Gossow et al. 1999). Higher predation rate of carp was observed during winter (Gossow et al. 1999; Kučerová and Roche 1999).

## *3.2 Quantification of Losses*

Besides economic parameters, the amount of damage should be a function of duration of otter presence in terms of otter-days presence at a pond and the proportion of commercial fish in the diet. We analyzed these two aspects and, in addition, investigated the potential impact of otters on fish in terms of secondary losses, since this is the central argument of fish farmers. Box 4 gives a short introduction to the methods and problems that occur when assessing the damage at a given pond.

**Box 4** Otter Damage Assessment

At fish farms, damage is usually defined as loss of stocked fish revealed when a pond is drained. Since a fish farming period expands usually over several months, the recorded losses at the moment of pond draining represent damage accumulated during a longer period, usually several months, sometimes years. At this time the causes of damage are not necessarily evident any more. A number of different reasons for losses (other fish predators, fish diseases etc., see also Fig. 3) exist and the otter is only one of them. Therefore, it is far from easy to properly assign the correct amount of damage to the different causes of loss.

The only link between otters and losses is circumstantial evidence: spraints (Fig. 6), tracks, and in some cases food remains found at the pond. It is assumed that if a pond is stocked by fish and otter signs are found on the bank of the pond next to the water, otters were hunting in the pond, and therefore there is a certain probability that damage has occurred in this pond. In cases when otters visited a pond and caused damage to fish stocked there long time ago, it may well happen that no signs of otter presence are visible anymore.

### 3.2.1 Relationship Between Otter Presence and Spraint Number

The relationship between utilization of a site by otter and number of spraints found was repeatedly discussed in the past (e.g., Kruuk and Conroy 1987; Macdonald and Mason 1987) with the result that the number of spraints should be used with caution as an index of otter numbers. A new method to estimate the utilization of a site was developed and tested (Gruber et al. 2008). "Visitation rate" (proportion of nights when at least one otter visits the site) of otters at particular ponds was identified by regular surveys of 55 ponds. These surveys were conducted in weekly intervals. Age (fresh, i.e. from previous night, versus old) of each spraint was recorded and all spraints were collected. The visitation of a particular pond was then calculated based on presence/absence of fresh/old spraints using a maximum likelihood approach. Subsequently, calculated visitation was compared with the number of spraints found at the pond. The comparison revealed a strong correlation between the visitation at each pond and the average number of spraints when counted every week. Weekly intervals for pond surveys are, however, not an option for damage quantification in praxis, since this would be too costly. Further simulations (sequential reduction of the number of surveys) revealed that at least two pond surveys within six months are necessary to get a meaningful correlation between the number of spraints and the visitation rate of otters during the whole period. A single survey during the period of several months does not bring reliable results because the number of spraints deposited by otters on a particular pond strongly varies during that time.

### 3.2.2 Utilization of Ponds by Otters

A high variation in visitation rate among different ponds was found. Some ponds were used intensively and others rather sporadically. On average, visitation rate was 20 %, which is equivalent to visits occurring every fifth day. Every monitored pond, where fish were stocked, was visited by otters to some degree during the vegetation period.

**Fig. 6** Otter spraint (*left*) with remaining fish bones and jelly (*right*), a greenish or brown secret from the anal glands. *Photo:* Lukáš Poledník

Data of four radio tracked otters in the same area provided a very similar visitation rate: 21 %. One otter had 18 ponds in its home range and an otter used on average three ponds per night (Poledník 2005).

Besides pond-to-pond differences in otter presence, spraint surveys and radio telemetry revealed an overall seasonal difference. During the summer period, all ponds are usually stocked and otters use all of them. In contrast, during winter some ponds are without water (about 23 %, Kranz et al. 2002) and most, which are filled with water, are at least for part of the time heavily frozen. Therefore, otters have to concentrate at running waters or at those ponds which are stocked and which provide some access to the water (Fig. 7).

### 3.2.3 Proportion of Commercial Fish in the Diet of Otters

The analysis of 2,265 spraints collected along the ponds surveyed for otter presence revealed a high variation (10–90 %) of commercial fish in otter diet. Amphibians, namely frogs (max. 49 % of diet), European perch, *Perca fluviatilis* (max. 49 %), roach, *Rutilus rutilus* (max. 31 %), and crayfish (max. 40 %) were the main alternative prey groups. Thus, high visitation rates at a particular pond and high numbers of otter signs do not necessarily imply high damage and vice versa.

**Fig. 7** Frozen pond in Austria. An otter has made the narrow pass around the sticks holding the wall of the pond overflow, the only place where open water is still accessible. *Photo:* Reinhard Klenke

### 3.2.4 Otter Impact on Fish Due to Consumption

Based on visitation and diet analyses, otter damage varied among ponds from 0.05 to 2.37 % in terms of stocked fish at 15 regularly surveyed ponds during the vegetation period (May to September). The average damage per pond was 0.9 %.

Based on the known number of otters (12 individuals) within an area of 10 × 10 km with about 120 ponds, an average food consumption of kg per day (Kruuk and Carss 1996), and the proportion of commercial fish in the diet (about 35 %), otters consumed about 630 kg of commercial fish in one vegetation period, which is equivalent to 1.1 % of stocked fish in that area.

Hence, two fundamentally different approaches provided very similar results (0.9 versus 1.1 %) and indicate that the overall impact of otter predation on commercial fish in fish ponds during the vegetation period is very low. This is true in particular when considering the overall level of natural mortality. In this area, fish farmers generally accept losses of up to 10 %.

Data on otter damage at ponds during the winter are insufficient, but the total number of commercial fish consumed by otters at ponds should not differ dramatically from the values calculated for the vegetation period, as the number of otters within the area remains the same. However, as the predation on commercial fish is not evenly distributed over the entire area but rather concentrated at fewer ponds (stocked and accessible), higher damage can be expected at those ponds where otters have access to fish.

#### 3.2.5 Otter Damage Caused by Disturbing Fish

The condition, health, growth, and survival rate of fish may be affected by stress, as frequently argued by fish farmers. Otters predating in fish ponds, especially during winter when fish are mostly inactive, may cause such stress to fish. In the experiments conducted as part of our project, common carp were stressed under controlled conditions by tame otters. The analysis of blood samples of experimental fish showed changes in nitrogen, glycid, and mineral metabolism, as well as levels of hormones and fat reserves in fish disturbed by otters (Poledník et al. 2008). However, these changes had no economic impact, since subsequent survival and growth rate of stressed fish did not decrease. Nevertheless, further experiments are recommended to better understand these complex relationships.

### *3.3 Fishermen and Damage Assessment*

The damage caused by otters assessed by regular monitoring of 55 ponds was compared with losses reported by fishermen. A strong discrepancy between these two data sets was found. There was no correlation between the damage perceived by fishermen and the damage as revealed by diet analysis and visitation rates of otters (Fig. 8).

### *3.4 Factors Influencing Damage*

In the light of a considerable variation of damage occurring at different ponds, we investigated small-scale landscape factors in order to identify those which make otter presence and damage more likely to occur at a given pond. Such findings would be a valuable piece of information for fish farmers and could be integrated into further development of the compensation scheme.

Ecological mitigation is also supposed to have an effect on the amount of damage, though some means may function more as a placebo by have an effect on the perception of fish farmers. Since little information is available on how effective traditional means of mitigation are, some of them were tested in a standardized way.

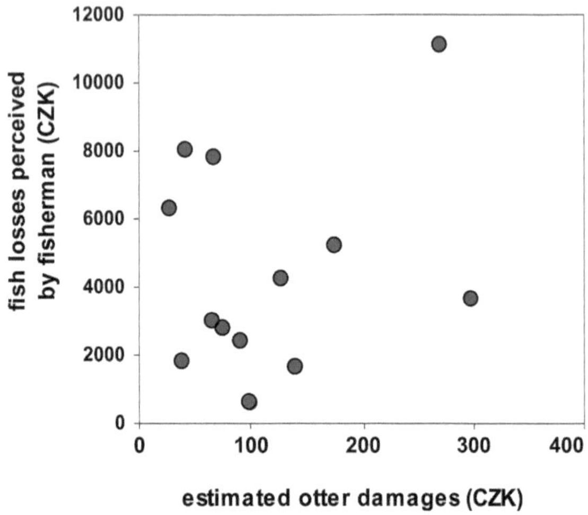


**Fig. 8** Comparison of fish losses perceived by fishermen at 13 different ponds and estimated otter damage revealed by diet analysis and visitation rates there

### 3.4.1 Damage and Small Landscape Factors

Apart from seasonal differences and otter densities, further characteristics of the ponds and their surroundings may have an influence on the extent of damage. Such characteristics may include size of pond, size (age) of carp, the amount of refuge habitat at a given pond and the distance to the nearest river, ditch, and pond. These landscape parameters were analyzed and it was found that their impact on the total damage at ponds was relatively low. Since such factors are complicated to quantify and their effects are interrelated, their use in damage compensation schemes appears problematic and of little added value (Santos-Reis 2006).

### 3.4.2 Evaluation of Existing Ecological Mitigation Measures

Fences, both normal and electric, are an effective means to keep otters from fish farms, if the banks allow otter-proof fencing. This is unfortunately often not the case at centuries old carp ponds (Bodner 1996). Electric fences depend on a proper power supply, may fail under certain weather conditions (wet, cold, high snow layers), and may cause barriers to other protected species, such as amphibians migrating to ponds. Winter ponds may be protected by making use of the ice cover and frost (protective measures at the inflow and outflow are required).

Scaring devices frequently used by fish farmers (fladry consisting of cloth pieces, human hair, and sheep wool) and additionally excrements of large carnivores were tested in paired ponds and were found to be inefficient (Kranz et al. subm).

Killing otters by shooting or trapping is at present an illegal measure. However, some aspects should be mentioned. First, it is necessary to distinguish between single killings, which have no effect on the population but may appease fish farmers, and regular culling, which may reduce the population. Both are ethically problematic, as females with dependent cubs might be killed. Culling would also have an effect on neighboring otter populations outside the fish farming area.

Translocation of otters causes considerable costs and may expose the translocated individual to high risks in the new habitat. It is also limited by the low acceptance of fishermen and anglers in the release areas (Kranz 1999; Conover 2002; Poledník et al. 2005; Hlaváč, pers. com.).

Deviation ponds offering alternative, more readily available food than commercial fish, may work when installed for short periods particularly critical for damage. Otherwise, they raise the carrying capacity of the otter habitat, resulting in more otters without any damage reduction. However, they may work well in combination with incentives for extensive fish farming as in Lower Austria (Bodner 1996).

# 4 Conclusion and Recommendations

## *4.1 Further Ecological Research*

Otter biology and the range of consumption of commercial fish by otters at a single pond are known with sufficient precision. This does not mean, however, that the damage occurring at a particular pond can be determined exactly or nearly exactly. Due to the variation in number, age, and sex[7] of otters visiting the ponds, the frequency of the visits, occurrence of alternative prey species, losses due to diseases, predation by other species, and other biotic and abiotic factors, the damage experienced by a single farmer can only be approximated. The uncertainty due to natural variation can be reduced modestly by more research.

Further research, however, can yield important data regarding the damage in extreme situations, such as surplus killing and injuries inflicted by otter. Finally, research concerning damages due to stress of fish is not sufficiently completed.

## *4.2 Continuous Assessment of the Compensation Scheme*

The effectiveness and acceptance of the compensation schemes in place, Act No. 115/2000, has been scarcely analyzed. Data about the claimed and compensated damage are not systematically collected[8]. A comprehensive assessment is, however, crucial and should include financial costs (payments to fish farmers and transaction costs), analysis of applicants (e.g., whether pond farming is their main

[7] Especially families (females with dependent cubs) do significant damage simply because they are more individuals together and because of learning to catch prey.

[8] The most comprehensive data about the claimed compensations are available in the Czech Otter Foundation Fund, which provided expert reports for most of claims.

occupation, geographical distribution), and stakeholders' perceptions of the compensation system. Similarly, it is important to collect information about all farmers eligible for damage compensation (both commercial and hobby farmers). An inventory of ponds used for carp farming and a statistical survey of all people engaged in pond farming are essential. Furthermore, the performance of various alternative measures (e.g., fencing the ponds, composition of fish stocks) and alternative compensation schemes need to be systematically compared.

## 5 Simplification and Differentiation of the Compensation Payments

Otters repeatedly visit ponds and other water bodies in their territory. In areas with numerous small ponds, such as the Czech-Moravian Highlands, where the otter is common, the assessment of actual damage represents a real challenge (see chapters Distribution and otter densities, Quantification of losses). The losses of fish due to otter predation are small on average, but they occur repetitively and vary in magnitude. The current compensation scheme (Act No. 115/2000) is associated with high transaction costs, especially for small and hobby farmers, and the administrative efforts connected with handling the numerous compensation claims are huge (see Box 2). As explained in previous sections, small and hobby farmers represent the group in which the acceptance of the otter and the success of the compensation payments are not satisfactory.

We propose to tackle these issues as follows:

- The current level of administrative burden to prove the damage should be relaxed. The compensation claims can be submitted to (and dealt with by) a single authority, either at the local or regional level [9]. The certification of otter presence at ponds is not necessary every time when damage occurs. In areas with permanent otter presence, otters are expected to visit each pond in the area. Reliable information about otter presence in the area can be obtained from other sources[10].
- Types of schemes differ widely in transaction costs, which are related to distribution of the damage (Schwerdtner and Gruber 2006). Due to distribution of otter damage, a lump-sum compensation scheme [11] is recommended. Compensation system based on case-by-case damage assessment has, in the case of otters, high transaction costs, which are in addition useless (inspections do not provide reliable information). The lumped-sum compensation scheme could be cheaper and moreover friendlier because it eases the burden of the damage proof.

[9] Currently, both local and regional authorities are involved in the review of the compensation claims, though at different stages (see Box 2).

[10] e.g., the otter surveys carried out for the Ministry of Environment.

[11] Lump-sum compensation is based on an estimation of the expected loss, independent of actual damage.

- The compensation of actual damage can be (at least partly) connected to or replaced by incentives fostering better prevention of damage and encouraging environmentally-responsible management techniques. These incentives can be connected to agri-environmental measures (e.g., Ring and Santos 2006). Examples of their successful employment are known from Saxony and Lower Austria (Thum et al. 2003; Myšiak et al. 2004). In the case of new ponds, the compliance with established standards, sound management practices, and acceptance of predators should be imposed in the permit to construct the pond.[12]
- The current damage compensation based on an extensive proof of the damage can be applied when the fish farmer believes that their actual damage is significantly higher than the lumped payments or the incentives. The evaluation of direct damage is basically a function of otter visitation rate, proportion of commercial fish in the diet of otters, and fish price. Reliable in-depth damage assessment based on these three components requires at least three separate surveys of pond (to collect enough spraints for diet analysis). The costs associated with in-depth damage assessment should be covered at least partly by the fish farmers themselves to prevent the misuse of the instrument.

## *5.1 Decentralization of the Compensations and Incentives*

To address regional differences exacerbating the conflict (see Box 1), the compensation schemes and incentives should be flexible enough to reflect the local actors' concerns and preferences. Such flexibility can be achieved, for example, by giving a discretion to the local authorities to adjust the rules for damage applications (e.g., deadlines for the applications, interval at which the applications are submitted), without compromising the aim of the compensations. In the case of environmental incentives, the discretion can involve setting priorities for environmental targets. Decentralized conflict management should foster attitude of shared responsibility and closer involvement of the relevant actors.

## *5.2 Joint Data Collection*

Currently, data about otter populations and the damage assessment are collected by various actors independently from each other. This practice results in different figures regarding otter population size and amount of damage, which instigates disagreements and conflicts. Joint data collection, including the collective

[12] Presently land use planning does not take the otter into account when new fish farms are built. The otter is not considered as a factor that affects future farming and some people, who build a fish farm, are not even aware that otters live in the area. In the course of land use planning and granting licenses to run a fish farm, mitigation measures and incentives for environmentally friendly management should be suggested respectively be a prerequisite.

definition of the applied methodology, requested precision, and time interval for collecting new data, can foster better communication and mutual understanding of the opposing parties. The joint data collection can be extended to other data necessary to assess the performance of the compensation schemes.

## *5.3 Support for Setting up an Organization/Association Representing the Concerns and Needs of, and Providing Practical Advice to Small and Hobby Pond Farmers*

Although the small and hobby farmers are formally represented in the Fish Farmers Association, their specific situation and needs are not sufficiently conveyed. We propose to establish an organization that will represent the interests of numerous small and hobby farmers. Such an organization will also be practical for dissemination of the relevant information, capacity building, or organization of training courses on how to prevent damage etc.

## *5.4 Involvement of a Wider Range of Stakeholders*

A wide range of stakeholders have been involved in the mitigation of the otter-related conflict in the Czech Republic, but the level of their involvement is far from satisfactory. A better involvement of all affected actors, applicable to almost all previous recommendations, can facilitate positive social responses and bolster legitimacy, acceptance, and satisfaction with the adopted policies.

**Acknowledgments** This work was financed by the EU 5th Framework Program (5th FP) Project "FRAP" (Development of a procedural Framework for Action Plans to Reconcile conflicts between large vertebrate conservation and the use of biological resources: fisheries and fish-eating vertebrates as a model case), contract number EVK 2-CT-2002-00142-FRAP.

# References

Baruš V, Zejda J (1981) The European otter (*Lutra lutra*) in the Czech Socialist Republic. Acta Sci Nat Brno 12:1–41

Bath AJ (2005) Human dimensions as a strategic tool for nature conservation: using examples from large carnivores and large herbivores. HD manual report for the large herbivore foundation (LHF) and the large carnivore Initiative for Europe (LCIE, http://www.lcie.org), pp 63

Bodner M (1996) Fischotter und Teichwirtschaft. Forschungsbericht Fischotter 3, WWF Österreich, vol 14. Vienna, pp 26–44

Brožová M (2005) Situační a výhledová zpráva. Ryby, Ministerstvo zemědělství ČR, Praha

Conover M (2002) Resolving human-wildlife conflicts. The science of wildlife damage management. Lewis Publishers, Boca Raton, p 418

Culková M (2004) Rybáři versus vydra říční. Bachelor work, Faculty of Science, Palacky University, pp 62
Gossow H, Kranz A, Knollseisen M (1999) Ecology of otters in a fish pond area and some of its socio-economic aspects. Final report for the Austrian science foundation, project No P10626-Bio, Vienna
Gruber B, Reineking B, Kranz A, Poledníková K, Poledník L, Klenke R, Valentin A, Henle K (2008) A new way of estimating visitation rates of cryptic animals via repeated surveys of indirect signs. J Appl Ecol 45:728–735
Hell P (1980) Die Situation des Fischotters in der ČSSR. In: Reuther C, Festetics A (eds) Der Fischotter in Europa, Verbreitung, Bedrohung, Erhaltung. Selbstverlag der Aktion Fischotterschutz e V, pp 195–198
Hlaváč V, Toman A, Bodešínský M (1998) Experimentální reintrodukce vydry v Jeseníkách. Bull Vydra Vol 8. pp 37–39
Kloskowski J (2005) Otter *Lutra lutra* damage at farmed fisheries in southeastern Poland, I: an interview survey. Wildl Biol 11(3):201–206
Knollseisen M (1995) Aspects of the feeding ecology of the Eurasian otter Lutra lutra L. in a fishpond area in central Europe (Austria and Czech Republic). Diploma thesis, university of natural resources and applied life sciences, Vienna, pp 59
Kranz A (1995) Verbreitung der bayerisch-böhmisch-österreichischen Otterpopulation (*Lutra lutra*) 1994 in Österreich. BOKU Rep Wildl Res Game Manag 9:1–25
Kranz A (1999) Fischotterschutz, Aktionsplan für Österreich. Schriftenreihe des Naturschutzbundes Niederösterreich—Band 2
Kranz A (2000) Otters (*Lutra lutra*) increasing in central Europe: from the threat of extinction to locally perceived overpopulation. Mammalia 64(4):357–368
Kranz A, Toman A, Roche K (1998) Otters and fisheries in Central Europe: what is the problem? BOKU Rep Wildl Res Game Manag 14:142–144
Kranz A, Toman A, Knollseisen M, Prášek V (2002) Fish ponds in central Europe: a rich but risky habitat for otters. IUCN Otter Spec Group Bull 19A:181–186
Kranz A, Poledník L, Poledníková K, Toman A (2007) Otters in central Europe—Status, habitats, and new conflicts. In: Conroy JWH, Gutleb AC, Ruiz-Olmo J, Yoxon GM (Eds) Proceedings of the European Otter Conference "Return of the otter in Europe—Where and how?". J Int Otter Surv Fund No 2
Kranz A, Poledník L, Poledníková K (subm.) Efficacy of otter (*Lutra lutra*) odour repellents at fish farms
Kruuk H (1995) Wild otters: predation and populations. Oxford University Press, Oxford
Kruuk H, Conroy J (1987) Surveying otter *Lutra lutra* populations: a discussion of problems with spraints. Biol Conserv 41:179–183
Kruuk H, Carss DN (1996) Costs and benefits of fishing by a semi-aquatic carnivore, the otter *Lutra lutra*. In: Greenstreet SPR, Tasker ML (eds) Aquatic predators and their prey. Fishing New Books, London, pp 10–16
Kučera L (1980) Bestandesentwicklung des Fischotters in Böhmen und Mähren (ČSSR). In: Reuther C, Festetics A (eds) Der Fischotter in Europa, Verbreitung, Bedrohung, Erhaltung. Selbstverlag der Aktion Fischotterschutz e V, pp 199–204
Kučerová M, Roche K (1999) Otter conservation in the Třeboň biosphere reserve and protected landscape. Area: scientific background and management recommendations. Internal document of the council of Europe. T-PVS (2000) 20, Strasbourg, 2000, pp 103
Kučerová M, Roche K, Toman A (2001) Rozšíření vydry říční (*Lutra lutra*) v České republice. Bull Vydra 11:37–39
Macdonald SM, Mason CF (1987) The use of spraints for surveying otter *Lutra lutra* population: an evaluation. Biol Conserv 41:167–177
Moravcová J (2002) Biologie a ekologie vydry říční (Lutra lutra), výchova a vzdělávání k její ochraně. Diploma work, pedagogical faculty, Charles University, Prague
Myšiak J, Schwerdtner K, Ring I (2004) Comparative analysis of the conflicts between carp pond farming and the protection of otters (*Lutra lutra*) in upper Lusatia and South Bohemia. UFZ

Discussion Papers 8/2004, UFZ-Umweltforschungszentrum Leipzig-Halle GmbH, Leipzig. http://www.frap-project.ufz.de/downloads/Schwerdtner_et_al_2004.pdf. Accessed 15 Oct 2007

Novotná E (1998) Lidé a vydra, závěrečná zpráva sociologického průzkumu. Jiho-česká sociologická agentura, Zahrádky, pp 24

Poledník L (2005) Otters (*Lutra lutra* L.) and fishponds in the Czech Republic: interactions and consequences. PhD thesis, Palacky University Olomouc, pp 109

Poledník L, Mitrenga R, Poledníková K, Lojkásek B (2004) The impact of methods of fishery management on the diet of otters (*Lutra lutra*). Folia Zool 53:27–36

Poledník L, Poledníková K, Roche M, Hájková P, Toman A, Culková M, Hlaváč V, Beran V, Nová P, Marhoul P (2005) Záchranný program—program péče pro vydru říční (*Lutra lutra*) v České republice v letech 2006–2015. AOPK Praha

Poledník L, Poledníková K, Hlaváč V (2007) Program péče o vydru říční a výsledky monitoringu vydry v roce 2006. Ochrana přírody 62(3):6–8

Poledník L, Řehulka J, Kranz A, Poledníková K, Hlaváč V, Kazihnitková H (2008) Physiological responses of overwintering common carp (*Cyprinus carpio*) to disturbance by Eurasian otter (*Lutra lutra*). Fish Physiol Biochem 34(3):223–234. doi:10.1007/s10695-007-9180-3

Ring I, Santos R (2006) Module 9: development of policy instruments. In: Klenke R, Ring I, Kranz A, Jepsen N, Rauschmayer F, Henle K (eds) Human-wildlife conflicts in Europe—fisheries and fish-eating vertebrates as a model case. Springer, Berlin Heidelberg

Roche K (2001) Sprainting behaviour, diet and foraging strategy of otters (*Lutra lutra*) in the Třeboňsko protected landscape area and biosphere reserve. PhD thesis, academy of sciences of the Czech Republic, pp 135

Roche M (2003) Výsledky mapování výskytu vydry říční v ČR a zkušenosti s náhradami ztrát způsobených vydrou. In: Sýkorová Z (ed). Rybářství a predátoři. Sborník referátů z Odborného Semináře, Praha 18.9.2003. Český rybářský svaz, pp 14–18

Roche M, Toman A (2003) Metodika stanovení výše náhrad škod pro vydru říční. Report for ministry of environment of the Czech Republic. Czech Otter foundation found and agency for nature conservation and landscape protection of the Czech Republic, Třeboň, pp 5

Santos-Reis M (2006) Module 2: wildlife abundance and habitat. In: Klenke R, Kranz A, Jepsen N, Rauschmayer F, Ring I, Henle K (eds) Human-wildlife conflicts in Europe—fisheries and fish-eating vertebrates as a model case. Springer, Berlin

Schwerdtner K, Gruber B (2006) A conceptual framework for damage compensation schemes. Biol Conserv 134(3):354–360

Spurný P, Mareš J, Kopp R, Fiala J (2003) Socioekonomická studie sportovního rybolovu v České republice. Český rybářský svaz, Brno, pp 24

Toman A (1992) První výsledky Akce Vydra. Bull Vydra 3:3–8

Toman A (1998) Otters (*Lutra lutra*) versus privatisation in the Czech Republic. BOKU Rep Wildl Res Game Manag 14:6–7

Thum R, Schwerdtner K, Ring I (2003) Artenschutz und Teichwirtschaft. Rechtliche und institutionelle Rahmenbedingungen in Deutschland unter besonderer Berücksichtigung des Freistaates Sachsen. UFZ-Diskussionspapiere 11/2003, UFZ-Umweltforschungszentrum Leipzig-Halle GmbH, Leipzig, http://www.frap-project.ufz.de/downloads/ThumSchwerdtnerRing_2003.pdf. Accessed 15 Oct 2007

# Otters in Saxony: A Story of Successful Conflict Resolution

**Reinhard A. Klenke, Irene Ring, Kathleen Schwerdtner Máñez, Randi Habighorst, Vivien Weiss, Heidi Wittmer, Bernd Gruber, Simone Lampa and Klaus Henle**

R. A. Klenke (✉) · B. Gruber · S. Lampa · K. Henle
Department of Conservation Biology, UFZ—Helmholtz Centre for Environmental Research, Permoserstraße 15, 04318, Leipzig, Germany
e-mail: reinhard.klenke@ufz.de

B. Gruber
e-mail: bernd.gruber@canberra.edu.au

S. Lampa
e-mail: simone.lampa@ufz.de

K. Henle
e-mail: klaus.henle@ufz.de

I. Ring · K. Schwerdtner Máñez
Department of Economics, UFZ—Helmholtz Centre for Environmental Research, Permoserstraße 15, 04318, Leipzig, Germany
e-mail: irene.ring@ufz.de

K. Schwerdtner Máñez
e-mail: kathleen.schwerdtner@zmt-bremen.de

R. Habighorst · H. Wittmer
Department of Environmental and Planning Law, UFZ—Helmholtz Centre for Environmental Research, Permoserstraße 15, 04318, Leipzig, Germany
e-mail: randi.habighorst@gmx.de

H. Wittmer
e-mail: heidi.wittmer@ufz.de

V. Weiss
Department of Urban and Environmental Sociology, UFZ—Helmholtz Centre for Environmental Research, Permoserstraße 15, 04318, Leipzig, Germany
e-mail: vivien.weiss@ufz.de

R. A. Klenke et al. (eds.), *Human-Wildlife Conflicts in Europe*,
Environmental Science and Engineering, DOI: 10.1007/978-3-540-34789-7_6,

**Abstract** An overview is given about the history of a conflict-laden relationship between Eurasian otters (*Lutra lutra*) and humans in Saxony for more than 100 years. The development of the conflict is described in front of the background of otter persecution, population development, social and economical factors as well as legal instruments in four historical time periods with different societal conditions that framed people's thoughts and actions. The time period from 1990 up to the presence was specifically investigated as part of the FRAP project. Here we made a detailed analysis of the ecological, economic, legal, and institutional as well as social conditions and elaborate the factors that contributed to successful conflict reconciliation.

## 1 Introduction

Most literature about human wildlife conflicts deals with problems and methods for conflict resolution (Breitenmoser 1998; Conover 2002; Woodroffe et al. 2005). If a conflict is not active but rather latent or even reconciled, it seems not worth writing about it. However, successfully reconciled conflicts and a reflection on the reconciliation history can tell us important lessons for conflict resolution.

In the case of Eurasian otters (*Lutra lutra*) in Saxony we have the unique chance to follow the history of a conflict-laden relationship between otters and humans for more than 100 years. There is well documented historical information about the persecution of otters by fishermen which has driven the population close to its extinction at the beginning of the 20th century. Over time, threats endangering the viability of the Saxon otter population have changed significantly, ranging from active persecution through hunting (Fiedler 1996; Kubasch 1987), to environmental pollution (Tschirch 1996), habitat destruction (Klenke 1996a, b), and traffic mortality, as well as other non-natural causes of death (Kubasch 1987; Zinke 1991, 1994, 1996).

We also can follow in detail the recovery of the population thanks to a long-term monitoring program of protected animals found dead that dates back to the late 1960s and was started by the taxidermist Rudolf Piechocki and the biologist Michael Stubbe at the Martin Luther University of Halle/Saale (Stubbe 1969, 1977, 1991; Stubbe and Heidecke 1991). In combination with this monitoring, several activities were started to involve volunteers in different aspects of otter conservation. Most of them were professionals in fields distant from biology but interested either in the animal or in the natural history of the region; among them—though few—fishermen with a strong empathy for the surrounding nature and landscape.

Further, there are four historical time periods with different societal and economic conditions that framed people's thoughts and actions:

- the early capitalism of the German Reich,
- the National Socialism during which the otter was protected for the first time by law,
- the socialism of the German Democratic Republic (GDR), and
- the social market economy after the political change of 1989 in Germany.

In the following, we summarize the development of the conflict and its dependence on societal conditions with a focus on the past 60 years. The time period from 1990 up to the presence was specifically investigated as part of the FRAP project. We present a detailed analysis of the ecological, economic, legal and institutional, as well as social conditions and elaborate the factors that contributed to successful conflict reconciliation.

## 2 The Geographical Region

Saxony is situated in the east of Germany and has common borders with the Czech Republic (to the south) and Poland (to the east). With a size of 18,416 $km^2$ and more than 4.2 million inhabitants it is one of the smaller but more densely populated federal states of East Germany. It shows a moderate north–south gradient of altitude from the glacially shaped lowlands to hilly landscapes and the mountains of the Saxon Switzerland and the Ore Mountains along the Czech border. Originally, otters probably were distributed more or less across the whole state area (Fiedler 1996). At present, the otter population is mainly concentrated in the eastern part of Saxony, the so-called Upper Lusatia, an area rich in ponds and watercourses (Fig. 1). The dominant land use in Upper Lusatia is agriculture followed by forestry. There are approximately 1,000 ponds in Saxony, covering more than 8,000 ha. With more than 5,000 ha pond area and 71 enterprises most of Saxony's fish farms are concentrated in Upper Lusatia (Usbeck et al. 2004). A more detailed description of the socio-economic characteristics of Upper Lusatia can be found in the chapter of Myšiak et al. (2013, in this book).

The development of pond fisheries in Upper Lusatia has a long tradition, starting in the 16th century (Myšiak et al. 2013). It was prompted by the suitable geological, geographical, and hydrological conditions of the region. The name "Lusatia" (originally "Łuža") means marsh- or swampland and was given to the region by the Slavic settlers because of its numerous moors and inland waters. Ponds were mainly built on regularly flooded moor land or sand and clay soils that are usually poor in nutrients and thus unsuitable for agriculture. Often bog iron ore was extracted by surface mining (e.g., Hartstock 2000). Additionally, there are numerous creeks and rivers, another important factor in the maintenance of ponds.

Due to its unique cultural and landscape value, characterized by a special mixture of landscape elements, traditional land-use forms, and handicraft, the central part of the Upper Lusatian lowlands has been designated a biosphere reserve "Upper Lusatia Heath and Pond Landscape" in 1994.

Several authors assume this area to host one of the highest densities of otters in Europe (Ansorge and Striese 1993; Grohmann and Klenke 1996), one reason for this concentration is the highly productive fish farming system. With only a few exceptions (Hertweck 1996a, b; Meyer 1994; Riebe 1994), most ecological otter research in the past was carried out in this area, funded mostly by regional or national authorities (e.g., Ansorge and Striese 1993; Hertweck et al. 2002;

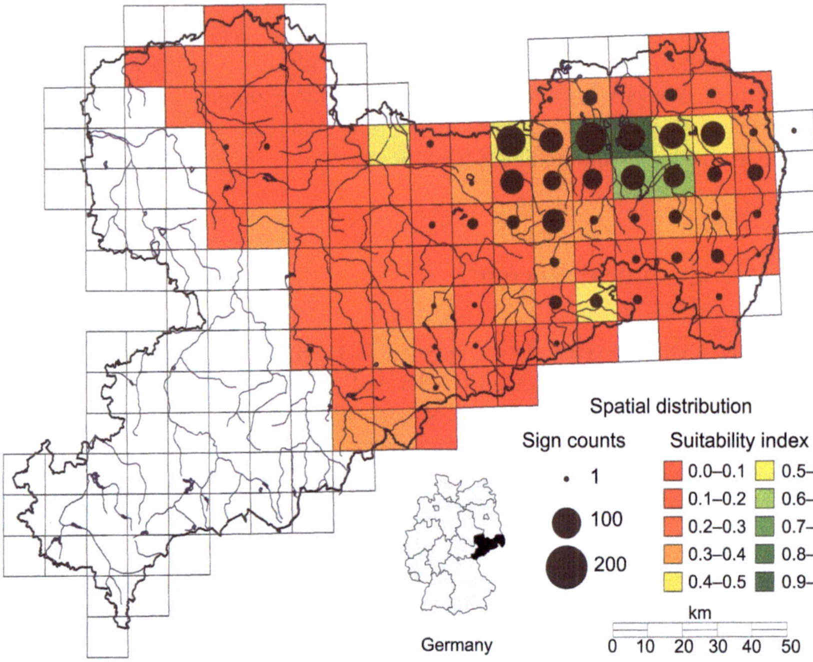


**Fig. 1** Density of otter signs and habitat suitability visualized in a grid of official topographical maps (10 min longitude, 6 min latitude, corresponding to about 11.65 × 11.14 km edge length) after Klenke (2002)

Kubasch 1984; Roth et al. 2000; Steffens 1996). As part of the FRAP project, ecological research focused on a small square study area of 625 km$^2$ north of Bautzen (Fig. 2). Socio-economic research concentrated on Upper Lusatia, but also included Saxon and German wide analyses for comparative purposes. Legal and institutional analysis further included European level framework regulation.

## 3 Otter Population Development

Detailed data about the eradication of the otter in Saxony were compiled by Fiedler (1996). The first published data of otter numbers in Saxony are available for the year 1884, published in "*Schriften des Sächsischen Fischereivereins* 1884–1921" (Fig. 3). This association of Saxon fishermen was founded in 1884 and one of its central aims was the eradication of the otter. Otters were hunted using different methods, mostly guns and traps. For each dead otter a premium bounty was paid and the best hunters were announced in published "boards of fame". The last payment for an otter dates from 1919.

Otter protection started on the 3rd of July 1934 with a closed hunting season for the whole year by the national hunting law of the Third Reich (Klueting 2003;

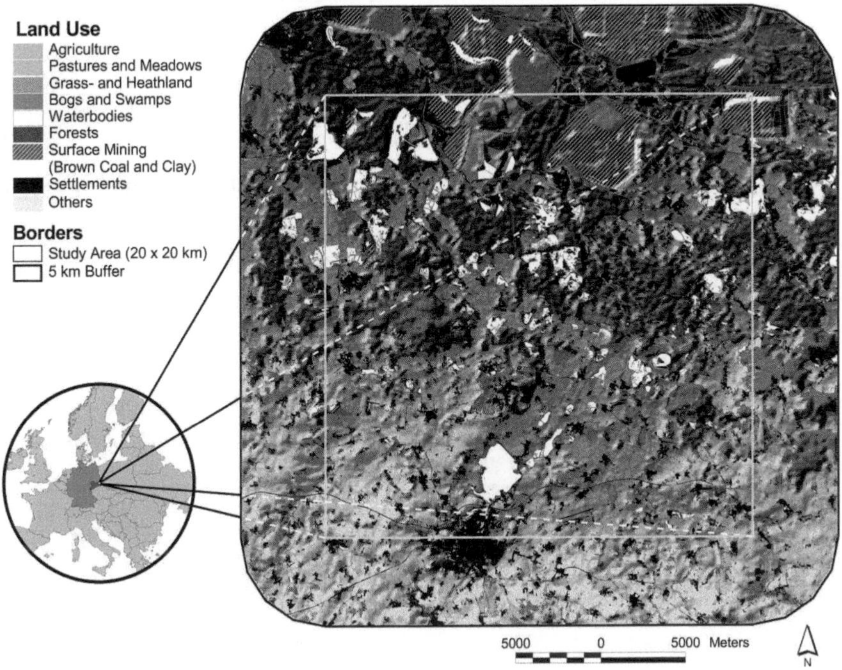


**Fig. 2** Map of the 25 × 25 km (625 km$^2$) study area surrounded by a 5 km buffer zoomed in from a nested map of Europe with Germany and the location of the study site in Upper Lusatia in Saxony. *Sources*: ATKIS (Land Survey Office Saxony perm. 515/04) and SRTM XSAR © DLR/ASI 2005

Lekan 2005). However, hunting was still allowed near ponds for many decades, until the otter was finally withdrawn from the list of huntable species by GDR regulation in the 1980s (Kubasch 1996b).

The development of the Saxon otter population based on death records and sightings is shown in Fig. 4. The data sources differ between time periods. For the first half of the reported time span, information about the decline of the otter population can only be derived from the reports on hunted otters. For the time between 1920 and 1950, published information is mainly available about scattered otter sightings or hunts, whereas later on an increasing part of otter reports originates from the monitoring of traffic accidents.

The inclusion of sightings from an unsystematic pool of different sources like newsletters, personal notes, or verbal reports for the time between 1920 and 1950 does not allow a rigid statistical analysis for this period, but shows that the otter was never extirpated in Saxony. Although hunting and mortality rate of otters most likely varied among years, requiring cautious interpretations, there seems to be a slow but continuous recovery of the population from the early 1950s onwards. Several consecutive sign surveys show a spread of the population into nearly all parts of Saxony that were occupied by the otter in the past (Klenke 2002; Kubasch 1996a).

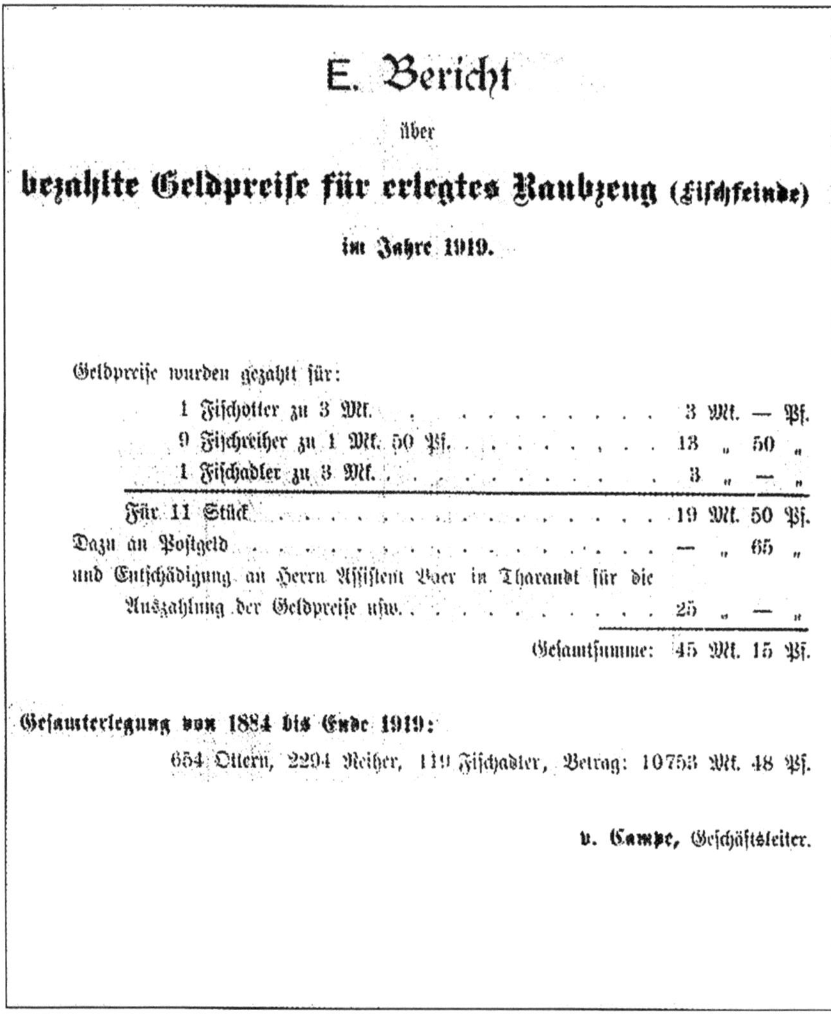

E. Bericht

über

bezahlte Geldpreise für erlegtes Raubzeug (Fischfeinde)

im Jahre 1919.

Geldpreise wurden gezahlt für:

| | |
|---|---|
| 1 Fischotter zu 3 Mk. | 3 Mk. — Pf. |
| 9 Fischreiher zu 1 Mk. 50 Pf. | 13 „ 50 „ |
| 1 Fischadler zu 3 Mk. | 3 „ — „ |
| Für 11 Stück | 19 Mk. 50 Pf. |
| Dazu an Postgeld | — „ 65 „ |
| und Entschädigung an Herrn Assistent Baer in Tharandt für die Auszahlung der Geldpreise usw. | 25 „ — „ |
| Gesamtsumme: | 45 Mk. 15 Pf. |

Gesamterlegung von 1884 bis Ende 1919:

654 Ottern, 2294 Reiher, 119 Fischadler, Betrag: 10753 Mk. 48 Pf.

v. Campe, Geschäftsleiter.

**Fig. 3** Last report about bounty payments for culled fish predators in *Schriften des Sächsischen Fischereivereins* 50/1920 (from Fiedler 1996)

# 4 Conflict History and Legal Frameworks

## *4.1 Historic Development*

The main reason for the fast decline of the otter population at the end of the 19th century in Saxony and probably all over Germany was active persecution, especially by fishermen. To analyze the reasons for the conflict, we depend on historical sources. There seems to have been mainly two reasons: the perceived

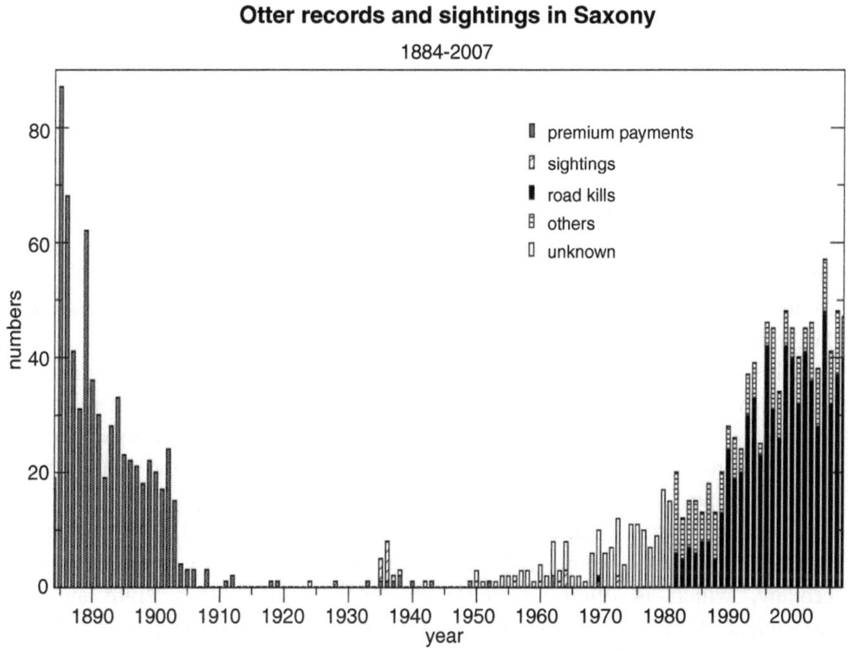


**Fig. 4** Decline and recovery of the otter population in Saxony documented indirectly on basis of death records and observations. Data from *Schriften des Sächsischen Fischereivereins* 1–51 (Fiedler 1996), published observations (Kubasch 1996b), and reported otters found dead since the 1950s (Kubasch 1996b; Zinke 1996), completed with unpublished data from Saxon museums for Natural History compiled by the Saxon State Office for the Environment and Geology Dresden

economic damage caused by the otter and the Zeitgeist. There is no information available on the economic relevance of the damage. However, there was a widespread societal agreement about the bad role of predators, such as Eurasian otters, great cormorants (*Phalacrocorax carbo sinensis*), grey herons (*Ardea cinerea*), or even ospreys (*Pandion haliaetus*), robbing biological resources (e.g., Brehm 1882). This situation has changed in the 1920s when voices came up also within the fisheries community to slow down the persecution of the otter (Fiedler 1996).

At this time a general change in the perception of nature and especially of such charismatic animals like the otter could be observed (e.g., Frohn 2005; Piechocki 2007). Eventually, it has become more widely accepted to preserve the otter, leading to its legal protection in 1934. After the Second World War hunting was generally prohibited by the Soviet occupying power until 1953 (Kubasch 1996b). The German Democratic Republic further provided very strong protection for the otter both by the hunting law (with the hunting season being closed for the whole year, VKdDDR 1953, 1985a) and the nature conservation law.

Until 1962 hunting was allowed only in special cases of hardship and only in accordance with the person responsible for nature conservation in the district.

From 1962 onwards hunting of otter was banned by law. In selected cases of hardship, permission for the shooting of single otters could be granted only by the highest hunting authority (*Oberste Jagdbehörde*) in Berlin (VKdDDR 1962). Nonetheless, we can find some cases of illegal hunting up to the late 1970s, especially in Upper Lusatia (Kubasch 1987).

In 1984 the legal status of the otter changed generally. From now on it was exclusively protected by nature conservation law (VKdDDR 1985b) and removed from the game species list. This is an important difference compared to the otter's status today because it was a clear decision for the strong protection of the otter without any doubts or consideration of traditional vested interests, for example related to hunting.

In this context it is worth mentioning that the social status of hunting has changed over the years. During GDR times, hunting activities were not related to personal income, landownership, or social status. Hunting was well organized and centrally controlled; furthermore scientific research about game management played an important role.

In contrast to the different situation in West Germany and whole Germany nowadays, there was neither a legal nor an illegal market for trophies in East Germany, which is often an important incentive for killing rare animals, independent of their legal protection status (Chardonnet et al. 2002; Tausch 2004). The reason for is also seen in the hard and direct punishment of such a delinquency, mostly already by the police in former time. A hunter would have lost his hunting permission immediately if it gets acquainted with having killed a strongly protected animal without permission. Comparable civil wrongs were handled more or less as a peccadillo by the courts in other countries and this is still the case nowadays in Germany (e.g., Reinhardt and Kluth 2007, p. 89). In addition, the professional training of taxidermists was tightly regulated and of comparable high quality (Anonymus 1978; Jungbluth 2004). They were trained and employed only at scientific institutes and museums; the founding of an own business was not allowed. Therefore, the workmanship of otter remains was controlled very strictly.

With the increase of the otter population well-known old problems surfaced again. However, contrary to the situation in the past, the increased abundance did not yet provoke strong conflicts. The problems were discussed early and constructively in stakeholder groups. It was commonly accepted that not only fish production was important but also the conservation of protected animals in a favorable status. So we can find some guiding documents aiming to reconcile conflicts between species conservation and hunting, forestry, agriculture, and fisheries using administrative top down approaches (e.g., RdB Dresden 1978). Very early in the stage of population increase of the otter, mitigation measures and compensation schemes were developed, especially for small aquaculture companies (RdB Dresden 1978, 1981). In cases of hardship, compensation payments were possible as well as state aid for technical mitigation measures, such as stationary fences. There were even technical standards, similar to industrial norms, for the protection of ponds against foxes and otters (VEB Meliorationsbau Dresden 1986). It is very important to note that under the socialist system economic damage caused by the otter was, with the exception of a few private owners, *per se* damage to society and not to individuals or companies.

Discussing these problems within several societal sectors in general and the administration in particular lead to an increasingly broad interest in the otter since 1970, not only by people interested in natural history and nature conservation, but also by fishermen, hunters, scientists, and regional authorities. Important topics were not only the damage caused by the otter, but also the development and viability of the otter population and the threats endangering it, such as landscape change by land improvement (Benndorf 1986; Bernhardt 1992), environmental pollution of rivers and other water bodies, and, especially in the case of the otter in Saxony, road kills. Despite these threats, the otter population recovery was on a good way at that time and the first positive results of the consequent protection were seen clearly.

The situation has completely changed after the German reunification in 1990. Due to rising car traffic (in Saxony: 224 cars per 1,000 inhabitants in 1988 vs. 470 cars per 1,000 inhabitants in 1997; Völlings 1999) the number of road kills increased dramatically. Four of five otters were killed by traffic in the early 1990s. Additionally, private ownership and market conditions started to play a very important role for the local economy. Now the fish losses caused by a predator directly lead to loss of private income and may even endanger the livelihood of fishermen. As a result, illegal shootings of otters started again (Zinke 1994). There was a strong need for new policies and fast actions with long lasting effects.

## 4.2 The Present Legal and Institutional Framework

Today, major European framework regulations relevant for the reconciliation of conflicts between otter conservation and fisheries consist in the EU Habitats Directive, the Common Agricultural Policy (CAP), the Common Fisheries Policy (CFP), and various European funds.

Regarding European funds, we analyzed Saxon support programs that were based on the European Agricultural Guidance and Guarantee Fund (EAGGF, Guidance Section) and the Fisheries Guidance (FIFG) for the period 2003–2006 (Similä et al. 2006). Since 2007, these funds have been replaced by the European Agricultural Fund for Rural Development (EAFRD) and the European Fisheries Funds (EFF), followed by amended support programs in Saxony. The European funds provide financial means to co-finance conservation, agri-environmental and aquaculture support programs in Saxony.

The EU Habitats Directive is the European Directive relevant for the protected otter and its habitats and has been implemented by national laws. The German legal system is characterized by its federal structure. For the relevant policy fields of nature conservation and hunting, there are federal framework laws, which provide a frame for the state laws, such as the Saxon Nature Conservation Act. The state laws implement the national law and may add further specifications. Saxony further issued regulation and administrative ordinances regarding nature conservation and biodiversity conflict mitigation, for example, the Saxon damage compensation scheme in cases of hardship.

Contrary to other European countries, species protection in Germany is mainly provided for by nature conservation law—as part of the Federal Nature Conservation Act—and only to a lesser extent by hunting law. The binding federal regulation of the Federal Nature Conservation Act is very detailed and, thus, state laws are largely uniform in this respect. In German conservation law, the otter is a strictly protected species, representing the highest category of protection (Thum et al. 2003). Although the otter is also subject to German and Saxon hunting law, there is no hunting season defined. Hunting of otters is neither allowed in Europe nor Germany due to Article 14 of the Habitats Directive that does not permit the use of the species (Similä et al. 2006).

## 5 The Species Action Plan for the Otter in Saxony

The Saxon State Office for the Environment and Geology (*Landesamt für Umwelt und Geologie*), founded after the reunification, is the statutory body for the development and implementation of action plans for endangered species in Saxony. The Eurasian otter was one of the first species for which an action plan was developed in Saxony (Klenke 1993; Steffens 1996), and it was the first for otters in one of the new federal states of reunified Germany.

This plan was built in a scientific process over a period of three years from 1993 to 1996 with scientists from local museums and further stakeholders from regional authorities responsible for road building, fisheries, nature conservation, the biosphere reserve administration "Upper Lusatia Heathland and Pond Landscape", and last but not least fishermen. The scientific results and recommendations were published in a small brochure (LfUG Sachsen 1996, Box 1). The action plan further contained practical pilot studies to test several ecological mitigation measures. Probably even more important for applied conservation management was the construction of a database containing 347 targets with a list of relevant activities to reduce hazard, improve the habitat quality or reconcile a conflict at a given location. This list was specifically addressed to the regional authorities with the aim to initiate actions and to solve local problems. The database was not meant as a static compilation, it rather was a start for actions following a general management strategy with emphasis on regional specifics (Zöphel et al. 1996). This strategy was flanked by the development of policy instruments, including economic incentives for conservation.

**Box 1** The Species Action Plan for the Eurasian Otter in Saxony

The species action plan for the otter in Saxony (LfUG Sachsen 1996) addresses the following issues:

1. History of otter conservation in Saxony.
2. Distribution and habitat of otters in Saxony, including the historical distribution and persecution until the beginning of the 20th century, the distribution of otters in Saxony in the recent past and current times as well as a habitat survey and habitat modeling results.
3. Ecology of the otter, including age structure and reproduction, sex ratio and reproductive status on basis of fecal steroids, spatial activity and diet studies in pond areas and freshwater habitats.
4. Population viability, including the temporal and spatial distribution of causes of death, the analysis of road kill hot spots, the role of environmental pollution, as well as economic damage and conflicts with fisheries.
5. Strategy of otter conservation in Saxony.
6. Ecological mitigation measures and policy instruments addressing endangered population viability due to road kills, environmental pollution, and habitat destruction, as well as for damage mitigation.
7. Target objects of action.

Besides the reduction of road kills, damage compensation, and testing of mitigation measures, the use of deviation ponds and additional stocking of wild fish have played an important role in the program. More than 75 % of the targets of action were in relation to road traffic and habitat loss. Considering that these were the most important threats for the otter in recent times, the concentration of the species action plan on appropriate mitigation measures against these threats is understandable. Only 44 actions (9 %) directly related to conflict mitigation and damage compensation measures were planned (Fig. 5).

Unfortunately, no detailed analysis exists on the general effects of this program. Only a small compilation of special prevention measures against traffic losses was made by Schreyer and Jahn (2006) for the area managed by the administration of the biosphere reserve. Probably most of the hot spots of otter road kills, with at least two or more victims, were defused in the past ten years in Saxony, mainly during the building or reconstruction of roads.

## 6 Characteristics of the Case Study

As part of the FRAP project, the conflict between species conservation and fisheries was analyzed more than ten years after the start of the species action plan for the Eurasian otter in Saxony. This action plan (see Box 1) and associated studies provided an excellent knowledge basis for further analysis:

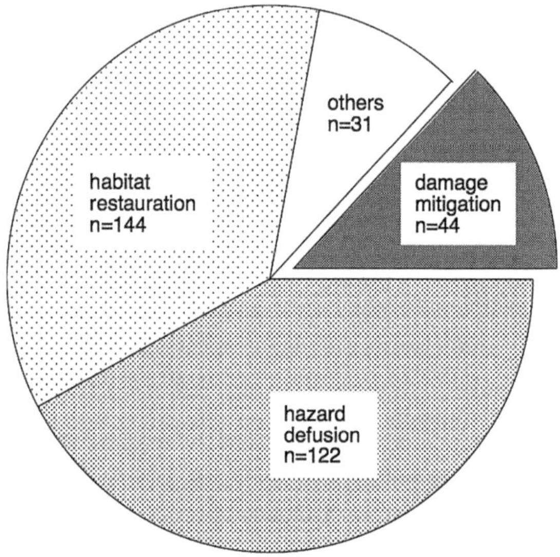


**Fig. 5** The Saxon otter action plan: proportion of measures categorized by mitigation type

1. Ecological studies about population viability, status of the population, the role of habitats, and measures to stop further decline were available in some detail and at least partly published in English (e.g., Ansorge et al. 1997; Geidezis 1999, 2002; Hertweck et al. 2002; Jurisch 1997; Jurisch and Geidezis 1997; Klenke 2002).
2. The species action plan also included additional elements, such as an analysis of the conflict history, basic information on the legal framework, and existing policy instruments.

Therefore, the species action plan provided many of the elements relevant for a successful reconciliation process as presented in this book with the generic framework for reconciling human-wildlife conflicts (Henle et al. 2013). While the ecological part of the species action plan was more or less detailed, the legal, economic, and social aspects, as well as their integration with ecological findings needed more attention. This is why in phase 2 of applying the generic framework the Saxon case study put a special focus on the analysis of the legal and institutional framework, the regional economic situation, the existing policy instruments and their implementation, as well as the perception of the conflict by various stakeholder groups. Regional policy analysis as well as stakeholder interviews showed, that no further participatory process was needed to support decision-makers in reconciling the otter-aquaculture conflict. Therefore, research in phase 3—Deriving and implementing resolutions—focused on improving the cost-effectiveness of existing compensation payments.

## 6.1 Regional Ecological Situation

Many important questions about the ecology of the otter in Saxony were already answered by the Saxon otter action plan and the studies initialized in this context. However, at least three questions were not answered in a satisfying manner:

1. Is there a relation between otter density and damage reared by the otters and how can we measure it?
2. How many otters live in Saxony or at least in the study area?
3. How is the abundance of the otter related to the landscape characteristics?

Thus we have concentrated our focus in the ecological modules of the conflict reconciliation framework (Henle et al. 2013) in Saxony on these questions. At first a detailed sampling approach was developed to gather fresh otter spraints in a systematic manner (Klenke et al. 2008). This approach was applied on ponds from which we also got data about the hatched fish and the surrounding landscape. On basis of these data, we have developed a method and a tool to estimate visiting rates of otters at the ponds (Gruber et al. 2008). The method is rather generic and can also be applied to other cryptic animals leaving detectable signs of presence. Because the question regarding otter density and reared damage is not only an ecological but rather an economic one, we will come back to the results in detail later in this chapter.

To answer the second question, we have tried to use the genetic information about an individuum contained in cells left on the surface of the spraints. Because age and handling of the spraints are important factors determining the results of the detection of an individuum, special attention was given on the improvement of the laboratory protocol (Lampa et al. 2008). The results of the genetical determination of individual otters were used to estimate the population density by a mark-recapture approach (Klenke et al. 2008; Lampa et al. 2011; Otis et al. 1978; White and Burnham 1999). The first results from data of the first year of our study give $n = 18$ ($se = 2.647$, 95 % confidence interval = 13–24, CMR model $M_{h\text{-misidentification}}$) corresponding to 0.4 otters per kilometer shoreline within the ecological study area.

It seems also that the landscape structure is an important factor determining otter densities as well as the damages reared by the otter. Multiple linear regressions with several variables related directly to the landscape structure in buffers of 250 and 1,000 m around the investigated ponds give hints that especially isolated ponds were under pressure by hungry otters (Klenke et al. 2005). Similar results were obtained by Polednik (2005) in the Czech Republic with the same methodological approach.

## 6.2 Regional Economic Situation

The political changes after the German reunification in 1990 had far-reaching consequences for all economic sectors in the former GDR, including the aquaculture sector in Saxony that mainly consists in carp pond farming (compare Myšiak et al. 2013, in this book). Two important aspects can be distinguished:

- a decreasing demand for freshwater fish and
- the privatization of ponds.

Both aspects had significant effects on the conflict between fish producers and stakeholders interested in otter conservation.

The decreasing demand for freshwater fish was a direct result of the opening of markets. With the large variety of sea fish as well as imported freshwater fish available, fewer consumers bought fish produced in Saxony. Pond fisheries underwent far-reaching production changes, resulting in a significant production decline. The total amount of fish produced dropped from 8,712 t in 1989 to 3,850 t in 1992 and reached its minimum in 2003 with only 2,866 t (SLfL2007). The production has only slightly increased since.

The decreasing economic gains combined with higher production costs created a situation, where the viability of pond fisheries as an economic activity is increasingly endangered (SLfL 2005a). From an economic point of view, an extension of the production intensity from currently about 600 kg/ha (SLfL 2005b) to approximately 1,000 kg/ha would be necessary to avert production abandonment (SLfL 2005a). Neither a higher carp production nor the abandonment would be in the interest of nature conservation.

On the one hand, the maintenance of ponds is crucial in this traditional cultural landscape, where species have become adapted to them as secondary habitats for centuries. Especially otters utilized the ponds as retreat areas, when the continuing pollution of rivers and streams made those water bodies unsuitable. Additionally, the ponds and their surrounding landscape provided a comparatively safe harbor as well as territories for females with cubs. Without cultivation, the shallow ponds would silt up and lose their habitat function. On the other hand, extensive production is one of the central conservation goals in this area, aiming at the preservation of natural structures in and around ponds as well as at minimizing the negative effects of intensive fish production such as the use of antibiotics and fertilizer.

The privatization of ponds and the changes in the employment structure resulted in a modified perception of otter damage. Formerly being paid independently from the actual production, fishermen's wages now depend on their economic gains. Damage caused by fish-eating species became much more significant, as it may endanger the economic viability of pond fisheries (compare also Kranz 2000). Thus, effective reconciliation strategies were needed.

## 6.3 Policy Instruments for Conflict Reconciliation

In response to solving the problems between nature conservation and fish production, the Saxon government implemented various policy instruments relatively soon after the political changes. The following empirical analysis of policy instruments reflects the status of the legal framework until 2007. A focus was put on economic instruments of nature conservation policy in the form of support measures and compensation payments. Two aims can be distinguished in this respect:

- the maintenance of production in an extensive way and
- the prevention and compensation of damage in order to raise acceptance for the otter and distribute the conservation costs more equitable.

### 6.3.1 Saxon Programs to Generally Support Environmentally Sound Pond Fisheries

Pond fisheries in Saxony were supported under a Saxon conservation contracting program from 1992 onwards. In the first years, the spatial focus was mainly on the biosphere reserve. Saxony also set up EU co-financed programs for environmentally sound agriculture. In 1993, the first Saxon agri-environmental program "Environmentally Sound Agriculture" (*Umweltgerechte Landwirtschaft*) came into force. It contained program No. 73/94 B, which explicitly supported pond fisheries. In 2000, the second Saxon agri-environmental program "Support of an Environmentally Sound Agriculture in Saxony"*(Förderung einer umweltgerechten Landwirtschaft im Freistaat Sachsen)* was implemented, which integrated all former conservation contracts. Part E of this program, "Conservation and Preservation of Cultural Landscapes" (*Naturschutz und Erhalt der Kulturlandschaft* or NAK) supported pond fisheries in general as well as "Environmentally Sound Pond Fisheries". Since 2007, the Saxon agri-environmental program has again been amended and enlarged, covering now environmentally sound agriculture (including environmentally sound pond fisheries) and the ecologically sound increase of forests (SMUL 2007). The measures related to pond fisheries have been transferred to the new program, although in a slightly changed manner (see below).

Under the former agri-environmental program, Saxon aquaculture was supported with 2.3–2.6 million € per year between 2000 and 2003, covering more than 95 % of Saxon pond area used for fish farming. NAK funds represented the major part of public support to aquaculture that summed up to about 20–30 % of the total gross income per hectare (Klemm 2001).

Under the NAK program, the measure "Pond Maintenance" could be applied for all ponds in Saxony and was supported with 150 € per ha and year. It aimed at a general continuation of pond cultivation without further specifications. In contrast, the support of "Environmentally Sound Pond Fisheries" aimed at the extensive production and maintenance of the structural diversity at ponds. Basic payments for the latter consisted in 200 € per ha and year, but could be extended by choosing additional measures such as no artificial feeding. A quarter of all NAK payments in Saxony were used for pond fisheries. In 2006, about 240 aquaculture companies and therefore approximately the entire Saxon pond area was supported under NAK (SLfL 2006). On average, about a quarter of the income in pond fisheries stemmed from public payments (Klemm 2001). Obviously, the NAK program provided a significant incentive for the continuation of pond cultivation and therefore the maintenance of ponds as otter habitats.

### 6.3.2 Policy Instruments Specifically Addressing the Otter-Aquaculture Conflict

There are three policy instruments directly relevant to the conflict between otter conservation and pond fisheries in Saxony (Similä et al. 2006). Most important is a specific support measure of the NAK program, aiming to provide a feeding habitat for protected species, next to a Saxon compensation scheme for damage caused by protected species. Last but not least, there is financial support for technical measures to avoid damage caused by protected species. These three instruments are each presented in more detail and later discussed in context. Only a good coordination of policy instruments allows for an effective reconciliation of the vested interests between otter conservation and pond fisheries, while saving public monies.

The specific measure of NAK "Support of a Feeding Habitat for Protected Species" is a typical payment for environmental services. Fish farmers who participated were paid 103 € per ha and year for an extra stocking of ponds with fish that is considered as food for otters. Therefore, this measure is sometimes called "otter bonus", although it is also applicable to other protected species. It is understood as a compensation for potential damage, a reason why this instrument can also be interpreted as a compensation for otter damage in advance (Schwerdtner and Gruber 2007; Schwerdtner 2008).

This measure was supposed to further create a deviation effect (LfUG Sachsen 1996a), so that otters spare the ponds and storing facilities used for keeping the fish for sale. Paying of the "otter bonus" depends on the presence of otters at the respective ponds, not on actual damage. Thus, it might as well become an income, turning the attendance of otters at a pond into something positive. In 2003, 2,687 ha pond area in the Upper Lusatia were supported with approximately 280,000 € as feeding habitat, corresponding to about half the pond area used for fish production in Upper Lusatia. Figure 6 gives an overview of the payments from 2000 to 2003 (Schwerdtner 2008).

Saxony is the only state in Germany that pays for damage caused by rare but locally abundant animals protected by law (Thum et al. 2003). Contrary to all other economic instruments mentioned, the damage compensation scheme is only funded from the budget of the State of Saxony without European co-funding. The legal basis for damage compensation is the "Compensation for Cases of Hardship Regulation". Since 1997, fishermen can file their application, if their total damage exceeds 1,000 € per ha per year. Signs for damage, such as fish remains or otter tracks, must be reported immediately to the authorities. Claims are made after the ponds have been drained. Therefore, this instrument is an *ex post* compensation scheme (Schwerdtner and Gruber 2007). A marginal return calculation is used, where a standardized "norm loss" is subtracted from the expected gain in order to

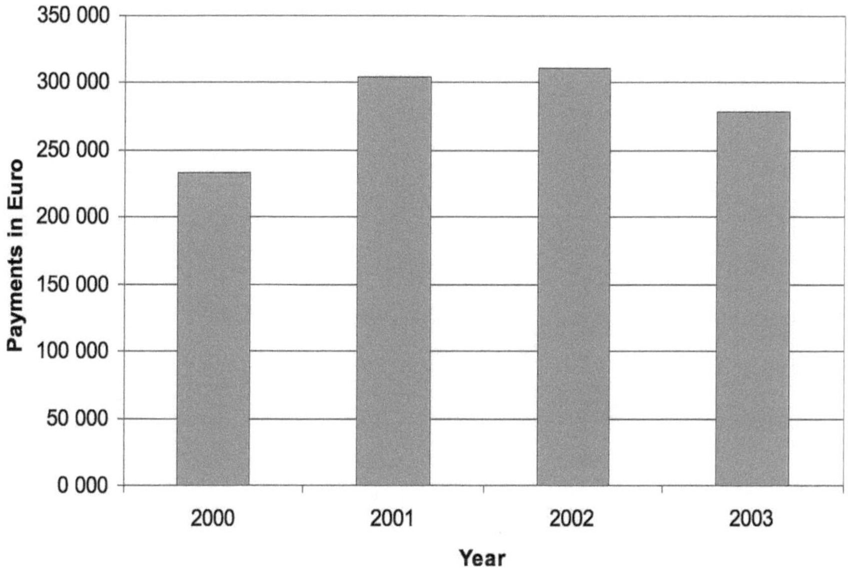


**Fig. 6** Payments for providing a feeding habitat for the otter in Upper Lusatia from 2000 to 2003. *Source*: Schwerdtner (2008) based on data from Saxon State Agency for Agriculture, Fisheries Authority 2003/2004

determine the amount of loss. As control mechanism, two expert opinions from fisheries and conservation authorities are necessary. If the claim is accepted, up to 80 % of the loss is compensated in the case of the strictly protected otter. Payments can fluctuate considerably from year to year, as can be seen from Fig. 7 (Schwerdtner 2008). Damage caused by other wild animals, e.g., cormorants, is only compensated to about 60 %.

Pond farmers sustaining small-scale damage from otters could obtain restitution in kind from some district authorities (Similä et al. 2006). The special guideline as the basis for this procedure is no longer in force. Therefore, local authorities are currently looking for a new solution to cover small-scale damage.

In Saxony, financial support is also provided for measures aimed at avoiding damage caused by protected species. The former Saxon support program for aquaculture was co-financed within the framework of the Financial Instruments for Fisheries Guidance (FIFG) of the European Union (Similä et al. 2006). As part of this program, technical measures were funded to protect fish stocks from cormorants, herons, and otters. Support was given as direct project support for 60 % of the total costs, although for Saxony, this incentive seemed to be too low for fish farmers to apply for this measure (Schwerdtner and Ring 2005). Only three applications were filed from 2001 to 2002 to build otter fences and financially supported with some 25,500 € . The type of equipment utilized is expensive and the remaining investment costs for the pond farmers are generally too high compared to the return of an average aquaculture company (Klemm 2001). This type of measures is still supported under

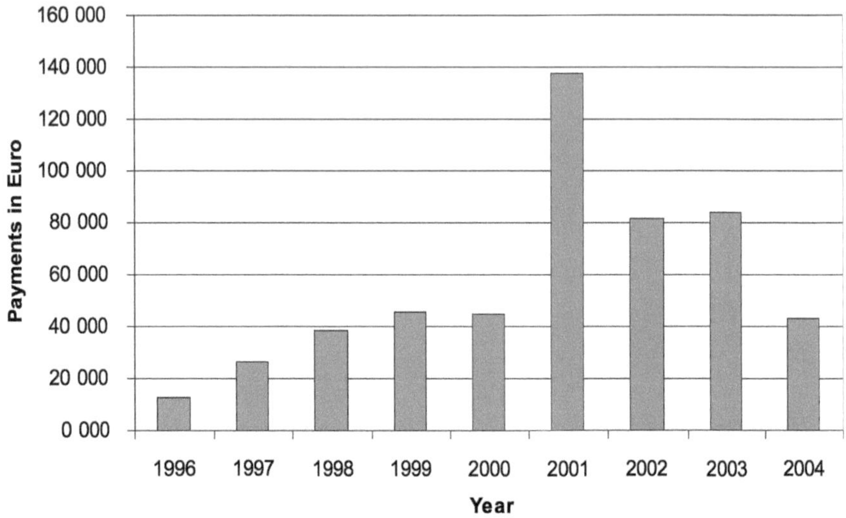


**Fig. 7** Payments under the "compensation for cases of hardship regulation" from 1996 to 2004 in Upper Lusatia. *Source*: Schwerdtner (2008) based on data provided by *Staatliches Umweltfachamt Bautzen* 2006

the amended Saxon program "Aquaculture and Fisheries", Saxony's implementing program of the new European Fisheries Fund (SMUL 2008).

For the otter, the previously mentioned "otter bonus" turns out to be the most important policy instrument, both in terms of financial support and pond area covered. However, the damage compensation scheme seems to be necessary for locally high damages. According to the regulation, damage compensation is only paid if measures to prevent damage have been taken beforehand (Schwerdtner 2008). In the case of the otter, the situation is different: the species depends on the utilization of ponds for food; therefore preventive measures would considerably decrease the otter's habitat quality. Consequently, the aquaculture support program is mainly used for fences to protect especially valuable fish, such as young fish or ornamental fish, and for storing facilities.

## 6.4 Stakeholder Perceptions and Participatory Processes

The core of a conflict consists in different interests and values among the main stakeholder groups. In Saxony, there were six groups identified as relevant for the topic of otter conservation. In short, the different positions could be distinguished into two main views: On the one side, fishermen and anglers complain about losses due to otters and call for financial compensation to keep their respective occupational or leisure activities. Nature conservationists on the other side aim at securing the abundance of otters by maintaining and improving their habitats.

It seems that the existing mitigation measures in Saxony, mainly in terms of policy instruments, have been capable to mitigate the conflict, which is perceived as reconciled by most of the stakeholders. A qualitative analysis of the discourses on the past and current state of the conflict was carried out to find out what the different perceptions of the stakeholders are and how they talk and argue on the topic of biodiversity and fisheries (Zwirner and Wittmer 2004; Behrens 2005). For this purpose 26 qualitative interviews (duration: 90–120 min) with 31 persons from different stakeholder groups were conducted between 2003 and 2004 in two areas of the Saxon study region: the Upper Lusatian Pond Area (*Teichlausitz*) and the Upper Lusatian Hill Country (*Lausitzer Hügelland*).

The main stakeholder groups interviewed consisted in:

- conservationists representing NGOs (8 interviews)
- fishermen (8 interviews)
- anglers (3 interviews)
- state authorities (4 interviews)
- specialists and scientific experts (3 interviews), and
- a representative of the tourism sector (1 interview).

The interviewees were asked to portray their respective view on the subject of the otter-aquaculture conflict. The situation in the Upper Lusatian Pond Landscape and the Upper Lusatian Hill Country was rather homogeneous, so the results will not be distinguished in the presentation. The sociological analysis of the interviews (Wilson 2004) revealed different storylines, which were and are decisive for the perception of the conflict.

In general, all stakeholder groups accepted the otter as part of the landscape, but they also admited that it can cause relevant damage under specific circumstances. Due to the historical situation, the otter has always been present in Upper Lusatia. Carp pond fishery has a long tradition in the region, and so there is a long tradition of dealing with the otter as an animal that is perceived to diminish the gain from aquaculture (Zwirner and Wittmer 2004). As shown before, interdependence between the abundance of otters and the maintenance of pond fisheries exists. Against this background, one can identify the factors, which have lead to the current reconciled situation.

The interviewed NGO conservationists perceived the level of overall payments as high enough to mitigate the conflict: "One can turn the pond fishery business into cash by obliging to nature conservation schemes", as an honorary conservationist stated. In the interviews some conservationists as well as some interviewees from conservation authorities claimed that the payments are (far) more generous than adequate and that the supervision by the authorities is not tough enough. Representatives from both the nature conservation and fisheries authorities generally agreed on the rules of the various compensations schemes, although they perceived it difficult to scrutinize the damage: "There is a tension between the reported damage and the real damage", as an interviewee from a regional nature conservation authority reported regarding the hardship regulation. Fishermen sometimes complained that the calculated direct loss is not fully compensated by this damage compensation scheme and indirect damage like injured fishes or loss of weight as consequence of frequent disturbance in wintering ponds is not included. "The loss of young fish is grave and 60 % of the occurred damage is compensated" said a local fisherman, although referring to compensation for cormorant damage. The interviews showed that stronger complaints related to the conflict about fish losses due to cormorants than due to otter.

That is in strong opposition to the findings of Kranz (2000) for Saxony, where the otter still seem to play the most important role causing damages in fish farms, closely followed by the cormorant. However, from the viewpoint of one of the authors, who coordinates the species conservation program, these findings are not conforming to the situation perceived in a lot of discussions with fishers and authorities. At these times the cormorant was in Saxony already in the main focus of the complaints expressed by the fisherman (see also Seiche and Wünsche 1996 and the following lines). One reason for this result is probably that only the frequency of complains was estimated, independent from the severity of the damages reared by a given species.

In general, fishermen relied on the compensation payments and "are willing to take them", as another fisherman conceded. The remarks above show that the design of compensation schemes and controls could be a further conflict issue, which could be mitigated by improving existing schemes (Kranz 2000; Schwerdtner 2008).

In Saxony, as was told in almost all interviews, most persons from nature conservation associations and authorities, fisheries authorities, and the aquaculture sector know each other already from GDR times and the times of political change in 1989/1990. As they have well established relations and communication patterns, they mainly resolve things cooperatively on a more or less informal level. Therefore, no conflict is salient. An aquaculture working group was set up, to bring together representatives from the fisheries sector, nature conservation associations, and the respective authorities. They met if problems or questions occurred, as a staff member of the Saxon State Ministry for the Environment and Agriculture reported, but nowadays cooperation is well enough established, so that many questions can be solved in bilateral dialogues.

Another factor is that, due to the political change and the German reunification, a window of opportunity emerged, which lead to new structures in the aquaculture sector: Before this time, fish farms in the GDR were state owned enterprises and financially supported. After the political change, fishermen were confronted with new challenges like economic pressure. However, ponds are perceived as important biotopes for endangered species by all stakeholder groups and there is an agreement about the maintenance of aquaculture. To support fishermen, different nature conversation programs helped to overcome the economic difficulties after the German reunification. For example an association, which aims at conserving the cultural and nature heritage of Saxony (*Landesverein Sächsischer Heimatschutz*), purchased a couple of ponds and fishermen became tenants with the obligation to perform ecologically sound aquaculture.

Recently the increasing cormorant population has become more and more problematic (Behrens et al. 2008; Seiche and Wünsche 1996; Thum 2004). This may be seen as a further factor why the otter is accepted. In comparison to the damage caused by cormorants, those caused by otters is less, and mitigation measures and compensation payments for otter damage are well established. The findings of our interviews reinforce earlier studies that already found stronger complaints by fishermen regarding cormorant damage (see for example, Seiche and Wünsche 1996). There is one study, however, indicating that the otter seemed to play the most important role in causing damage to fish farms, although closely followed by the cormorant (Kranz 2000). One reason for this may be that the latter study only estimated the frequency of complaints independent of the severity of the damage reared by the species in concern.

To sum up, almost all stakeholders agreed throughout the interviews that the conflict between otter conservation and pond fisheries is not salient due to various compensation payments, partly co-financed by the European Union, and the historical relationship of aquaculture and nature conservation in Saxony. As these factors are relatively stable, one can assume that the conflict will not become salient in the near future.

## *6.5 Increasing the Cost-Effectiveness of Compensation Payments*

Despite having no active conflict, it is in the interest of the state and the tax payer that public monies are invested in a cost-effective way. Especially regarding the combination of the *ex ante* compensation in the form of the "otter bonus" and the *ex post* damage compensation scheme, a correct calculation of total potential damage caused by otters is highly relevant.

Unfortunately, a correct calculation of the amount of damage, i.e. fish eaten by otters, seems highly difficult for the following reasons (see also Kranz 2000; Myšiak et al. 2013, and Poledníková et al. 2013 in this book):

- Ponds are a "black box". The losses can only be estimated after the draining (Fig. 8).
- In most cases, it is not possible to distinguish between damage causing animals. Damage might also be caused by great cormorants, grey herons, white-tailed sea eagles (*Haliaeetus albicilla*) and ospreys, or other fish-eating species.
- Beside direct losses, damages can also rising by indirect effects like stress and loss of fitness as an effect of the disturbance of wintering ponds
- The responsible authorities depend on information from fish farmers: Losses can also be the result of bad management practices.

A basic question is therefore: Do we have any indication on the amount of total damage? As a very rough basis, we can base such a calculation on the following data: After Ansorge et al. (1997), approximately 500 individuals including 20 %

**Fig. 8** A good catch. The stock of fish in ponds can be reduced by parasites, fish diseases, or even due to direct (small photo) or indirect losses (stress related factors) caused by predators like the Eurasian otter. *Photo:* André Künzelmann

offspring live in Saxony, most of them in Upper Lusatia. The necessary daily amount of food is about 0.577 kg for adult animals and about 0.289 kg for offspring (Reuther 1993). Following Geidezis and Jurisch (1996), about 90 % of otter food in Upper Lusatia is fish. The possible annual damage for each age class can be estimated with the following equation:

$$D_{age\,class} = F_{daily} \cdot N \cdot 365 \tag{1}$$

where

$D_{age\,class}$ = maximum amount of food for a particular age class,
$F_{daily}$ = daily amount of food in this age class,
$N$ = number of individuals in this age class.

The maximum possible damage $D_{max}$ is the sum of the age class specific damage $D_{juv}$ for juvenile and $D_{ad}$ for adult animals:

$$D_{\max} = D_{juv} + D_{ad} \tag{2}$$

Inserting the daily amount of food for juvenile and adult otters and assuming $N = 100$ for juvenile and $N = 400$ for adult otters, we get an estimated maximum amount of food eaten per year by all otters in Saxony of:

$$maxD = 10{,}549\,\text{kg} + 84{,}242\,\text{kg} = 94{,}791\,\text{kg} \quad (3)$$

Because approximately 90 % of the eaten food is fish (see citations above) the possible maximum amount of damage is about 85 t (85,312 t) of fish per year. Compared with an average annual production of 3,124 t of adult fish for the years 2000–2003, a maximum of 2.7 % of the fish production can get lost due to otter predation. However, otters mostly eat fish that is offspring older than one year. To estimate the monetary value of this loss, the maximum possible loss $D_{max}$ is reduced by 10 % (see above) and multiplied with the average price for carp in 2000–2003, using a mixed price based on different marketing forms such as direct sale, wholesale, or sales to angling associations.

$$85{,}312\,\text{kg} \times 2.08\,€ = 177{,}411.52\,€ \quad (4)$$

Both the physical damage estimation and its monetary value are only very crude estimates[1]. Yet, when comparing the monetary value of damage estimated here to the expenditure of the two compensation payment schemes for the years from 2000 to 2003, at least some overcompensation seems to take place (Schwerdtner 2008, Fig. 9)[2]. Compensation payments reach more than twice the estimated monetary damage for three consecutive years. The maximum estimated damage is thus effectively compensated; however, a cost-effective compensation of otter damage at minimum costs does not seem to take place (Schwerdtner 2008). An indispensable prerequisite for more correct figures is an otter monitoring with spatially differentiated population size estimation. This would allow controlling both the claims for damage compensation and represent a sound basis for the "otter bonus".

Studies about the influence of otter densities on damage have disclosed discrepancies between the estimated visiting rates of otters at wintering ponds and the amount of damage that was claimed by the fishermen according to the hardship regulation and registered by the administration (Fig. 10). Though these studies neglected payments for the "otter bonus", which ideally should have been considered as well, a correlation between estimated visiting rates and amount of damage compensation payment can still be expected. The damage compensation scheme is explicitly meant for large-scale damage, and the latter is to be expected where high otter visiting rates occur.

---

[1] Compare Schwerdtner (2008, p. 161ff) for a critical discussion of the estimated monetary value of otter damage.

[2] Readers might question why we consider both the hardship regulation and the otter bonus in this context, because the latter has been introduced as a payment for environmental services, and thus, payments are usually based on conservation benefits rather than damage costs. However, the official NAK measure is called "Providing a feeding habitat for protected species" and when deciding upon the per hectare payments with regard to the otter, Saxon authorities estimated the numbers of otters and the average amount of food eaten by otters per year. Fish farmers are thus paid for extra carp to add to the ponds that then is supposed to be food for otters. This is the positive way of framing the situation; the negative way perceives any fish eaten by otters as damage costs.

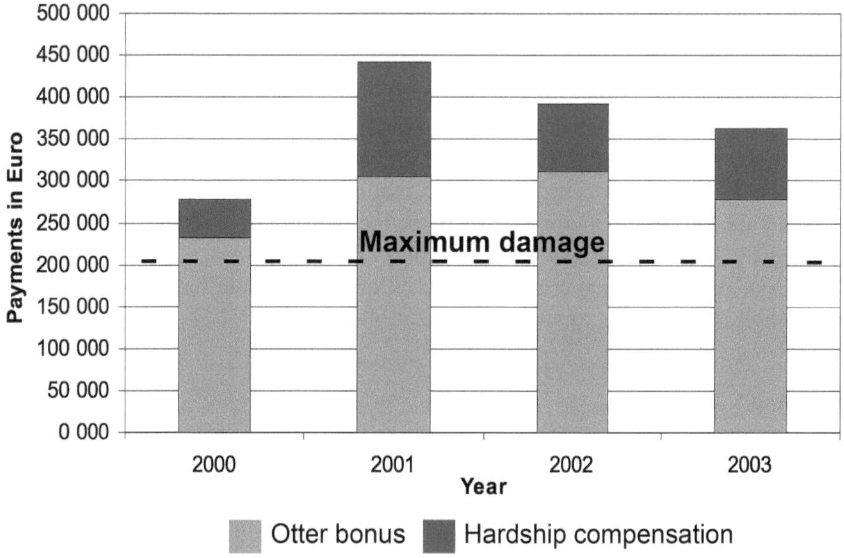


**Fig. 9** Overall payments for the NAK support measure feeding habitat ("otter bonus") and the damage compensation scheme (hardship regulation) from 2000 to 2003, compared to the estimated monetary value of maximum potential otter damage (*dashed line*). *Source*: Schwerdtner (2008)

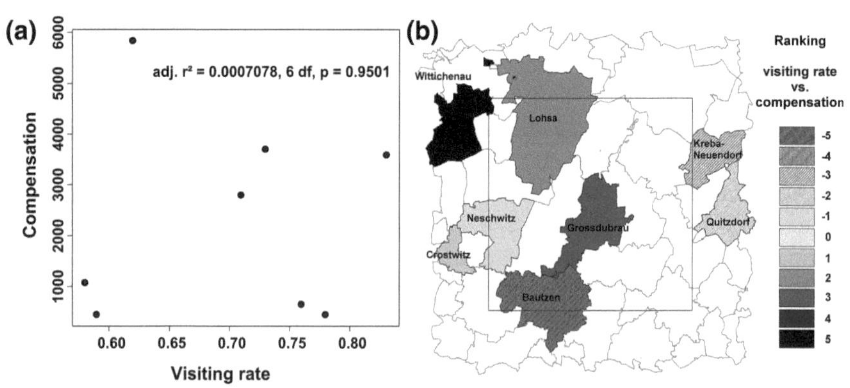


**Fig. 10** **a** Correlation between visiting rates and damage compensation payments (hardship regulation). **b** Spatial distribution of ranks obtained from a comparison of visiting rates versus damage compensation payments. *Source* ATKIS (Land Survey Office Saxony perm. 515/04)

The visiting rates were estimated at selected ponds with a method described in Gruber et al. (2008). They show a high dependency on landscape parameters in a multiple linear regression model (Adjusted $R^2 = 0.6397$, 6 *df*, $p = 0.0007$) that was used for the prediction of visiting rates in each municipality (Klenke et al. 2005). Regarding damage claims, data were unfortunately not available for all municipalities, whereas data about landscape parameters of ponds were available for the whole study site. Where data on damage claims existed, the requested damage claims were low in municipalities with higher predicted visiting rates of otters but high in those with lower predicted visiting rates. On the one hand, there are individual fish farmers in the area that are known for complaining loudly about otter damage and regularly make use of the hardship regulation. On the other hand, many fish farmers are satisfied with the *ex ante* compensation scheme, i.e. the NAK measure for feeding habitat. The latter provides lower monetary compensation, but also involves lower transaction costs both for applicants and authorities (compare Schwerdtner and Gruber 2007).

Damage calculation on basis of the current market conditions can lead to to distortions and incomparability of data from different periods. As an alternative, information about damage could be obtained in a standardized way being rather based on physical measures in the form of losses of number and tons of fishes than on actual market prices. However, abandoning a monetary estimate of potential otter damage makes it difficult to investigate the cost-effectiveness of a damage compensation scheme that compensates in monetary terms.

The NAK measure of providing a feeding habitat for the otter offered another option for improving cost-effectiveness of compensation payments. Although conceptualized as a payment for ecological services provided by fish farmers, the amount of the payment was based on the potential damage to be expected by protected species. For many years, fish farmers were paid 103 € per ha pond area, irrespective of the size of the ponds. However, otters do forage close to the borders of a pond (compare Poledníková et al. 2013 in this book). Therefore, damage from otters is relatively higher on small and medium-size ponds compared to large ponds. Owners of large ponds were thus paid relatively too much compensation. The German FRAP team recommended that the Saxon State Agency for Environment and Geology should in the future base this measure on the circumference of the ponds rather than on their overall area and provided options to change the calculation procedure for compensation payments. Fortunately, the NAK program underwent an amendment phase during that time, and therefore, the policy window was open to consider a more cost-effective procedure to calculate the compensation payments.

Under the new agri-environmental support program as of November 2007 (SMUL 2007), the amended compensation measure is now called "Environmentally Sound Pond Management Including Conservation Measures for Species/Biotic Communities of Ponds" and provides 232 € for an additional stocking of ponds with fish to prevent damage from protected species for the first 3 ha per pond and 207 € per ha up to a pond size of 20 ha. This measure is the direct follow-up of the former "otter bonus" and now reflects much better the costs related to fish lost through otters. Any hectare beyond a pond size of 20 ha is not considered any more, saving the state a considerable amount of money, whereas payments for small ponds are highest and medium-sized ponds still get payments higher compared to the former regulation.

### 6.5.1 Reasons for Successful Conflict Reconciliation in the Past: A Case for Best Practices in the Future?

As we can see from this chapter, the most important factors for the successful reconciliation of the otter-aquaculture conflict in Saxony were:

1. The recovery of the otter population took place in a period characterized by a changed societal understanding of the value of rare species and ecological mechanisms.
2. Because population development was very slow, society had time for habituation combined with an intensive dialogue starting very early among the conflict parties.
3. The relations between economic effort and private income were relaxed in socialist time.
4. Under new market conditions, and very soon after the political changes, various aquaculture support programs and compensation schemes were set up, being most important for conflict reconciliation in the recent past.
5. At the same time, a communication process with stakeholders was established very early by the administration, by setting up working groups that included all relevant stakeholder groups.
6. The otter is a very charismatic species for which most people feel sympathy.

The general framework conditions after the political change were organized by politics. In Saxony we can see that the responsible persons at different administrative levels realized the necessities to act quickly and effectively.

An important step in this context was the design of the species action plan for the otter where otter conservation changed its quality from more or less isolated actions driven mostly by individual persons at the regional level to a concerted action led and coordinated by the administration at the state level. Already in the conceptual phase of the program, the following questions were asked:

- Which information is needed about the history of the conflict, the ecology of the otter, and damages caused by the species?
- Which stakeholders are crucial and what is necessary to keep the stakeholders in dialogue?
- Are there existing technical mitigation measures or is it necessary to newly develop or improve such measures?
- How successful can technical measures, economic incentives, institutional or legal responses, social processes, or information and education campaigns be to reconcile the conflict?
- What is the relevant level of decision-making and what is the time horizon to be addressed with a conflict reconciliation process?

This phase therefore represented a screening of the conflict to find appropriate ways and the right measures to solve it. On the basis of the results of this screening process, further research, development, testing, and implementation measures were designed as part of the species action plan. Nevertheless, we can find still some gaps regarding the implementation and the evaluation of this species action plan. After the scientific phase, the implementation of the measures to resolve the targeted problems was more or less accidental. A general evaluation of the implemented measures is still lacking until today. Only the development of economic incentives was tackled purposefully.

The agri-environmental measure supporting an extra stocking of ponds with fish for otters rewards ecological services provided by fish farmers to nature conservation. This approach is to be favored both from an ecological and an economic perspective compared to an *ex post* damage compensation (Simila et al. 2006; Schwerdtner and Gruber 2007; Schwerdtner 2008). *Ex post* damage compensation schemes involve higher transaction costs. Correct damage calculation is always more complicated and thus often leads to dissatisfaction of some stakeholders. In contrast, payments for ecological services do not need to be based on an exact calculation of damage; they may just reward the presence of the protected species concerned. These views are shared in other conflicts (compare Suvantola 2013, in this book; Zabel and Holm-Müller 2008). Notwithstanding, rewarding land users for the presence of protected species requires species monitoring but synergies with the monitoring duties according to the EU Habitats Directive could be used.

Although the various compensation schemes turned out to be essential for conflict reconciliation in the Saxon otter-aquaculture conflict, the communication patterns among stakeholders are just as important. This is not only apparent from the analysis of stakeholder perceptions in Saxony but also in the German state of Brandenburg, where this strategy also achieved good results (Zöphel 2008, personal communication). Too strong a focus on a number of economic instruments can even bear the danger of a continuous discussion of the amount and type of payments and this can be a disincentive to further creative responses (Wilson 2004).

Despite the success in Saxony to reconcile the otter-aquaculture conflict, open questions remain. We know little about the economic and societal drivers that can lead to a change in the societal perceptions of the role of rare species even though their ecological status has not changed. A protected species may turn into a scapegoat for environmental or societal problems that have nothing to do with the species as such (Bell 2004). We also know little about the transfer of societal perceptions from a fast growing population of one conflict species, such as recently the great cormorant, on other species that feed on the same resources like the otter. The otter-aquaculture conflict in Saxony is presently reconciled but it still needs careful monitoring to detect in due time and risks that may turn it from a latent into a salient conflict in the future.

**Acknowledgments** This work would not have been possible without the support of private fishermen, voluntary people, conservtionists, as well as of several state authorities (State Ministry for Environment and Agriculture, State Office for the Environment, Agriculture and Geology, Biosphere Reserve Administration Upper Lusatia Heathland and Pond Landscape) in Saxony. It was financed by the EU 5th Framework Program (5th FP) Project "FRAP" (Development of a procedural Framework for Action Plans to Reconcile conflicts between large vertebrate conservation and the use of biological resources: fisheries and fish-eating vertebrates as a model case), contract number EVK 2-CT-2002-00142-FRAP. For geographical analyses we used selected data from ATKIS® under permission 515/04 from Land Survey Office Saxony *(Landesvermessungsamt).*

## References

Anonymus (1978) Präparatoren ausbildung in der DDR. Der Präparator 24:292

Ansorge H, Schipke R, Zinke O (1997) Population structure of the otter, *Lutra lutra*. Parameters and model for a Central European region. Mamm Biol/Z Säugetierkd 62:143–151

Ansorge H, Striese M (1993) Zum Bestand des Fischotters in der östlichen Oberlausitz. Abh Ber Naturkundemus Görlitz 67:14–19

Behrens V (2005) Soziale Konflikte im Arten- und Naturschutz. Diplomarbeit, Universität Hannover 129 p

Behrens V, Rauschmayer F, Wittmer H (2008) Managing international 'problem' species: Why pan-European cormorant management is so difficult. Environ Conserv 35(1):55–63

Bell S (2004) Integrated management of European wetlands. Final project report. Available at http://www.dur.ac.uk/imew.ecproject/. Accessed 2 Mar 2009, p 241

Benndorf J (1986) Erhaltung und Wiederherstellung naturnaher Fließgewässer als Voraussetzung für ihre Mehrzwecknutzung. Naturschutzarb Sachsen 28:21–32

Bernhardt H (1992) Anthropogene geoökologische Veränderungen der kleinen Offenland-Hohlformen am Erzgebirgsrand im 20. Jahrhundert. In: Billwitz K, Jäger K-D, Janke W (eds) Jungquartäre Lebensräume. Springer, Berlin, pp 272–291

Brehm AE (1882) Brehms Thierleben. Zweite umgearbeitete und vermehrte Auflage. Kolorierte Ausgabe, Verlag des Bibliographischen Instituts, Leipzig

Breitenmoser U (1998) Large predators in the Alps: the fall and rise of man's competitors. Biol Conserv 83:279–289

Chardonnet P, Bd Clers, Fischer J, Gerhold R, Jori F, Lamarque F (2002) The value of wildlife. Rev Sci Tech Off Int Epiz 21:15–51

Conover M (2002) Resolving human-wildlife conflicts: the science of wildlife damage management. Lewis Publishers (CRC Press LLC), Boca Raton, London

Fiedler F (1996) Verbreitung und Lebensraum des Fischotters in Sachsen—Abriß der historischen Verbreitung bis zum Anfang des 20 Jahrhunderts. In: Sächsisches Landesamt für Umwelt, Geologie (ed) Artenschutzprogramm Fischotter in Sachsen. Mater Natursch Landschaftspfl. Freistaat Sachsen, Radebeul, pp 7–9

Frohn H-W (2005) Busy for nature—seit fast einem Jahrhundert. Nat Landschaft 80:278–283

Geidezis CL (1999) Food selection of Eurasian otters (*Lutra lutra*) in a fish pond area. Studies in the Oberlausitz pondland, Germany. PhD Thesis, Friedrich-Alexander-University Erlangen-Nürnberg, Department of Zoology I, Erlangen, pp 1–172

Geidezis L, Jurisch C (1996) Beiträge zur Ökologie des Fischotters–Nahrungsuntersuchungen: Ergebnisse aus dem Oberlausitzer Teichgebiet. In: Sächsisches Landesamt für Umwelt, Geologie (ed) Artenschutzprogramm Fischotter in Sachsen. Mater Natursch Landschaftspfl. Freistaat Sachsen, Radebeul, pp 39–50

Geidezis LC (2002) Diet of otters (*Lutra lutra*) in relation to prey availability in a fish pond area in Germany. In: Dulfer R, Conroy JH, Nel J, Gutleb AC (eds) Otter conservation—an example for a sustainable use of wetlands. In: Proceedings 7th international otter colloquium, 14–19 Mar 1998, Trebon. IUCN/SSC Otter Spec Group Bull 19A (spec. Edition), pp 72–76. http://otterspecialistgroup.org/Bulletin/Volume19A/Vol19_A_Index.html. Accessed 13 Dec 2008

Grohmann O, Klenke R (1996) Beiträge zur Ökologie des Fischotters—Methodische Aspekte und praktische Beispiele zur Raumnutzung: Farbmarkierte Nahrung. In: Sächsisches Landesamt für Umwelt, Geologie (ed) Artenschutzprogramm Fischotter in Sachsen. Mater Natursch Landschaftspfl. Freistaat Sachsen, Radebeul, pp 30–32

Gruber B, Reinking B, Calabrese JM, Kranz A, Poledníková K, Poledník L, Klenke R, Valentin A, Henle K (2008) A new method for estimating visitation rates of cryptic animals via repeated surveys of indirect signs. J Appl Ecol 45:728–735

Henle K, Ring I, Klenke R, Jepsen N, Kranz A, Neßhöver C, Rauschmayer F (2013) A generic framework for the development of conflict reconciliation action plans. In: Klenke R, Ring I, Kranz A, Jepsen N, Rauschmayer F, Henle K (eds) Human wildlife conflicts in Europe—fisheries and fish-eating vertebrates as a model case. Springer, Berlin Germany

Hartstock E (2000) Entstehung und Entwicklung der Teichwirtschaft Moritzburg. Schriftenr Sächsische Landesanst Landwirtsch 5:1–107

Hertweck K (1996a) Beiträge zur Ökologie des Fischotters—Methodische Aspekte und praktische Beispiele zur Raumnutzung: Ergebnisse von Geländeuntersuchungen in der sächsischen Schweiz. In: Sächsisches Landesamt für Umwelt, Geologie (ed) Artenschutzprogramm Fischotter in Sachsen. Mater Natursch Landschaftspfl. Freistaat Sachsen, Radebeul, pp 37–38

Hertweck K (1996b) Beiträge zur Ökologie des Fischotters—Nahrungsuntersuchungen: Ergebnisse aus der Sächsischen Schweiz. In: Sächsisches Landesamt für Umwelt, Geologie (ed) Artenschutzprogramm Fischotter in Sachsen. Mater Natursch Landschaftspfl. Freistaat Sachsen, Radebeul, pp 50–53

Hertweck K, Klenke R, Henle K (2002) Estimating the density of otter *Lutra lutra* populations using individual analysis of tracks. In: Dulfer R, Conroy JH, Nel J, Gutleb AC (eds) Otter conservation—an example for a sustainable use of wetlands. In: Proceedings 7th international otter colloquium 14–19 Mar 1998, Trebon. IUCN/SSC Otter Spec Group Bull 19A (spec. Edition), pp 115–119. http://otterspecialistgroup.org/Bulletin/Volume19A/Vol19_A_Index.html. Accessed 13 Dec 2008

Jungbluth JH (2004) Die "Arbeitsgruppe Kuratoren" in der Gesellschaft für Biologische Systematik [GfBS]/the "curators working group" within the Gesellschaft für Biologische Systematik [GfBS]. Decheniana (Bonn) 157:131–134

Jurisch C (1997) Quantifizierung von aufgenommenen Beutefischen aus den Faeces (Spraints) des Eurasischen Fischotters *Lutra lutra*, Linne 1785 (*Mustelidae*). In: Toman A, Hlavac V (eds) Neprodejne, Praha, pp 39–41

Jurisch C, Geidezis CL (1997) Minimum passage rate of fishes through the digestive tract of otters *Lutra lutra* Linne, 1758 (*Mustelidae*). Mammalia 61:123–126

Klemm R (2001) Bericht zur Wirtschaftlichkeit sächsischer Teichwirtschaften 1996/97 bis 1999/2000. Sächsische Landesanstalt für Landwirtschaft, Dresden, p 27

Klenke R (1993) Das Artenschutzprogramm Fischotter des Freistaates Sachsen. Naturschutzarb Sachsen 35:31–36

Klenke R (1996a) Handlungsempfehlungen für ausgewählte Maßnahmekomplexe—Maßnahmen zur Minderung des Gefährdungspotentials. In: Sächsisches Landesamt für Umwelt, Geologie (ed) Artenschutzprogramm Fischotter in Sachsen. Mater Natursch Landschaftspfl. Freistaat Sachsen, Radebeul, pp 67–73

Klenke R (2002) Habitat suitability and apparent density of the Eurasian otter (*Lutra lutra*) in Saxony (Germany). In: Dulfer R, Conroy JH, Nel J, Gutleb AC (eds) Otter conservation—an example for a sustainable use of wetlands. Proceedings 7th international otter colloquium 14–19 Mar 1998, Trebon. IUCN/SSC Otter Spec Group Bull 19A (spec. Edition), pp 167–171. http://otterspecialistgroup.org/Bulletin/Volume19A/Vol19_A_Index.html. Accessed 13 Dec 2008

Klenke R (1996b) Bestandserfassung und Lebensraumkartierung des Fischotters. In: Sächsisches Landesamt für Umwelt, Geologie (ed) Artenschutzprogramm Fischotter in Sachsen. Mater Natursch Landschaftspfl. Freistaat Sachsen, Radebeul, pp 24–27

Klenke R, Gruber B, Schwerdtner K, Vallentin A (2005) Influence of landscape structure in Upper Lusatia (Saxony, Germany) on damage in fish ponds caused by otters. In: 5th European vertebrate pest management conference 5–8 Sep 2005, Abstracts, Budapest, p 79

Klenke R, Tschierschke A, Gruber B, Lampa S, Hempel U, Geissler S, Helm C, Hofmann C, Kalbitz J, Kaulfuss J, Liu Z, Henle K (2008) Räumliche und zeitliche Aktivitätsmuster von

Fischottern (*Lutra lutra* Linnaeus, 1758) anhand von frischen Markierungsfunden in einer Teichlandschaft. Säugetierkdl. Inf 6(37):229–246

Klueting E (2003) Die gesetzlichen Regelungen der nationalsozialistischen Reichsregierung für den Tierschutz, den Naturschutz und den Umweltschutz. In: Radkau J, Uekötter F (eds) Naturschutz im nationalsozialismus. Campus, Frankfurt/M, New York, pp 77–106

Kranz A (2000) Otters (*Lutra lutra*) increasing in Central Europe: from the threat of extinction to locally perceived overpopulation? Mammalia 64:357–368

Kubasch H (1984) Zum Vorkommen des Fischotters, *Lutra lutra* (L. 1758) im Bezirk Dresden. Veröff Mus Westlausitz Kamenz 8:15–28

Kubasch H (1987) Über Fischotter-Verluste im Bezirk Dresden zwischen 1945 und 1985. Veröff Mus Westlausitz Kamenz 10:45–49

Kubasch H (1996a) Verbreitung und Lebensraum des Fischotters in Sachsen—Die Verbreitung von 1930 bis 1993. In: Sächsisches Landesamt für Umwelt, Geologie (ed) Artenschutzprogramm Fischotter in Sachsen. Mater Natursch Landschaftspfl. Freistaat Sachsen, Radebeul, pp 9–12

Kubasch H (1996b) Zur Geschichte des Fischotterschutzes in Sachsen. In: Sächsisches Landesamt für Umwelt und Geologie (ed) Artenschutzprogramm Fischotter in Sachsen. Mater Natursch Landschaftspfl. Freistaat Sachsen, Radebeul, pp 5–7

Lampa S, Henle K, Klenke R, Gruber B (2011) Analysing population dynamics ofEurasian otters (*Lutra lutra*) using non-invasive genetic capture-mark-recapture (CMR):pitfalls and solutions. In: Proceedings of the IUCN XI International Otter Colloquium.Hystrix It J Mamm (n.s.) Supp. (2011): 103.http://www.italian-journal-of-mammalogy.it/article/view/4744/4547

Lampa S, Gruber B, Henle K, Höhn M (2008) An optimisation approach to increase DNA amplification success of otter faeces. Conserv Genet 9:201–210

Lekan TM (2005) "It shall be the whole landscape!" the reich nature protection law and regional landscape planning in the third reich. In: Cioc M, Brüggemeier F-J, Zeller T (eds) How green were the nazis? Nature, environment, and nation in the third reich. Ohio University Press, Athens, pp 73–100

Sachsen LfUG (ed) (1996) Artenschutzprogramm Fischotter in Sachsen. Mater Natursch Landschaftspfl, 1st edn. Freistaat Sachsen, Radebeul

Meyer M (1994) Einige Bemerkungen zum Vorkommen des Fischotters in Nordwestsachsen, Veröff. Naturkundemuseum Leipzig, pp 27–35

Myšiak J, Schwerdtner-Manez K, Ring I (2013) Comparative analysis of the conflicts between carp pond farming and the protection of otters (*Lutra lutra*) in Upper Lusatia and South Bohemia. In: Klenke R, Ring I, Kranz A, Jepsen N, Rauschmayer F, Henle K (eds) Human wildlife conflicts in Europe—fisheries and fish-eating vertebrates as a model case. Springer, Berlin Germany

Otis DL, Burnham KP, White GC, Anderson DR (1978) Statistical inference from capture data on closed animal populations. Wildl Monogr 62:1–135

Piechocki R (2007) Genese der Schutzbegriffe 6. Artenschutz Nat Landschaft 82:286–287

Poledník L (2005) Otters (*Lutra lutra* L.) and fishponds in the Czech Republic: interactions and consequences. PhD thesis, Palacky University Olomouc, pp 109

Poledníková K, Kranz A, Poledník L, Myšiak J (2013) Otters causing conflicts - The fish farming case of the Czech Republic. In: Klenke R, Ring I, Kranz A, Jepsen N, Rauschmayer F, Henle K (eds) Human 1003 wildlife conflicts in Europe—fisheries and fish-eating vertebrates as a model case. Springer, Berlin Germany

RdB Dresden (1978) Schutzmaßnahmen für die vom Aussterben bedrohten Tierarten Seeadler, Schwarzstorch, Uhu, Sperlingskauz, Kranich und Elbebiber. In: Weisung Nr. 1/1978. Rat des Bezirkes Dresden, Stellvertreter des Vorsitzenden für Land- Forst- und Nahrungsgüterwirtschaft, Mitglied des Rates für Umweltschutz und Wasserwirtschaft, pp 1–4

RdB Dresden (1981) Zum Vorkommen und zur Bearbeitung von Schäden durch den Fischotter (*Lutra lutra* L.). Rat des Bezirkes Dresden, Abteilung Forstwirtschaft, Bezirksjagdbehörde/ Bezirksnaturschutzorgan, pp 1–2

Reinhardt I, Kluth G (2007) Leben mit Wölfen. Leitfaden für den Umgang mit einer konfliktträchtigen Tierart in Deutschland. BfN-Skripten, vol 201, p 181. http://www.bfn.de/fileadmin/MDB/documents/service/skript201.pdf. Accessed 27 Dec 2008

Reuther C (1993) *Lutra lutra*—Fischotter. In: Stubbe M, Krapp F (eds) Raubsäuger—*Carnivora (Fissipedia)*. Teil II: *Mustelidae 2, Viveridae, Herpestidae, Felidae*. In: Niethammer J, Krapp F (eds) Handbuch der Säugetiere Europas, vol 5. Aula, Wiesbaden, pp 907–961

Riebe H (1994) Der Fischotter *Lutra lutra* L. 1758 in der Sächsischen Schweiz. Nationalpark Sächsische Schweiz 2:63–72

Roth M, Walliser G, Henle K, Hertweck K, Binner U, Waterstraat A, Klenke R, Hagenguth A (2000) Habitatzerschneidung und Landschaftsstruktur—Auswirkungen auf populationsökologische Parameter und das Raum-Zeit-Muster marderartiger Säugetiere. Laufener Seminarbeitr 2(00):47–64

Schreyer RM, Jahn A (2006) Erfahrungen mit Fischotterquerungshilfen im Biosphärenreservat Oberlausitzer Heide- und Teichlandschaft. Naturschutzarb Sachsen 48:50–64

Schwerdtner K, Gruber B (2007) A conceptual framework for damage compensation schemes. Biol Conserv 134:354–360

Schwerdtner K, Ring I (2005) Development of policy instruments in Germany. In: FRAP project report (unpublished), Helmholtz Centre for Environmental Research (UFZ), Leipzig, pp 1–44

Schwerdtner MK (2008) Zur Umsetzung von Artenschutz. Eine ökologisch-ökonomische Analyse. Thesis (Dr. agr.). Naturwissenschaftliche Fakultät III, Institut für Agrar- und Ernährungswissenschaften, Martin-Luther-Universität, Halle/Saale, pp 1–205

Seiche K, Wünsche A (1996) Kormoran und Graureier im Freistaat Sachsen. In: Sächsisches Staatsministerium für Umwelt und Landesentwicklung (ed) Materialien zu Naturschutz und Landespflege, Dresden, 1/1996, pp 1–102

Similä J, Thum R, Varjopuro R, Ring I (2006) Protected species in conflict with fisheries: the interplay between European and national regulation. JEEPL 5:432–445

SLfL (2005a) Ökonomie der Karpfenteichwirtschaft. Schriftenr Sächsische Landesanst Landwirtsch 1(10):1–76

SLfL (2005b) Zahlen zur Binnenfischerei im Freistaat Sachsen. In: Sächsische Landesanstalt für Landwirtschaft (ed) Jahresbericht 2004. Fischereibehörde, Königswartha, pp 1–46

SLfL (2006) Zahlen zur Binnenfischerei im Freistaat Sachsen. In: Sächsische Landesanstalt für Landwirtschaft (ed) Jahresbericht 2005. Fischereibehörde, Königswartha, pp 1–41

SLfL (2007) Zahlen zur Binnenfischerei im Freistaat Sachsen. In: Sächsische Landesanstalt für Landwirtschaft (ed) Jahresbericht 2006. Fischereibehörde, Königswartha, pp 1–48

SMUL (2007) Richtlinie des Sächsischen Staatsministeriums für Umwelt und Landwirtschaft zur Förderung von flächenbezogenen Agrarumweltmaßnahmen und der ökologischen Waldmehrung im Freistaat Sachsen (Förderrichtlinie Agrarumweltmaßnahmen und Waldmehrung—RL AuW/2007) vom, 13 Nov 2007. Sächsisches Amtsblatt 49:1694–1734

SMUL (2008) Richtlinie des Sächsischen Staatsministeriums für Umwelt und Landwirtschaft zur Förderung der Aquakultur und der Fischerei (Förderrichtlinie Aquakultur und Fischerei—RL AuF/2007) vom, 1 Nov 2007. Sächsisches Amtsblatt 5:213–217

Steffens R (1996) Artenschutzprogramm Fischotter in Sachsen—Einführung. In: Sächsisches Landesamt für Umwelt und Geologie (ed) Artenschutzprogramm Fischotter in Sachsen. Mater Natursch Landschaftspfl. Freistaat Sachsen, Radebeul, pp 4–5

Stubbe M (1969) Populationsbiologische Untersuchungen an Mustela-Arten. Hercynia (NF) 6:306–318

Stubbe M (1977) Der Fischotter, *Lutra lutra*, in den Südbezirken der DDR. (Ergebnisse der "Aktion Fischotter 1972", Nr. 9). Abh Ber Naturkundemus Görlitz, vol 51 (5), pp V/1–V/19

Stubbe M (1991) Dokumentation der wissenschaftlichen Arbeiten an bedrohten Tierarten und im Rahmen der Todesursachenforschung am zoologischen Institut der Martin-Luther-Universität Halle-Wittenberg. In: Brandenburg L (ed) Schutz des Fischotters—Dokumentation der ersten Fachtagung im Land Brandenburg. Landesumweltamt Brandenburg, Potsdam, p 43

Stubbe M, Heidecke D (1991) Die Verbreitung des Fischotters in den östlichen Ländern Deutschlands und Ergebnsse der interdisziplinären Totfundanalytik. In: Brandenburg Landesumweltamt (ed) Schutz des Fischotters—Dokumentation der ersten Fachtagung im Land Brandenburg. Landesumweltamt Brandenburg, Potsdam, pp 4–10

Suvantola L (2013) The Golden eagle compensation scheme in Finland as an example of incentive measures — Potential for conflict management? In: Klenke R, Ring I, Kranz A, Jepsen N, Rauschmayer F, Henle K (eds) Human 1003 wildlife conflicts in Europe—fisheries and fish eating vertebrates as a model case. Springer, Berlin Germany

Tausch, F (2004) Panoptikum der Glasaugen: Kuckuckseier in Schachteln und grüne Mambas im Keller—in Deutschland blüht ein schwunghafter Handel mit geschützten Tieren, Sz-online/ Sächsische Zeitung, 18 Feb 2004. http://www.sz-online.de/nachrichten/artikel.asp?id=575700. Accessed 14 July 2009

Thum R, Schwerdtner K, Ring I (2003) Artenschutz und Teichwirtschaft. Rechtliche und institutionelle Rahmenbedingungen in Deutschland unter besonderer Berücksichtigung des Freistaates Sachsen. UFZ Diskussionspapier, Nr. 11/2003, Leipzig, p 48

Thum R (2004) Rechtliche Instrumente zur Lösung von Konflikten zwischen Artenschutz und wirtschaftlicher Nutzung natorlicher Ressourcen durch den Menschen am Beispiel Kormoranschutz und Teichwirtschaft. Natur und Recht 9: 580–587

Tschirch W (1996) Gefährdung des Fischotters—Gefährdung durch Schadstoffe. In: Sächsisches Landesamt für Umwelt, Geologie (ed) Artenschutzprogramm Fischotter in Sachsen. Mater Natursch Landschaftspfl. Freistaat Sachsen, Radebeul, pp 62–63

Usbeck H, Ebert A, Usbeck F, Schwerdtner K, Ring I (2004) Socio-economic report of the German model region "Oberlausitz-Niederschlesien". In: FRAP project report, Usbeck GmbH and UFZ Centre for Environmental Research, Leipzig, Germany

VEB Meliorationsbau Dresden (1986) Binnenfischerei: Schutzanlagen gegen Wildtiere. Pat. MELIOR 12–050 00 01. Direktor des VEB Meliorationsbau Dresden, German Democratic Republic, pp 1–9

VKdDDR (1953) Gesetz zur Regelung des Jagdwesens vom 25 Nov 1953 (GBl. I Nr. 125 S. 1175), Volkskammer der Deutschen Demokratischen Republik. Staatsverlag der Deutschen Demokratischen Republik, Berlin

VKdDDR (1962) Achte Durchführungsbestimmung zur Regelung des Jagdwesens vom 14.02. 1962, Volkskammer der Deutschen Demokratischen Republik. In: Ministerium für Land-, Forst- und Nahrungsgüterwirtschaft. In: Jagdbehörde O (ed) Gesetzblatt I.II. Staatsverlag der Deutschen Demokratischen Republik, Berlin, p 255

VKdDDR (1985a) Gesetz über das Jagdwesen der Deutschen Demokratischen Republik—Jagdgesetz—vom 15. Juni 1984 (GBl. I Nr. 18 S. 217), Volkskammer der Deutschen Demokratischen Republik. In: Ministerium für Land, Forst- und Nahrungsgüterwirtschaft – Oberste Jagdbehörde (ed) Jagdrecht. Textausgabe. Staatsverlag der Deutschen Demokratischen Republik, Berlin, pp 17–23

VKdDDR (1985 b) Erste Durchführungsbestimmung zur Naturschutzverordnung—Schutz von Pflanzen und Tierarten—(Artenschutzbestimmung) vom 1. Oktober 1984 (GBl. I Nr. 31 S. 381), Volkskammer der Deutschen Demokratischen Republik. In: Ministerium für Land-, Forst- und Nahrungsgüterwirtschaft—Oberste Jagdbehörde (ed) Jagdrecht. Textausgabe. Staatsverlag der Deutschen Demokratischen Republik, Berlin, pp 177–185

Völlings A (1999) Umwelt und Verkehr—Chancen für die eigene Mobilität. In: Sächsiches Landesamt für Umwelt und Geologie, Stabsstelle 1, Öffentlichkeitsarbeit (ed). http://www.smul.sachsen.de/umwelt/download/klima/UmweltVerkehr.pdf. Accessed 13 Oct 2010

White GC, Burnham KP (1999) Program MARK: survival estimation from populations of marked animals. Bird Study, 46 Supplement, pp 120–138. http://warnercnr.colostate.edu/~gwhite/mark/euring.PDF. Accessed 11 June 2011

Wilson DC (ed) (2004) Discourse analysis. In: FRAP project report. UFZ Helmholtz Centre for Environmental Research, Leipzig, p 355

Woodroffe R, Thirgood S, Rabinowitz A (2005) People and wildlife—conflict or coexistence?. Cambridge University Press, Cambridge 497

Zabel A, Holm-Müller K (2008) Performance payments for carnivore conservation in Sweden. Cons Biol 22(2):247–251

Zinke O (1991) Die Todesursachen der im Museum der Westlausitz Kamenz von 1985–1991 eingelieferten Fischotter Lutra lutra (L., 1758). Veröff Mus Westlausitz Kamenz 15:57–63

Zinke O (1994) Ein aktueller Nachweis illegaler Bejagung des Fischotters. Veröff Mus Westlausitz Kamenz 17:90–92

Zinke O (1996) Gefährdung des Fischotters—Gefährdungsschwerpunkte und Verlustursachen aus der Totfundanalyse. In: Sächsisches Landesamt für Umwelt und Geologie (ed) Artenschutzprogramm Fischotter in Sachsen. Mater Natursch Landschaftspfl. Freistaat Sachsen, Radebeul, pp 53–59

Zöphel U, Steffens R, Klenke AR (1996) Strategien und Grundsätze des Fischotterschutzes in Sachsen. In: Sächsisches Landesamt für Umwelt und Geologie (ed) Artenschutzprogramm Fischotter in Sachsen. Mater Natursch Landschaftspfl. Freistaat Sachsen, Radebeul, pp 64–67

Zwirner O, Wittmer H (2004) Germany. In: Wilson DC (ed) Discourse analysis. FRAP project report. UFZ Helmholtz centre for environmental research, Leipzig, pp 125–155

Fishermen in South Bohemia hauling the net through the pond. *Photo:* Lukáš Poledník

# Comparative Analysis of the Conflicts Between Carp Pond Farming and the Protection of Otters (*Lutra lutra*) in Upper Lusatia and South Bohemia

**Jaroslav Myšiak, Kathleen Schwerdtner Máñez and Irene Ring**

**Abstract** Protection of the Eurasian otter (*Lutra lutra*) has been successful in recent years but is increasingly running into conflict due to the damage caused on the fish stock in ponds aquaculture. In this chapter we compare the conflicts in two regions with a long history of carp farming—Upper Lusatia in Saxony (Germany) and South Bohemia in the Czech Republic, teasing out the factors which amplify or attenuate the conflicts. We show that financial compensation for the damage occurred is insufficient to mitigate the conflict or stop it from becoming worse. To succeed in long term, a set of mitigation measures, both financial and non-financial, should be deployed. These measures perform best when they are spatially differentiated and tailored to the size of farms and farming practices.

J. Myšiak (✉)
Fondazione Eni Enrico Mattei and Euro-Mediterranean Center for Climate Change, Castello 5252, 30122, Venice, Italy
e-mail: jaroslav.mysiak@cmcc.it

K. Schwerdtner Máñez
Department of Social Sciences, Leibniz Center for Tropical Marine Ecology, Fahrenheitstrasse 6, 28359, Bremen, Germany
e-mail: kathleen.schwerdtner@zmt-bremen.de

I. Ring
Department of Economics, UFZ—Helmholtz Centre for Environmental Research Leipzig, Permoserstr. 15, 04318, Leipzig, Germany
e-mail: irene.ring@ufz.de

R. A. Klenke et al. (eds.), *Human-Wildlife Conflicts in Europe*, Environmental Science and Engineering, DOI: 10.1007/978-3-540-34789-7_7,

# 1 Introduction

The fish-pond landscapes in Central and Eastern Europe have immense historical and cultural value, and are highly significant as a habitat for numerous endangered species. As typical examples of cultural landscapes, their maintenance depends on fish-pond farming. Although the protection of endangered piscivorous predators in these landscapes, such as the Eurasian otter (*Lutra lutra*), has been successful in recent years, it is increasingly running into conflict due to the damage caused by this species in the fish ponds.

Formerly widespread throughout Europe, the otter underwent a rapid decline in numbers during the twentieth century. For centuries otters were regarded as pest whose damage to fish populations caused high losses in aquaculture. As a result of persecution, otter populations dwindled and became endangered. Under strict species protection policies starting in the second half of the twentieth century, the persecution and killing of otters were outlawed, allowing otter populations to slowly regenerate. When the EU Habitats Directive entered into force in 1992, the otter finally became a "strictly protected species of common interest" (Council of the European Community 1992), meaning that catching, killing and disturbing otters as well as damaging and disturbing their habitats became strictly prohibited in all European member states. Yet, the slow regeneration of otter populations in Central Europe was accompanied by complaints about otter damage, especially by carp farmers (Kranz 2000; Bodner 1998).

We present a comparative analysis of conflicts between otter protection and aquaculture in two regions with a long history of carp-farming—Upper Lusatia in Saxony (Germany) and South Bohemia in the Czech Republic (Fig. 1).[1] Whereas the previous two chapters go into much more detail regarding the specifics of each of the conflicts (compare Klenke et al. 2013 and Poledníková et al. 2013, in this book), the purpose of this chapter is to tease out the similarities and differences between these two cases. A critical comparison lends itself to the identification of factors which amplify or attenuate conflicts. Both areas are characterized by numerous artificial ponds for carp farming (mostly *Cyprinus carpio*), allowing for viable otter populations, spreading also into neighboring habitats. Furthermore, both areas experienced substantial structural changes, resulting in new political, economic and social conditions after the political upheaval in 1989. But this alone is not sufficient to explain the recent developments in the respective conflicts.

We first briefly introduce the two study areas in terms of their natural and socio-economic features. Secondly, the fisheries sectors are compared, after which the otter populations in the two areas are characterized. In the third step, the respective compensation schemes as a means to contribute to conflict mitigation are presented and analyzed. Combined with information on the perception of the conflict

[1] The analysis reflects the period up to 2004. More recent developments of the conflicts reported here and the attempts to cope with are not addressed in this article.

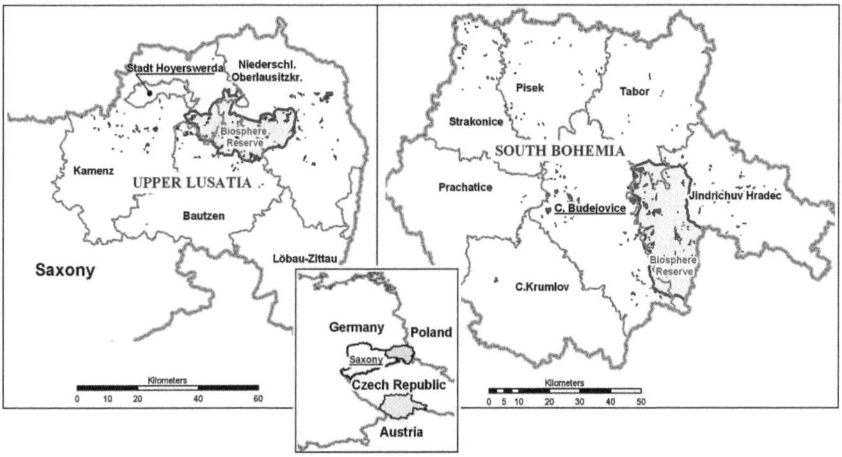


**Fig. 1** Geographical overview of the study areas with major ponds: *Left*—Upper Lusatia situated in Saxony (Germany), *right*—South Bohemia (Czech Republic)

and the respective compensation schemes by relevant stakeholder groups, recommendations for the improvement of such payments are presented.

# 2 Natural and Socio-Economic Characteristics of Upper Lusatia and South Bohemia

## *2.1 Location, History, and Land Use*

Upper Lusatia belongs to the state of Saxony and is situated in the south-east of Germany close to the border with the Czech Republic (to the south) and Poland (to the east). The South Bohemian pond region is located in the southern part of the Czech Republic (CR) on the border with Austria (to the south) and Bavaria/Germany (to the south-west).

The development of pond fisheries in both Upper Lusatia and South Bohemia was prompted by their suitable geological, geographical, and hydrological conditions. Ponds were mainly built on regularly flooded moor land or sand and clay soils that are usually poor in nutrients and therefore unsuitable for agriculture. Both areas have numerous creeks and rivers, another main factor for the existence of ponds. Due to their unique cultural and landscape value, parts of both regions have been declared as biosphere reserves.

In Saxony, the biosphere reserve "Upper Lusatia Heath and Pond Landscape" was designated in 1994, whereas the biosphere reserve "Třeboň Basin" in South Bohemia has existed since 1977. The name "Lusatia" (originally "Łuža") means marshland or swampland and was given to the region by the Slavic settlers because

**Table 1** Selected socio-economic descriptors of the compared regions (2002)

| | South Bohemia | Czech Republic total | Upper Lusatia | Saxony | Germany total |
|---|---|---|---|---|---|
| Total area (km$^2$) | 10,056 | 78,868 | 4,400 | 18,413 | 357,027[a] |
| Numbers of inhabitants | 625,000 | 10,206,000 | 607,000 | 4,345,000 | 83,000,000 |
| Changes between 1990 and 2002 | +2,200 | −155,600 | −78,000 | −381,900[b] | +2,823,000[b] |
| Population density (persons/km$^2$) | 62 | 129 | 137 | 236 | 231 |
| Share of agricultural/ forestry land use (%) | 52/30 | 54/33 | 47/34 | 56/27 | 55/30 |
| Number of employees in agriculture, forestry, and fisheries | 26,900 | 194,000 | 8,600 | 52,560 | 929,000 |
| Unemployment (%) | 6.65 | 9.81 | 22.5[c] | 19.2 | 10 |
| GDP per capita (EUR) | 5,852 | 6,195 | 14,300 | 16,900 | 25,400 |

*Sources*: Data from Czech Republic—Český Statistický Úřad (2003), from Germany—Usbeck et al. (2004), and [a] Statistisches Bundesamt Deutschland (2004), [b] Data available only up to 2000, [c] data from 2001

of its numerous moors and inland waters. Similarly, the river, which supplies most of the numerous ponds in the Třeboň Basin, is called the Lužnice. The Upper Lusatia Heath and Pond Landscape Biosphere Reserve encompasses 301 km$^2$ and is inhabited by 12,800 people. The Třeboň Basin Biosphere Reserve covers an area of 700 km$^2$ and has a total population of 28,500.

Upper Lusatia comprises the districts of Kamenz, Bautzen, and Niederschlesischer Oberlausitzkreis, as well as the town of Hoyerswerda. The district of Löbau-Zittau is also included in the analysis, a highland area with fewer and smaller ponds. The total area encompasses 133 municipalities on 4,400 km$^2$. With just over 10,000 km$^2$, the pond region in South Bohemia is twice as big as Upper Lusatia (Table 1). South Bohemia encompasses seven districts (České Budějovice, Český Krumlov, Jindřichův Hradec, Písek, Prachatice, Strakonice, Tábor) and 623 municipalities. Nevertheless, the two areas have approximately the same number of inhabitants (about 607,000 in Upper Lusatia and 625,000 in South Bohemia). Both study areas are considered rural regions because of their low population densities.

The dominant land use in Upper Lusatia is agriculture (47 %), followed by forestry (34 %). There exist about 1,000 ponds in Saxony, covering more than 8,000 ha. More than 5,000 ha pond area and 71 enterprises are located in Upper Lusatia, revealing that much of Saxony's pond fisheries are concentrated in this area (Usbeck et al. 2004). The political upheaval and German reunification in 1990 led to lasting structural changes in the area. Many industrial enterprises and large agricultural cooperatives collapsed, causing high unemployment which in turn led to high migration losses, especially among younger people. In fact since 1990, Upper Lusatia has lost approximately 78,000 inhabitants or 11.4 % of its total population (Table 1). At the same time, the area lost about 40,000 jobs; the share of people working in the primary sector (agriculture, forestry, and fisheries) dropped from 7.5 to 3.6 % (Usbeck et al. 2004). The monthly net income in the region is significantly below the average in Saxony, while the unemployment rate in 2001 of 22.5 % was more than twice the German average (StaLA Sachsen 2001). GDP is 14,300 € per capita, just 57 % of the German average.[2]

The prevailing forms of land use in South Bohemia are agriculture (52 %) and forestry (30 %). Numerous large and small ponds cover 25,000 ha, about 2.5 % of the region's area. South Bohemia does not rank among the country's key industrial areas; in 2001 it accounted for just 5.1 % of the Czech Republic's industrial turnover total, although some 11 % of total agricultural output is produced here. The political and economical changes after 1989 did not affect the population, which by 2002 had slightly increased. The average gross wage in 2001 was ca. 443 €[3] per inhabitant, 88.5 % of the Czech average. While in other sectors the average gross wage is lower than the national average, it is slightly higher for the primary sector. Registered unemployment in late December 2002 was 6.65 %, South Bohemia ranking second best after the capital Prague. Although the area's GDP accounts for only 5.4 % of that of the Czech Republic, converted to GDP per inhabitant (5,852 €), it amounts to just 87.8 % of the national average and ranks fourth in the Czech Republic.

## *2.2 Carp Pond Farming*

Both study areas have a long tradition of carp-pond farming going back to the thirteenth century. Driven by the continuing land settlement and high population growth, fish-pond construction boomed in the fourteenth and fifteenth century. At that time the total pond area in both countries was as much as twice that nowadays. The high popularity of fish-pond farming was due to the higher profitability of fish ponds compared to cropping on low fertility soil (Hartstock 2000). The 30 Years' War (1618–1648) precipitated a decline from which fish-pond farming in Central and Eastern Europe never completely recovered. Since the eighteenth and nineteenth centuries many ponds have been transformed into fields and in many places were replaced by sugar beet plants. Towards the end of the nineteenth century, the systematic use of additional feeding (Benecke 1885 in Thiem 2002; Vogel 1928 in

[2] For the German case study, socio-economic indicators are reported for both the study area Upper Lusatia and the state of Saxony as a whole. This is because a number of indicators – especially those related to the fisheries sector presented later – are unavailable for just the study area itself.

[3] We adopted the exchange rate of 1 Euro (EUR) = 0.85797 US Dollar (USD) and 31.68 Czech Krone (CZK) as of 02.01.2002.

**Table 2** Comparison of main production characteristics of the two fish-pond areas

| | South Bohemia | Czech Republic total | Upper Lusatia | Saxony | Germany total |
|---|---|---|---|---|---|
| Total fish production (tons) | >10,000[a] | ~20,000 | n.a. | 2,931[g] | ~36,000[e] |
| carp (tons) | ~9,000[a] | ~18,000 | n.a. | 2,620[g] | ~11,000[e] |
| Fish-pond area (ha) | ~25,000[a] | ~52,000 | 5,016[d] | 8,419[g] | 30,000[f] |
| Number of ponds | ~7,000 | ~50,000 | n.a. | ~1,000[j] | n.a |
| Number of employees | n.a. | ~2,000[b] | n.a. | ~600[h] | n.a |
| Stock density (kg/ha) | n.a. (in some cases more than 1,200[c]) | ~600[b] | ~600[i] | ~600[i] | n.a |

*Sources*: [a] Český Statistický Úřad (2003); [b] CFFA (2003), [c] Bureš (2000), Faina (2000), Kranz and Knolleisen (1998), [d] Usbeck et al. (2004), [e] FEAP (2004), [f] Wedekind et al. (2001), [g] SLfL (2002), [h] StaLa Sachsen (1995), [i] Thiem (2002) and [j] SMUL (2004)

Thiem 2002) and fertilizers (Demoll 1925 in Thiem 2002) allowed increased pond productivity. This marked the beginning of intensification and mechanization in aquaculture (Thiem 2002, Table 2).

Under the communist system installed in both countries, fish-pond farms were nationalized and merged to form large state-owned enterprises. Fish production was carried out intensively, often involving high additional feeding, fertilizing and stocking rates. Following the political changes in 1989 the fish ponds were returned to their previous owners or privatized. While in Saxony the carp production has diminished ever since (from 6,686 tons in 1989 to 2,620 tons in 2002; SLfL 2002), production in South Bohemia remained stable or even increased. The production decrease in Saxony is mainly a result of the changed economic conditions, including reduced demand. Additionally, the fact that most ponds are included in one or more conservation support programs has led to a generally more extensive production scheme (Thiem 2002). Due to private data protection and the lack of periodical statistical surveys, only little statistical information about current fish-pond farming in terms of number of employees, turnout, and profit is available at a regional and local level.

In Saxony, some 8,419 ha of pond area exists today, almost all of which is used for carp production. This makes the state of Saxony the second biggest carp producer in Germany, following Bavaria. Large companies dominate carp production in Saxony, the 15 biggest operating on 56 % of the total pond area. Out of the total of 170 companies, 55 work on primary occupation, 99 on secondary occupation, and 16 producers are angling or conservation associations (SLfL 2002). Annual fish production varied between 3,351 tons in 2001 and 2,931 tons in 2002 (BfLuE 2002). The average pond area is 153 ha; however, in Upper Lusatia the average pond area per company varies between 218 ha in the district of Bautzen and 2.4 ha in the district of Löbau-Zittau, where for most of the fisheries employees this is a second occupation.[4] Saxony's average production is about 600 kg/ha and therefore rather low compared to 2,000 kg/ha during GDR times (Thiem 2002). Some 89 % of all the fish produced is carp. In 1994, 622 people worked in aquaculture in Saxony (StaLa Sachsen 1995).

In South Bohemia ca. 7,000 fish ponds have a total area of about 25,000 ha, almost half the total pond area in the Czech Republic (ca. 52,000 ha). Carp dominates production (ca. 87 %), with other species, such as salmonids, tench (*Tinca tinca*), and whitefish, having less importance. More than half the fish produced in the region is exported (CFFA 2003). All in all, more than half the Czech Republic's output of fish is produced here (ČSÚ 2003). In terms of company size, 124 of the 131 companies have fewer than 10 employees; most of them work on a part-time basis (ČSÚ 2003). By contrast, only nine companies have more than 10 employees, and just one has over 100 employees. A small number of companies owns most of the fish ponds. One single company operates on ca. 400 fish ponds mostly located in the Třeboň Basin Biosphere Reserve. The total area of these ponds amounts to 7,000 ha fish ponds (~30 % of the region's pond area), 1,213 ha of which are situated in nature reserves. The company, which is the biggest single carp producer in Europe, produces ca. 3,000 tons of fish annually—about the same as the total production of Saxony. According to the CFFA (2003), average production in fish ponds in the Czech Republic accounts for ca. 450 kg/ha. However, in many ponds the stocking density exceeds 1,000 and in some cases even 1,200 kg/ha (Bureš 2000; Faina 2000; Kranz et al. 1996).

## 2.3 The Otter Populations in Saxony and South Bohemia

The Eurasian otter is naturally distributed among inland waters all over Europe, its populations stretching from the Iberian Peninsula to Northern Siberia and from Scandinavia to South India. In Central Europe, the western border of the current otter distribution covers extensive distances across Germany. To the east and south, German, Polish and Czech otter populations are more or less connected (Reuther et al. 2002).

Persecuted for centuries, the study areas with their numerous fish ponds played a crucial role as refuges for otter populations from which, having been protected, they then spread to neighboring states. There are numerous references to the severe persecution of otters in the past. In Saxony, one of the central aims of the Saxon Fishery Association after its foundation in 1884 was the persecution of the otter. Due to intensive hunting, the population collapsed in 1903 (Fiedler 1996).

Under the German Federal Conservation Act, the otter is a specially protected species. It is also protected under the Hunting Act, which guarantees a closed season throughout the year. In the Czech Republic the otter is listed under the Act No. 114/1992 on the Protection of Nature and the Landscape and Directive 395/1992 as a

[4] Data gathered from the Saxon State Office for Agriculture, Department of Fisheries in 2004, own calculations.

severely endangered species. In addition, the otter is covered by the Hunting Act. Although the Hunting Act allows an open season for the otter throughout the year, it may only be stalked by hunters who have obtained a special permit from the conservation authorities beforehand.

The vital otter population in the East of Germany (Saxony, Brandenburg, Mecklenburg-Western Pomerania) is of special importance for the conservation and spread of the species into adjacent areas of Germany (Lower Saxony, Schleswig-Holstein, Saxony-Anhalt, Thuringia) and for connectivity with populations in Bohemia and Bavaria (Reuther 1999). Until the end of the nineteenth century, otters were present on all suitable inland waters and wetlands in Germany. By the mid-twentieth century, however, the species had disappeared in most areas in western and central Germany. Since the early 1980s, there have been clear indications that otters have started to recolonize their former habitats. Nowadays, approximately 20 % of German territory is populated by otters again (Reuther 1999). The German study area Upper Lusatia hosts one of the most abundant otter populations in central Europe (Ansorge and Striese 1993). However, there is high uncertainty about otter numbers. Ansorge et al. (1997) estimate 400 adult animals for the area, Klenke (1996) mentions a population size of 200 (−100, +200) for the Upper Lusatia and East Saxony. Densities may vary considerable: from 3–6 otters per 100 $km^2$ (Ansorge and Striese 1993) to 30 animals per 100 $km^2$ (Grohmann and Klenke 1996).

In the Czech Republic, in the early 1990s the otter was distributed on only 25–30 % of the area in three isolated groups (Toman and Kadlečík 1992). Later on (1998–2001), an increase in the South Bohemian population was observed. The entire Czech population currently numbers ca. 800–1,100 individuals distributed over 40 % of the country's area (Roche 2003). Recent mapping of otter distribution suggests a continuous spread and increase in numbers mainly in southern Bohemia and the Czech-Moravian Highlands. The main cause of the present expansion is probably the decrease in water pollution together with more intensive fish farming following the restitution program in 1989 (Toman 1998a; Kučerová 2000). The study area contains the largest and most stable otter population in the Czech Republic, which extends across the South Bohemian fish ponds (Třeboň basin), the Šumava Mountains, and the Czech-Moravian Highlands. This population is connected to the remaining populations in the Bavarian forests and the Austrian Waldviertel .

## 3 Compensation Schemes as a Means to Conflict Mitigation

### *3.1 Rationale for Compensation Schemes in Biodiversity Conservation*

From an economic perspective, biodiversity (or the services it provides) represents a public good. It does not have the characteristics of private, marketable goods. Therefore, it is the state that has to create adequate framework conditions for sufficient biodiversity conservation. The conservation of protected species is considered to be a responsibility that must be shared by the whole society, since the entire society may benefit from it (Fourli 1999; Hampicke 2005). Conservation measures, such as species protection or the establishment of nature reserves, are often associated with additional costs for land users, who are restricted in their management choices. Depending on property and use rights of the country concerned, at least part of these costs should be compensated (Bromley and Hodge 1990; Hanley et al. 1998; Ring 2004).

In our case, fish farmers may be restricted in production methods especially when located in a protected area. If they adopt environmentally sound land-use practices, these practices usually involve less intensive production methods and thus, less economic profit. Costs are also caused by protected species that feed on commercial fish. Due to the protection status of species, fish farmers are limited in their possibilities to avoid such damage. This holds even more for fish farms located in protected areas, where the use of technical mitigation measures can be restricted or forbidden.

To justify public expenditure for conservation projects, non-marketable benefits of biodiversity conservation are frequently estimated in terms of the public's willingness to pay (WTP). The WTP for conservation programs targeted at threatened and endangered species, investigated by contingent valuation studies, expresses the perceived use (or usefulness) of the protection of these species (Loomis and White 1996; MacMillan et al. 2004). There are few studies quantifying the WTP for otters and none of them applies to the case study areas compared. For example, White et al. (1997) investigated the willingness to pay for otter protection in North Yorkshire. The mean WTP obtained from his survey amounted for £11.91. A similar study exists for the sea otter (*Enhydra lutris*) carried out by Hageman (1985) in which the WTP amounted to US$29. Albeit the problems related to the transferability, these data give a certain evidence on the society's willingness to pay for otter protection. In any case for Germany, the estimated willingness to pay for national species protection programs was found to be substantially higher than their actual costs (Hampicke et al. 1991) and biodiversity conservation is notoriously undersupplied, receiving too low financial resources (BfN 2002).

Especially the primary sector including agriculture, forestry, and fisheries is of outstanding importance for the overall success of biodiversity policies. In the Central European cultural landscape, many species depend on secondary habitats created by human land uses. Much of the biodiversity-rich land in the EU depends on low-intensity farming and other traditionally extensive land uses (Wiseman and Hopkins 2001). Although not competitive under present economic constraints, the extensive production methods are indispensable for the success of biodiversity conservation.

Theoretically, compensation to fish farmers may be paid in two forms: Either as payments for ecological services, often on the basis of agri-environmental schemes (Wätzold and Schwerdtner 2005) or as damage compensation for losses due to predation by protected species. As shown in the following section, their practical implementation may be less distinctive due to the large uncertainties associated with the estimation of damage caused on carp ponds by otter predation.

### *3.2 Compensation Payments in the Otter-Aquaculture Conflict*

The compensation payments in both study areas differ considerably in the instruments applied, assumptions made, and associated transaction costs.[5]

## 4 Damage Compensation

In Saxony compensation for damage caused by wild animals that are not part of the hunting law is paid under a program for "cases of hardship".[6] The program is based on §38 of the Saxon Nature Conservation Act and in place since 1995. It also compensates for otter damage on a voluntary basis, as otters are part of the hunting law. In aquaculture, the total financial loss must exceed 1,000 € per year. Damage is calculated by subtracting a standard loss from the expected fish production. The standard loss differs among age classes of fish and between summer and winter ponds. It varies between 12 and 50 %, with an average of 28 %. The standard loss is supposed to cover all natural losses due to fish mortality, diseases, and a certain extent of predation (Langner, personal communication). If losses exceed the standard loss, fish farmers can claim a hardship and ask for compensation. The compensation is bound to concrete evidence of the damage but real inspections rarely take place. Compensation is paid on the base of expert reports from fishery and conservation authorities. This makes the damage assessment to differ from real damage. Very likely, predation by other species such as cormorants or herons, poor water quality or improper management practices lead to losses which are indistinguishable from damage caused by otter. On the other side, the advantage of the calculation methods is that it partly covers the secondary damage, such as serious injuries caused by otters.

From 1998–2003, fish farmers in Upper Lusatia received an average damage compensation of about 58,000 € per year. There is no legal entitlement to such payments—compensation is paid as long as public funds are available. Otter damage is usually compensated to 60–80 %, depending on the time of loss (summer or winter) and the age class of fish. However, especially small-scale fish farmers often fail to reach the minimum loss of 1,000 € per year. To overcome this problem and to raise the acceptance of the otter, local authorities in two districts initiated an additional damage compensation scheme and provided living fish for those farmers. Ever since the latter schemes were stopped, attempts have been made to include small-scale farmers in the program for "cases of hardship".

---

[5] Compare Klenke et al. (2013) and Poledníková et al. (2013) for further information on the respective programs.

[6] Economic losses in agriculture, forestry, and fisheries caused by protected species are understood as 'hardship' when they exceed a certain amount.

In the Czech Republic compensation for damage caused by protected species is imposed by the Law 115 (and its later revisions) and is in place since 2000. The species, whose damage is compensated are explicitly listed in §3 and include otter, the great cormorant (*Phalocrocorax carbo sinensis*), brown bear (*Ursus arctos*), wolf (*Canis lupus*), moose (*Alces alces*), beaver (*Castor fiber*), and Eurasian lynx (*Lynx lynx*). Regarding protected piscivorous species, only damage caused to artificial fish ponds is compensated; the damage in streams and rivers is not object of the compensation. To compensate the damage caused by otter, a proof of otter presence in the pond is required. The affected fish farmers have to report the damage (within 48 h) to the responsible local authority, which inspects the fish pond and confirms the damage. Applications for damage compensation are submitted in 6 months cycles and must be complemented by an independent expert report assessing the number of otters visiting the pond and damage arisen.

The methodology applied to assess the damage extent was developed by Czech Otter Foundation and the Environmental Protection Agency (Roche 2003). It distinguishes between detailed assessments techniques, applied only exceptionally, and simplified techniques, applied more commonly. The detailed assessment bases on regular monitoring of water quality in the pond, climate factors, fish diseases, and population of other piscivorous predators. It has not yet been applied due to its high costs. The simplified techniques base on assumptions and available evidence about the daily consumption of otter, number of otters visiting the pond and their visiting frequency, and the market price of the farmed fish. In case of a single pond or a small complex of ponds, the pond area is considered. The damage is assumed higher in small ponds. Therefore, the estimated damage is increased by 20 % for ponds smaller than 2 ha and decreased by 20 % for ponds larger than 5 ha. Damage assessment covers the direct damage (fish actually eaten by the otter), since secondary damage (caused by injuring or stressing the fish in winter) is not regulated. The methodology does not consider damage caused by other piscivorous species such as cormorants or grey herons (*Ardea cinerea*), or by low water quality.

Scant information is available about the actual payment at the national and regional level. Šilhavý (2003) reports that in 2002 ca. 410,350 € was paid for damage caused by both the otter and the cormorant to the members of the Czech Fish Farmers' Association, which represents 60 fish producers managing ca. 85 % of the fish-pond area in the Czech Republic. According to Roche (2003), by September 2003 the Czech Otter Foundation, the institution which examines approximately 90 % of all compensation claims for otter damage in South Bohemia, had registered 160 claims amounting approximately 205,180 €.

## 5 Compensation for Ecological Services

In Saxony, incentives for extensive fish farming are paid under the "NAK"[7] program based on Council Regulation (EC) No. 1257/99. It is partly designed to support environmentally sound aquaculture and to maintain the historical pond landscape.

Taking the form of voluntary contracts for a duration of 5 years, fish farmers can choose from various measures besides the general maintenance of ponds such as extensive production with no additional feeding, no additional stocking or the support of wild fish stocks. There are also payments for an extra stocking to create feeding habitats for endangered species, which are mostly used for otters and unofficially called "otter bonus". In 2003, 99.8 % of Saxony's pond area was supported under the NAK program (SMUL 2004). In Upper Lusatia, 53 % of the pond area cultivated under NAK is also used for the "otter bonus" and supported with 103 € per ha and year. In 2003, some 280,000 € have been paid as "otter bonus" in this area.

In addition, there is an aquaculture program supporting the protection of fishing stocks against piscivorous predators by technical mitigation measures such as fencing and wires, under which 25,500 € were spent in 2001–2002 for pond fencing in Upper Lusatia.

In Czech Republic the agri-environmental schemes (AES) alongside with subsidies for less favored areas (LFA) were introduced by the Horizontal Plan of Rural Area Development, based on the Council Regulation (EC) No. 1257/99. They are available only since 2004. The only measure applicable in our context is construction of bio-corridors, which, if planted along the water courses, may improve the habitat for both fish and otter populations. The LFA measures compensating farmers for the restriction due to imposed extensive management regimes in protected areas generally do not apply to fish ponds.

More relevant measures for the otter-fish farming conflict have been introduced by the program Rural Development and Multifunctional Agriculture, also supported by the EU structural funds. Fish farming represents one of the main areas addressed by the program, including measures for fish processing, aquaculture (especially increasing production capacity and modernization of existing facilities) and consultation activities. These measures are generally not associated with obligation to pursue environmentally friendly production techniques but their provision can be made depending on compliance with environmental legislation.

In 2003, a compensation program (up to 32/ha €) from national sources was introduced, which rewards protection of littoral zones, reduction of fish feeding, fertilization of the fish-pond bottom, and other measures aimed at environmental protection in ponds larger than 5 ha. Tolerating protected piscivorous species is

[7] NAK stands for "Naturschutz und Erhalt der Kulturlandschaft" – nature protection and conservation of cultural landscapes and is part of the general program for environmentally sound agriculture.

not addressed by the program. An additional subsidy program is aimed at supporting fish farmers to tackle the increasing siltation of fish ponds.

## *5.1 Conflict Development, Perception of Conflict and Compensation Schemes*

The compensation schemes as described in the previous section have a different ability to mitigate conflicts between the involved actors, such as fish farmers and anglers, nature conservationists, and hunters, to name but a few. A conflict is a social construct characterized by disagreement, interference, and negative emotion (Barki and Hartwick 2004). Rooted in divergence of values, needs, interests, opinions or goals, existence of conflicts, and the strategies employed to mitigate them may significantly influence the success of compensation schemes and subsequently the protection of threatened and endangered species, such as the otter. Therefore, we have analyzed the recent conflict evolution in both study areas, and especially, how suitable the different compensation schemes are to mitigate the conflicts at hand.

Before the compensation payments were introduced in 2000, the conflicts seemed to escalate in South Bohemia. As a result more than 100 otters a year were thought to be illegally killed (Kranz et al. 1998). This was because the previously state-owned ponds were privatized and the damage was no longer accepted as part of naturally occurring fish losses (Toman 1998a; Roche 2003; Samek and Dušek 2003; Kučerová 2000). Unlike Saxony, the production intensity in the fish ponds remained stable or even increased in some areas. The negative trend in both market prices for carp and the demand for carp[8] may also have contributed to the growing conflict. Additionally, thanks to improved water quality, the otter population gradually grew and reoccupied areas from which it had been eliminated since the 1950s.

The damage compensation schemes implemented in 2000 and subsequently reformed in 2002 and 2003 satisfied (at least to some extent) the large fish-pond enterprises. However, the owners of small fish ponds complain about high transaction costs associated with the compensation claim. Especially in the Czech-Moravian Highlands with suboptimum climatic (e.g., long snowy winters, cold water) and geological (acid soil) conditions for carp farming (Kranz et al. 1998) the conflict seems to persist. The low-nutrient ponds predominating here are also smaller than those in lowland areas around the Třeboň and are owned by small-scale farmers. Therefore, the otter causes higher relative damage up to the complete depletion of the fish stock (Toman 1998b). Although between 2000 and 2003 exponential growth was reported in the number of applications for damage

[8] Carp prices dropped (after constant growth until 1997) in the period 1997–2000 by 30 %, stabilizing at ca. 83 % of the 1997 price level.

compensation (Roche 2003), in relative terms the owners of small fish ponds are less represented. Another interest group left out of the compensation schemes are the (hobby) anglers utilizing water courses. The South Bohemian Anglers' Association sued the Czech government to be considered for damage compensation in future.

The conflict seemed to be escalating again in 2004 with the implementation of the NATURA 2000 program. Under this program, special areas have been dedicated to protected species (including the otter) listed in Annex II of the Habitats Directive. At many places, this requires the restriction of production intensity in fish ponds.

Although the otter population in the Czech Republic and South Bohemia is well documented, the estimated population size differs considerably among the conflicting parties (e.g., Adámek et al. 2003; Hanzal and Havránek 2000). For example, the Czech Anglers Association (Sýkorová 2003) reports more than 1,400 otters in the Czech Republic (some 800 otters in the study area), the Association of Hunters (2003) ca. 1,300, while the monitoring program established by the Nature Protection Agency and the Czech Otter Foundation assesses the population size at approximately 800–1,000 individuals (Roche 2003). Strong disagreement also exists about the damage caused by otters due to (1) the different assessment of the population size and (2) different assumptions about daily consumption and secondary damage. Assessments of otter consumption per day vary between approximately 0.5 kg/day (Toman 1998b; Kranz et al. 2004) and more than 1 kg/day (e.g., Sýkorová 2003). The secondary damage is assessed by a factor of 1.9 by fish farmers and anglers but neglected by the compensation scheme. The average price of preyed fish varies between 4.2/kg € (according to the anglers) and 1.6/kg € (fish farmers). Subsequently, the damage assessments reported by the Czech Anglers Association (based on an assumed population size of ca. 1,460 otters, daily consumption of 1 kg fish at a price of 4.2/kg €, and secondary damage of a factor of 1.9) amount to 4.16 millions € (Sýkorová 2003). The corresponding assessment (based on the daily consumption of 0.75 kg/otter and a fish price of 1.6 €) by fish farmers amounts to 694,444 € for 2002 (Šilhavý 2003).

In Saxony, conflicts raised again with the increasing otter population in the 1960 and 1970s. Several applications for killing permits (possible until 1984, when the otter became totally protected under the conservation law of the former GDR) were made but mostly not accepted. This resulted in some illegal killings which were strictly punished (Kubasch 1996, compare also Klenke et al. 2013 in this book). In 1978, a fish farm management plan for the district of Dresden was agreed upon. Intensive carp production was restricted to a third of the pond area, while leaving the rest to traditional extensive or semi-extensive management. Otters were tolerated by the fish farmers who also bore the costs for technical mitigation measures. An information campaign started in 1986 raising the acceptance for otter protection even more (Kubasch 1996).

The political changes in 1989 and the following privatization of pond farms raised the conflict again. Fishers were no longer state paid and depended on fish production. Otters were again perceived as a problem species that endangered an effective production (Kubasch 1996). The early introduction of agri-environmental schemes combined with a decreasing demand for carp had very positive influences on the conflict development. By being paid for ecological services, fishers do less depend on production. Most measures under the NAK program for pond farming effectively support habitat quality for otters (Schwerdtner and Ring 2005).

The Department of Fisheries of the Saxon State Office for Agriculture assumes that each otter causes an annual damage of 500 € (Langner, personal communication). Using estimated number of 400 adult otters in the area (Ansorge et al. 1997), annual damage of about 200,000 € would occur. Summing payments for damage compensation and the otter bonus for the year 2003 in the Upper Lusatia, results in total compensation payments of 364,000 €. This exceeds by far the estimated financial damage to pond farmers. This indicates that either otter numbers are underestimated, requiring an update by an otter monitoring program, or overall compensation payments are too high, demanding an improvement of the single schemes respectively a better coordination of both compensation schemes.[9]

All Saxon stakeholders agree that one of the main reasons for the absence of a conflict is the existence of the compensation payments (Zwirner and Wittmer 2004). Though, slight disagreements exist whether payments are adequate. Whereas fishers consider them as adequate, authorities criticized payments being used as compensation for losses not connected to otter predation and for providing an additional (alternative) income.

In combination, agri-environmental schemes and compensation payments for otter-related costs mitigate the conflict about otters to a great extent. They are no longer considered as a problem species but perceived as part of the landscape (Zwirner and Wittmer 2004). Additionally, damage by cormorants and grey herons have outweighed the otter in the negative perceptions of fishing personnel (SLfL 2002; Zwirner and Wittmer 2004).

A persistent problem is the fish loss of small-scale farmers and hobby producers. Adequate compensation is necessary for those farmers whose losses may cover a considerable amount of their production. Damage compensation is important in this respect to raise the acceptance of otters in the area, especially in the case of isolated ponds that play an important role for the distribution of the species (Rothmann, personal communication).

In terms of stakeholder participation, the importance of the fisheries council has to be highlighted, which is legally based in the Saxon Fisheries Act. In this council, various stakeholder groups such as scientists, conservationists, fisheries and angling associations meet to discuss fisheries related problems such as damage by fish-eating species. Furthermore, there is a working group pond farming that is assembled at times when coordination is required specifically related to pond

[9] Compare Klenke et al. (2013, in this book) for a damage assessment based on otter numbers and daily food uptake resulting in approximately 17,000 € annual damage to fish farms. Although the rule-of-thumb assessment by the Department of Fisheries leads to slightly higher annual damage estimates, both approaches finally lead to the same conclusion regarding a potential overcompensation of damage, if both compensation schemes are considered together.

farming. These two stakeholder forums are relevant for regular communication among different stakeholder groups and for consensus-driven decision-making. The long tradition of communication between stakeholders since GDR times has led to the opinion of most stakeholders that conflicts are solved cooperatively (Zwirner and Wittmer 2004).

## 6 Comparative Analysis of Conflicts in Relation to Compensation Schemes

In the previous sections we analyzed factors governing the conflicts, either directly or through their manifold interactions, in the examined regions. The problem complexity increases if the broader context is considered in which the conflicts take place: the decline or improvement of water quality, insufficient availability of natural prey, the changes to river morphology and the subsequent loss of riverine habitats, and humankind-driven changes of fish populations in water courses. Each conflict is unique in terms of its perception (conceptualization), related drivers, and their relevance. That makes it difficult to generalize experience gained from analyzing different cases, which is of crucial importance for designing sustainable mitigation strategies.

The study areas compared bear important similarities: (1) they both have a long history of fish farming in ponds going back to the thirteenth century, (2) they have comparable geographical, hydrological, and geological conditions; and at least partly (3) they underwent similar political and societal developments after the Second World War. Despite these substantial similarities, there are also important differences. Since the political changes in the 1990s, Upper Lusatia suffered substantial population decline due to migration to the rest of Germany, whereas the political changes and economic reforms in South Bohemia were applied in a different context and similar migration did not take place. Unemployment rates are significantly higher in Upper Lusatia, both in comparison to the national average and to South Bohemia. Perhaps most importantly, the transformations regarding the pond fisheries were very different in both study areas, favoring transition to an extensive farming management in Upper Lusatia while encouraging stable or even higher fish stocks in South Bohemia.

An important limitation for designing mitigation measures is the lack of consensus, shared among the parties involved in the conflict, about the population size of species in question and about the actual damage caused. Even if monitoring campaigns are put in place, the confidence intervals of their results allow interpretations, which may equally sustain different positions. In our case, expert assessments point out that on average otter density in Upper Lusatia might be higher than in South Bohemia (Andreas Kranz, Reinhard Klenke; personal communications). Local densities can of course vary considerably, especially taking into account different geomorphologic conditions and pond number and average size in both study areas. While in Upper Lusatia the medium-sized ponds prevail, in South Bohemia very large pond complexes stand for a significant proportion of total pond area. Numerous small ponds with higher pressure by otter are located in territories (e.g., the Czech-Moravian Highlands) from which otters disappeared in the past due to severe persecution and poor water quality and which are now progressively reoccupied.

Despite less favorable economic conditions and a high otter pressure, the conflict between otter protection and pond farming in Upper Lusatia seems to be more relaxed than in South Bohemia. Although otters feed on commercial fish resources, concerned stakeholder groups do not perceive the otter as creating a real conflict with fish farming. The conflict potential seems to have been paid high attention early enough in order to avoid conflict escalation as known from South Bohemia. In the early 1990s, Saxony implemented voluntary compensation schemes as part of the agri-environmental program for less intensive production methods. The reduction in market demand for carp, experienced in that period, was mitigated by these schemes as well. Later, these early measures evolved to the current schemes. Using compensation as an incentive for more extensive farming methods, better suitable for the preservation of a historically valuable cultural landscape, is the key feature of the mitigation measures in Saxony.

To the contrary, in South Bohemia, the problematic perception of otter persists among various stakeholder groups despite the damage compensation scheme introduced in 2000. The mitigation measure has been introduced late, only after the extent of illegal killing already achieved considerable level. Lower confidence in the measure, especially among small farmers, and high uncertainty about how long the compensations will go on may be a lasting consequence of it. How fragile the mitigation measures are has been demonstrated during the designation of NATURA 2000 sites. In addition, the damage compensation scheme does not relate to other support schemes aimed at conserving unique landscape or pursuing more environmentally sound production techniques.

In both study areas, the compensation schemes apply different methodologies to deal with or assess the damage and exercise different rules for compensation. In Saxony, payments for ecological services in the form of providing a feeding habitat for the otter are by far more important than actual damage compensation payments. Regarding the hardship regulation in Saxony, applicants are not automatically entitled to damage compensation and the budget available is limited. Since there is less stringent obligation for proving the physical extent of the damage, the scheme allows to some extent for compensation for secondary damage due to injury or stress suffered by fish. In the Czech Republic, compensation is guaranteed by law and its calculation requires reliable assessment of the damage provided by a recognized expert. Since it bases on assumption about the daily consumption by the otter, it does not cover secondary damage. These differences do not yet explain the effectiveness of conflict mitigation. Apparently, there is a trade-off between the transaction costs associated with the compensation measures and the level of scientific accuracy surrounding damage assessment. Although the lesson learned from comparing the study areas suggests that successful conflict mitigation strategies require compensation schemes to address different objectives, acknowledging the scale of damage and further requirements are important.

Concerning the scale of damage, spatially differentiated and flexible schemes, able to pursue different measures for different extents of damage, are generally more likely to succeed. The extent of damage in the case of large pond farms, situated in lowlands in both Upper Lusatia and South Bohemia, are likely to exceed thresholds imposed by the compensation schemes. In Saxony, the threshold is explicitly defined by the hardship regulation, while in South Bohemia it is the perception of administrative obstacle, which is better coped with by larger companies than by small farmers. In Upper Lusatia, small and medium-sized farmers, which fail to reach the threshold, can still apply for the "otter bonus" in context of the agri-environmental program. This measure rewards the presence of the otter (as strictly protected species) as an ecological service. At the same time, the whole program provides an incentive for farmers to apply extensive forms of fish farming, reducing the potential for large-scale damage and by doing so, taking off the edge of the conflict. Small-scale damage used to be compensated in a non-monetary way with living carps. After the end of this measure, the "otter bonus" remains the only possibility for compensation. In South Bohemia, a similarly differentiated approach to damage compensation has yet to be found. The need for such an approach seems to be well justified in the light of the persistent conflict especially among the small farmers.

Even though, compensation schemes are an essential feature of conflict mitigation strategies from an economic point of view, often they are not sufficient to actually reconcile an existing conflict (Montag 2003). Apart from economic aspects, ecological and social factors are crucial in conflict mitigation. Mitigation measures formulated in a participatory process, open to relevant stakeholders, and based on trust and understanding of positions of opposing parties, are more likely to succeed. There is little information about the processes behind the design process of compensation schemes in both study areas. Indirectly, the success of such processes can be assessed by the level of agreement achieved in the compensation schemes. In this context, the interests of small-scale farmers and anglers have been marginalized in the Czech Republic.

## 7 Conclusions

The protection of the otter has been reported as a success story throughout Europe (Kruuk 2002). The return of otters to their previous habitats, however, is accompanied by conflicts due to the increased damage to fish stocks in fish ponds and water courses. The acceptance of the otter by fish farmers also declines with factors such as (1) growing market pressure (resulting in low carp prices), (2) the increasing siltation of the ponds, and (3) growing pressure of other piscivorous predators (e.g., great cormorant), to name but a few. In addition, the recent EU enlargement is suspected by fish farmers to increase the market pressures and to restrict fish production to comply with European legislation (e.g., Habitats Directive).

In this chapter we analyzed conflicts emerging between otter protection and fish farming in two study areas—Upper Lusatia (Germany) and South Bohemia (Czech Republic). When comparing the conflicts and the compensation schemes existing in both study areas, it may be concluded that designing economic instruments aimed at conflict mitigation is no easy task for several reasons:

1. Actual damage assessment is difficult and needs to address different scales of damage (and thus satisfying the parties affected). Indeed, especially in the Czech Republic, the conflict is characterized by serious disagreement about the facts.
2. Financial compensation has limited scope to solve the conflict. Monetary benefit is not the exclusive reason for fish-pond farming, especially in small ponds. Instead, people often breed fish as pets, as a pastime, or to share a good meal with friends—things that can hardly be compensated for in monetary terms. In such a case, material compensation with fish losses being substituted by replacement fish, as it used to be carried out in some districts in Upper Lusatia, is likely to be more successful.
3. Successful conflict mitigation strategies must involve relevant stakeholder groups in decision-making. A stakeholder forum, such as the fisheries council and the working group "pond farming" in Saxony, can effectively contribute to consensus-finding about important facts relating to the conflict and thereby mitigate it. Depending on the seriousness and extent of the conflict, it might be relevant not only to include professional fishermen, but also non-professional groups, such as anglers and their associations.

A proper set of mitigation measures should aim at changing the attitude of all or at least the most affected actors and mitigate the conflict or at least stop it from becoming more serious. The financial compensation of damage, if not supported by other mitigation measures, is unlikely to mitigate the conflict in the long-term. In order to be successful, a set of mitigation measures has to (1) address ecological, economic, and social aspects of the conflict; (2) be spatially differentiated, considering for example the differences between fish farming conditions in lowlands and in highlands; (3) be differentiated regarding the size and ownership of the fish ponds; (4) acknowledge the social needs of the various stakeholder groups involved; and (5) be composed by a number of measures, both financial and non-financial, which, besides compensating for the damage, help increasing the acceptance of the endangered and threatened piscivorous predators and preventing higher damage.

**Acknowledgments** This work was kindly supported by the European Commission under contract no. HPMD-CT-2001-00117 and under the 5th Framework Program (5th FP) Project "FRAP" (Development of a procedural Framework for Action Plans to Reconcile conflicts between large vertebrate conservation and the use of biological resources: fisheries and fish-eating vertebrates as a model case), contract number EVK 2-CT-2002-00142-FRAP.

## References

Adámek Z, Kortan D, Lepič P, Andreji J (2003) Impacts of otter (*Lutra lutra* L.) predation on fishponds: a study of fish remains at ponds in the Czech Republic. Aquacult Int 11:389–396

Ansorge H, Schipke R, Zinke O (1997) Population ecology of the otter *Lutra lutra*—parameters and model for a central European region. Mamm Biol/Z Säugetierkd 62:143–151

Ansorge H, Striese M (1993) Zum Bestand des Fischotters in der östlichen Oberlausitz. In: Abh Ber Naturkundemus Görlitz 67(5):13–19

Barki H, Hartwick J (2004) Conceptualizing the construct of interpersonal conflict. Int J Conf Man 15(3):216–244

BfN—Bundesamt für Naturschutz (2002) Daten zur Natur 2002. Landwirtschaftsverlag, Bonn

Bodner M (1998) Damage to stock in fish ponds as a result of otter (*Lutra lutra* L.) predation. BOKU Rep Wildl Res Game Manag 14:106–117

Bromley DW, Hodge I (1990) Private property rights and presumptive policy entitlements: reconsidering the premises of rural policy. Eur Rev Agr Econ 17:197–214

BfLuE (2002) Karpfenbericht 2002. Bericht über die Marktversorgung und die Außenhandelssituation von Karpfen und anderen Süßwasserfischen in der Bundesrepublik Deutschland. Bundesanstalt für Landwirtschaft und Ernährung, Referat 521—Fischwirtschaft. Hamburg

Bureš J (2000) Vývoj zonace a sítě maloplošných zvláště chráněných území Třeboňska. In: Pokorný J, Šulcová J, Hátle M, Hlásek J (eds) Třeboňsko 2000: Ekologie a ekonomika Třeboňska po dvaceti letech. Sborník příspěvků ze stejnomenné conference 12-14.4. 2000, Třeboň, pp 40–43

CFFA (2003) Czech Fish Farmers Association

ČSÚ (2003) Statistical yearbook of the Jihočeský region. Czech statistical office, regional office České Budějovice, ref.no. 61/2003, České Budějovice

Faina R (2000) Alternativy k tradičnímu pojetí rybářské intenzifikace na rybnících v CHKO Třeboňsko a na rybničných rezervacích. In: Pokorný J, Šulcová J, Hátle M, Hlásek J (eds) Třeboňsko 2000: Ekologie a ekonomika Třeboňska po dvaceti letech. Sborník příspěvků ze stejnomenné conference 12-14.4. 2000, Třeboň, pp 192–196

Fiedler F (1996) Verbreitung und Lebensraum des Fischotters in Sachsen. Abriß der historischen Verbreitung bis zum Anfang des 20. Jahrhunderts. In: Sächsisches Landesamt für Umwelt und Geologie (eds) Artenschutzprogramm Fischotter in Sachsen. Mater Natursch Landschaftspfl. Feistaat Sachsen, Radebeul, pp 7–9

Fourli M (1999) Compensation for damage caused by bears and wolves in the European Union. Office for Official Publications of the European Communities, Luxembourg

Grohmann O, Klenke R (1996) Beiträge zur Ökologie des Fischotters. Methodische Aspekte and Praktische Beispiele zur Raumnutzung—farbmarkierte Nahrung. In: Sächsisches Landesamt für Umwelt und Geologie (eds) Artenschutzprogramm Fischotter in Sachsen. Mater Natursch Landschaftspfl. Freistaat Sachsen, Radebeul, pp 30–32

Hageman R (1985) Valuing marine mammal populations: benefit valuations in a multi-specied ecosystem. Administrative report LJ-85-22. Southwest Fisheries Center, National Marine Fisheries Service, La Jolla

Hampicke U, Horlitz H, Kiemstedt K, Tampe D, Timp D, Walters M (1991) Kosten und Wertschätzung des Arten- und Biotopschutzes. Erich-Schmidt-Verlag, Berlin

Hampicke, U. (2005) Naturschutzpolitik. In: Hansjürgens B, Wätzold F (ed) Umweltpolitik und umweltökonomische Politikberatung in Deutschland. Z Angew Umweltforsch (Sonderheft) 15:162–177

Hanley N, Kirkpatrick H, Simpson I, Oglethorpe D (1998) Principles for the provision of public goods from agriculture: modelling moorland conservation in Scotland. Land Econ 74:102–113

Hanzal V, Havránek F (2000) Vydra říční—škůdce, nebo klenot naší přírody? In: Sborník Referátů, Predátoři v Myslivosti, 1-2. 9. 2000, Česká lesnícka společnost, Hranice. pp 58–59

Hartstock E (2000) Entstehung und Entwicklung der Oberlausitzer Teichwirtschaft. Schriftenr Sächsische Landesanst Landwirtsch 5, Dresden

Klenke R (1996) Ergebnisse der Erfassung von Fischotternachweisen von 1993–1995. In: Sächsisches Landesamt für Umwelt und Geologie (eds) Artenschutzprogramm Fischotter in Sachsen. Mater Natursch Landschaftspfl. Freistaat Sachsen, Radebeul, pp 12–17

Klenke RA, Ring I, Schwerdtner Máñez K, Habighorst R, Weiss V, Wittmer H, Gruber B, Henle K (2013) Otters in Saxony—a story of successful conflict resolution. In: Klenke RA, Ring I, Kranz A, Jepsen N, Rauschmayer F, Henle K (eds) Human-wildlife conflicts in Europe—fisheries and fish-eating vertebrates as a model case. Springer, Heidelberg

Kranz A (2000) Otters (*Lutra lutra*) increasing in central Europe: from the threat of extinction to locally perceived overpopulation? Mammalia 64:357–368

Kranz A, Knolleisen M (1998) How many otters live here? a discussion about counting otters. BOKU Rep Wildl Res Game Manag 14:120–125

Kranz A, Poledník L, Toman A (2004) Der Fischotter in Österreich. Der Anblick. Z Jagd, Fischerei, Jagdhundewes Natursch 2:12–15

Kranz A, Toman A, Knolleisen M, Prasek V (1996) Fish ponds in Central Europe—a rich but risky habitat for otters. In: Dulfer R, Conroy JH, Nel J, Gutleb AC (eds) Otter conservation—an example for a sustainable use of wetlands. Proceedings 7th International Otter Colloquium March 14–19, 1998, Trebon. IUCN/SSC Otter Spec Group Bull 19A (spec. Edition):181–186. http://otterspecialistgroup.org/Bulletin/Volume19A/Vol19_A_Index.html. Accessed 13 Dec 2008

Kranz A, Toman A, Roche K (1998) Otters and fisheries in Central Europe—What is the problem? BOKU Rep Wildl Res Game Manag 14:142–144

Kruuk H (2002) Hunters and hunted—relationship between carnivores and people. Cambridge University Press, Cambridge

Kubasch H (1996) Zur Geschichte des Fischotterschutzes in Sachsen. In: Sächsisches Landesamt für Umwelt und Geologie (ed) Artenschutzprogramm Fischotter in Sachsen. Mater Natursch Landschaftspfl. Freistaat Sachsen, Radebeul, pp 5–7

Kučerová M (2000) Ochrana vydry říční jako příklad ochrany vlajkového druhu. In: Pokorný J, Šulcová J, Hátle M, Hlásek J (ed) Třeboňsko 2000: Ekologie a ekonomika Třeboňska po dvaceti letech. Sborník příspěvků ze stejnomenné conference 12-14.4. 2000, Třeboň, pp 78–80

Loomis JB, White DS (1996) Economic benefits of rare and endangered species: summary and meta-analysis. Ecol Econ 18:197–206

MacMillan D, Hanley N, Daw M (2004) Costs and benefits of wild goose conservation in Scotland. Biol Conserv 119:475–485

Montag J (2003) Compensation and predator conservation: limitations of compensation. In: Angst C, Landry JM, Linnell J, Breitenmoser U (eds) Carnivore damage prevention News 6:2–5

Poledníková K, Kranz A, Poledník L, Myšiak J (2013) Otters causing conflicts—the fish farming case of the Czech Republic. In: Klenke R, Ring I, Kranz A, Jepsen N, Rauschmayer F, Henle K (eds) Human-wildlife conflicts in Europe—fisheries and fish-eating vertebrates as a model case. Springer, Heidelberg

Reuther C (1999) Otter 2000. Eine Vision für den Otterschutz in Deutschland. In: Reuther C (ed) Otterschutz in Deutschland. Arbeitsberichte der Aktion Fischotterschutz e. V., Hankensbüttel. Habitat 7:85–92

Reuther C, Dolch D, Drews A, Ehlers M, Heidemann G, Klaus S, Mau H, Sellheim P, Teubner J, Wölfel L (2002) Fischotterschutz in Deutschland—Grundlagen für einen nationalen

Artenschutzplan. Arbeitsberichte der Aktion Fischotterschutz e. V., Hankensbüttel. Habitat 14:1–159

Ring I (2004) Naturschutz in der föderalen Aufgabenteilung: Zur Notwendigkeit einer Bundeskompetenz aus ökonomischer Perspektive. Nat Landschaft 79:494–500

Roche M (2003) Výsledky mapování výskytu vydry říční v ČR a zkušenosti s náhradami ztrát způsobených vydrou. In: Sýkorová Z (ed) Rybářství a predátoři. Sborník Referátů z Odborného Semináře, 18.9.2003, Český rybářský svaz, Praha, pp 14–18

SLfL— Sächsische Landesanstalt für Landwirtschaft (2002) Zahlen zur Binnenfischerei im Freistaat Sachsen. Jahresbericht der Sächsischen Landesanstalt für Landwirtschaft, Dresden

Samek R, Dušek M (2003) Problematika rybích predátorů v aktuálním pohledu ochrany přirody. In: Sýkorová Z (ed) Rybářství a predátoři. Sborník Referátů z Odborného Semináře, 18.9.2003, Český rybářský svaz, Praha, pp 10–11

Schwerdtner K, Ring I (2005) Development of policy instruments (Germany). Practice and potential. FRAP project report. UFZ Centre for Environmental Research, Leipzig. 44 p

Šilhavý V (2003) Problémy způsobené komerčnímu rybářství neadekvátní ochranou predátorů. In: Sýkorová Z (ed) Rybářství a predátoři. Sborník Referátů z Odborného Semináře, 18.9.2003, Český rybářský svaz, Praha, pp 7–9

SMUL—Sächsisches Ministerium für Umwelt und Landwirtschaft. Official communication of 22.04.2004, Dresden. http://www.smul.sachsen.de, Accessed 20 March 2005

StaLa Sachsen—Statistisches Landesamt des Freistaates Sachsen (1995) Statistische Berichte. Die Teichwirtschaft im Freistaat Sachsen. Ergebnisse der Binnenfischereierhebung 1994. Kamenz

StaLa Sachsen—Statistisches Landesamt des Freistaates Sachsen (2001) Available on internet under: http://www.statistik.sachsen.de, Accessed 24 April 2005

Statistisches Bundesamt Deutschland (2004) Available on internet under: http://www.destatis.de, Accessed 24 April 2005

Sýkorová Z (2003) Vpliv predačního tlaku vydry, kormorána, volavky popelavé a dalších predátorů na rybí spoločenstva vodných toků v roce 2003. Český rybářsky svaz, Praha

Thiem A (2002) Naturschutzfachliche Grundsätze zur Bewirtschaftung von Karpfenteichen in Sachsen. Mater Natursch Landschaftspfl. Sächsische Druck- and Verlagshaus AG, Dresden

Toman A (1998a) Otters versus privatisation in the Czech Republic. BOKU Rep Wildl Res Game Manag 14:6–7

Toman A (1998b) Otter damage compensation in the Czech Republic. BOKU Rep Wildl Res Game Manag 14:118–119

Toman A, Kadlečík J (1992) Bull Vydra 3. Český ústav ochrany přírody, Praha, pp 3–59

Usbeck H, Ebert A, Usbeck F, Schwerdtner K, Ring I (2004) Socio-economic report of the German model region "Oberlausitz-Niederschlesien". FRAP project report. Usbeck GmbH and UFZ Centre for environmental research, Leipzig

Wätzold F, Schwerdtner K (2005) Why be wasteful when preserving a valuable resource? a review article on the cost-effectiveness of European biodiversity conservation policy. Biol Conserv 123:327–338

Wedekind H, Hilge V, Steffens W (2001) Present status, and social and economic significance of inland fisheries in Germany. Fish Manag Ecol 8:405–414

White PCL, Keith WG, Lindley PJ, Richards G (1997) Economic values of threatened mammals in Britain: a case study of the otter *Lutra lutra* and the water vole *Arvicola terrestris*. Biol Conserv 82:345–354

Wiseman R, Hopkins L (2001) Sowing the seeds for sustainability: agriculture, biodiversity, economics and society: proceedings of the 8th Interactive Session held at the Second IUCN World Conservation Congress, Amman, Jordan, 7 October 2000. IUCN, Gland, Switzerland and Cambridge, UK. pp 133. http://cmsdata.iucn.org/downloads/2001_049.pdf. Accessed 30. May 2011

Zwirner O, Wittmer H (2004) Germany. In: Wilson DC (ed) Discourse analysis. WP 6—Local mitigation efforts and local stakeholder analysis. FRAP project report, Institute for fisheries management and coastal community development, Hirtshals, pp 125–155

Great cormorant (*Phalacrocorax carbo sinensis*). *Photo:* André Künzelmann

# Cormorants in Denmark

## Re-enforced Management and Scientific Evidence

**Niels Jepsen and Thomas Olesen**

**Abstract** In Denmark, the conflict between fisheries and bird protection has been well known for decades. The conflict is based on the fact that the numbers of cormorants, feeding from the Danish coastal and freshwater areas have dramatically increased since protection measures in the early 1980s. Both commercial and recreational fishing organizations claim that the high density of cormorants is a threat not only to their fishing activities, but for the whole ecosystem. The mitigation of this conflict is partly institutionalized, because a group of stakeholders, managers and experts regularly meet to debate and agree on national cormorant management plans. In the ongoing debate, scientific evidence has been shown to play an important role. In this chapter, the background, the legislation, the economics and the politics concerning the management of this conflict is described and discussed.

## 1 Introduction

The Danish cormorant experience is interesting for several reasons both from a national as well as from a European perspective. Denmark was among the first European countries to experience the explosive cormorant population growth that

N. Jepsen (✉)
National Institute of Aquatic Resources, Section of Freshwater Fisheries Ecology, Technical University of Denmark, Vejlsøvej 39,
8600 Silkeborg, Denmark
e-mail: nj@aqua.dtu.dk

T. Olesen
Innovative Fisheries Management, An Aalborg University Research Centre, Willemoesvej 5, 9850 Hirtshals, Denmark
e-mail: dw@ifm.aau.dk

R. A. Klenke et al. (eds.), *Human-Wildlife Conflicts in Europe*,
Environmental Science and Engineering, DOI: 10.1007/978-3-540-34789-7_8,

has occurred throughout Europe, a consequence of the 1979 EEC Birds Directive proscribing to protect the cormorant. Denmark was also among the first countries to produce a cormorant management plan in an attempt to reconcile the ensuing conflicts. However, after 18 years of management plans the conflict is basically still the same. Management plans and stakeholder participation are no guarantee of reconciling the conflict.

The findings from the research of the European FRAP project carried out in Denmark are important lessons for other attempts to reconcile nature-human conflicts. Most important is that no easy solutions exists. The dynamics of such conflicts are complex and, in order to understand why this conflict has continued for years, one has to look beyond cormorant-eat-fish arguments. The drivers of the conflict are much more than the damage inflicted by cormorants. It is also a matter of conflicting perceptions of nature use as well as a clash between research-based knowledge and experience-based knowledge.

## 2 Evolution of the Conflict

The protection of cormorants following the EEC Birds Directive (European Commission 1979) initiated a rapid increase of the cormorant population in Denmark growing from 2,000 breeding pairs in 1980 to 38,000 breeding pairs in 1994. Since 1994, the size of the breeding population has stabilized (Eskildsen 2005). The level and intensity of the conflict has fluctuated over the years but conflicts have erupted as the population growth has resulted in cormorants settling or becoming regular visitors in new places. Moreover, stakeholders have become aware that cormorants might affect their own interests negatively. The first conflicts appeared in the pound net fishery. Pound nets are fixed circular net entrapments that are easily accessible for cormorants and therefore the easiest food source available. Soon after the population growth had started, pound net fishers began to complain about loss of catches, not only because of cormorants eating fish from the live hold but also fish being damaged by bite marks. The predation in pound nets takes place in the entire pound net season from early spring to late autumn. Due to the spread of colonies to most parts of Denmark with the existence of a large population of mobile young cormorants, which are not nesting, today, all pound net fishers are affected. The conflict has expanded into the recreational fisheries and, during the last years, urban areas have experienced cormorant settlements as well. Consequently, the need to balance conflicting interests is increasing rather than decreasing.

For the first two decades, the authorities turned down complaints from pound net fishers as groundless. A lack of scientific documentation of fisheries' claims gave authorities and environmental stakeholders no reason to change position on the cormorant management despite increasing complaints. On the contrary in 1992, the conflict resulted in the first management plan (Miljøministeriet 1992) with the objective to ensure that the population was managed in accordance with the protection status and that population growth could continue. However, the conflicts continued and in 1994, elements of population control were introduced in the administration of the management plan after political pressure in parliament. This was possible because the management plan is a framework allowing adaptations to ongoing population and conflict developments. Additional efforts into developing technical mitigation measures did not result in an effective reconciliation and the conflict continued. In 2002, the management plan was revised and the objective changed towards a more population control oriented management.

# 3 Assessment and Analysis

## *3.1 Ecological Background*

The great cormorant (*Phalacrocorax carbo sinensis*) is a large and conspicuous colonial water bird with a wingspan of up to 1 m and a body mass up to 3 kg. (It is often seen standing with its wings held out to dry.) Regarded by some as black, sinister, and greedy, cormorants are supreme fishers, which bring them into conflict with fisheries, and they have been persecuted in the past. Cormorants are mainly migratory, nesting in the northern parts and wintering further south. Cormorants breed in colonies, ranging from a few nests to several thousand. The colonies are usually situated in trees, but recently many ground-based colonies have been established (Fig. 1). Colonies are also known from man-made structures like lighthouse, power lines and old fishing vessels. Cormorant colonies in northern Europe range from one single pair to over 8,000 pairs, with a median size of 50–500 pairs. Tree colonies have a density of 200–750 nests/ha and ground colonies can reach densities up to 3,500 nests/ha. Cormorants start breeding at the ages of 2–7 years, where they will usually lay 3–6 eggs each year. The cormorants are rather long-lived and can reach ages of 15–20 years. Cormorants are highly mobile predators and they can fly up to 40 km from their nesting or roosting site for foraging. Nesting birds are stationary and will primarily be found within a radius of 20 km from the colony, whereas young non-nesting birds are more free-ranging and can be found almost anywhere. The preferred habitats are shallow coastal areas, lagoons, estuaries, and fiords, but cormorants also use freshwater areas (lakes, rivers, and reservoirs) for foraging and nesting.

## *3.2 Population, Habitat, and Landscape Factors*

It is relatively easy to observe cormorants and thus, much information is available about their distribution and abundance. As they nest in colonies, the population is usually monitored by counting the number of nests. The population structure of

**Fig. 1** Cormorant colony in Ringkøbing Fjord, Denmark. *Photo:* Niels Jepsen

cormorants (long-lived birds with a high number of non-breeding individuals) means that they are rather insensitive to population control by "random killing". Even if high numbers are killed, little effects are seen on total number of birds (Carrs and Marzano 2005; Alberti personal communication). Apparently, the availability of undisturbed nesting areas and food, limit the number of cormorants. Recent analyses of the development of the Danish and German colonies show that availability of foraging area within 10 km radius from the colony is the most important factor (Klenke et al. in prep.). In Denmark, the total number of nesting birds seems to have stabilized since the early rapid growth and during the last 12 years, around 40,000 nesting pairs have been counted. The total number of individual cormorants in Denmark at the same time is difficult to estimate because of emigration and immigration, but a late summer maximum of 300,000 birds is realistic. In winter, only a fraction of this number can be found here.

### 3.3 Diet

Adult cormorants eat from 250–1450 g of fish/day (Carss 2003). Cormorants have not only a wide species range of prey, but also a wide size range. Cormorants eat very small items, such as shrimps and large fishes up to 1.5 kg, with the most commonly foraged species usually being around a mean of 15 cm. There is, however, a significant temporal and spatial variation in the mean size of each prey species. The issue of cormorants showing distinct prey preferences is very difficult to investigate, but a review of published diet studies gives the impression that these birds can and will eat everything they can catch and swallow. The size distribution of prey fish mainly reflects whatever is available for the cormorants to catch.

It is clear that cormorants can have an impact on coastal fisheries. The damages done to pound-net fisheries have been documented and are widely accepted (Dieperink 1995). It is not really known if cormorant predation can have negative impact on whole fish populations in coastal areas. It can be calculated that, during summer time, Danish cormorants eat approximately 70 tons of fish per day. Another problem is the damage of fish in pound nets. When the birds use these "open-top" nets as preserved tables, they not only eat much of the catch, but also damage many of the larger fish, making these worthless (see Box 1. Indirect effects). Few studies have been carried out but according to an investigation from 1998, an average of 30 % of the catch had bite marks (Bildsø and Bundgaard 1998).

Most studies have based their estimate of cormorant diet solely on analyses of hard parts from cormorant pellets, collected in colonies. Such analyses give a good qualitative estimate of the diet composition on the day of collection. However, the lack of temporal resolution and methodological problems of different representations of these hard parts from different fish species makes quantitative analyses very weak (McKay et al. 2003). The FRAP study used a combination of methods to estimate the effect of cormorant predation on fish populations in a large, shallow Danish estuary. Our results indicate that the cormorants ate a significant part of the recruitment (young fish) of several important species. Thus, micro-tagging (Fig. 2) experiments showed that 40–50 % of small tagged eel were eaten in one year, 25–40 % of tagged salmon (*Salmo salar*) smolts were eaten during the smolt-migration in April/May, and nearly all the tagged juvenile flounders (*Platichthys flesus*) were eaten within a very short time after tagging (Jepsen et al. 2010). These results were based on certain assumptions and are subject to variation and uncertainty with rather wide statistical ranges, but, along with previous studies, they point in one direction: fish populations, already under significant pressure from fishing and habitat degradation, may decrease dramatically because of predation by large numbers of cormorants. More experiments using recovery of tags from pellets should be carried out in other areas to provide a wider basis for general conclusions about cormorants' effect on wild fish populations.

The general conclusions from the FRAP tagging studies are that a high number of cormorants can exert a significant impact on local fish populations and that this predation is the main population-regulating factor for several important fish species. In areas where cormorants have this kind of influence, technical mitigation measures, such as protecting the fishing gear or scaring the cormorants away, will not help reducing the conflicts. Other studies from other areas have not found the cormorants to have much effect on fish populations (Engstrøm 2001). Consequently, stakeholders from both sides can find scientific documentation supporting their opinion. The fact is, like in most cases concerning ecosystems, that in some situations the cormorants can fit into the system without bringing about much (undesired) change, whereas in other situations they may have a (negative) impact on the whole system.

**Fig. 2** Small individually or batch coded pieces of steel wire (coded wire tags) were used to mark fish. After release the proportion of fish eaten by cormorants can be estimated by counting the tags in regurgitated pellets by the use of a special metal detector. *Photos* **a–c**: Northwest Marine Technology Inc., **d**, **f**: Niels Jepsen, **e**: Per Huniche Jensen

Questions like: "How much damage do a given number of cormorants do?" are impossible to answer generally because the answer depends solely on the context.

Thus, a major challenge here is to inform/convince the stakeholders about the "real" situation, which is far from being black or white, but more like a spectrum of effects.

**Box 1** Indirect Effects

*Predation*

In the process of evaluating the impact of predation on a given resource, the experience from our case studies have shown that besides the direct effects of the predators eating prey, indirect effects can be of major importance. Such observed indirect effects include:

- Cormorants damaging fish in pound nets in addition to what they eat
- Seals only biting out chunks of salmon
- Otters inducing severe stress to carp held in wintering ponds in high density
- Cormorants often catch more fish than they eat, because gulls steal the prey.

These are examples of situations, where a direct measurement of the predators' food intake will underestimate the amount of fish actually "removed from the stock". There are also situations where the predator can fuel a conflict by other means than just by eating, killing or damaging.
Such situations include:

- Seals damaging valuable fishing gear and tearing holes in net-pens holding farmed salmon
- Cormorants entering fyke-nets, damage the net and drowned birds block the entrance for fish
- Otters visiting several fish farms/ponds can spread serious diseases.

As can be seen from the examples given here, the indirect effects can be much more severe than the direct, so it is rarely enough to describe/estimate the damage just by evaluating the consumption of the predators.

*Mitigation measures*

When management measures are taken, the direct effect on the predators is usually evaluated, but also in these cases, several types of indirect effects must be considered. As examples from the research done in FRAP, can be mentioned:

- Use of cover nets excludes cormorants from entering pound nets from above, but increases the number of birds diving into the net, increasing the number of birds drowning in the net. Cover nets can also reduce the catch of several pelagic species.
- Use of scaring devices (like gas canons) in most cases will also have negative effects on other animals and people.
- Fencing-in fish ponds to keep out otters will exclude wildlife that use the ponds for drinking.

When evaluating the effect of a management measure, such indirect ecological effects must be carefully judged to avoid the creation of new conflicts.

## *3.4 Economics*

The development in the pound net fishery has been negative for years and it has been negative in areas with and without cormorants. The implication of this is that other factors than the cormorant population size influence the state of the fish stocks, e.g., environmental conditions, such as recruitment problems for eel (*Anguilla anguilla*), eutrophication as well as the pattern of exploitation of fish stocks in general. However, for the individual pound net fisher, it is beyond discussion that cormorant predation poses a problem and the fishers interviewed in the FRAP project were convinced that the predation from cormorants is the main explanation for a declining profitability of the fishery.

The economic implication of the conflict is primarily a concern of pound net fishers. In terms of total catch and economic value, the contribution of the pound net fishery is insignificant compared to the Danish fisheries sector except of the eel fishery where pound net fishers in 1997 caught 84 % of the total eel catches (Danmarks Fiskeriforenings Bundgarnsundersoegelse 1997). The magnitude of damage is correlated to the number of pound net fishers. There is no record of the exact number of pound net fishers due to the way official statistical data are compiled. In 2004, the Association of Pound Net Fishers assessed that there were approximately 50 full-time pound net fishers and 150 part-time pound net fishers, of which many are trawl or gillnet fishers as well (Buch, pers.com.).

The economic implication for the angling sector is difficult to quantify. Anglers and recreational gear fishers have financed a number of habitat restorations and restocking programs through a mandatory license fee. In the FRAP project, it has been documented that, as cormorants eat a high percentage (up to 50 %) of stocked fish, with annual stocking programs worth 2–2.4 million €, cormorants do inflict costs on angling/recreational fishery. It is important to mention, though, that a number of successful projects improving the condition of local fish stocks have been possible in areas with a large population of cormorants. The value of angling/recreational gear fishing tourism is significant but there are currently no indications of the sector suffering economic damage due to cormorant predation.

It is clear that the costs of the environmental success of regenerating the cormorant populations are borne by the private sector: pound net fishers, fishers in general, and anglers. There is no desire among stakeholders to introduce compensation schemes. The Danish FRAP results could indicate that the scope of problems goes beyond the realm of compensation schemes because of the possible ecosystem effects of cormorant predation.

The Forest and Nature Agency, being the responsible authority, considers cormorants to be one of the species that are costly to manage. According to the management plan (see below) egg culling (by spraying the eggs with paraffin oil) and destruction of new colonies are the main actions. There is no official calculation of the costs of managing cormorants but an estimation of the cost of culling by oiling and deterrence activities, which constitute the major part of the daily management, have been calculated. The cost estimates are probably underestimates in the range of 10–20 %. In 2002, there were oiling and deterrence activities in 10 colonies. The total costs were 6,800 € (DKK 51,000) and 66 % of the costs were related to oiling. (Bregnballe and Olesen 2004).

### *3.5 Legal Framework and Management Plan*

The EEC Birds Directive limits the possibilities of managing the cormorants. Its article 9 provides the possibility to adopt certain management options in order to avoid damages despite the protection status. This article is taken up in the Danish Hunting and Game Management Act (Skov og Naturstyrelsen 1997, 2003).

The Danish management plan for cormorants has been specifically formulated as a conflict mitigation plan. The management plan has provided guidelines for management and frames the conflict politically, outlining the possibilities and limits to a regulation. This is an important function in a case like the management of cormorants, which has been the subject of intense public and political debates. Furthermore, the management plan serves as a political signal to the public that the issue is being addressed.

The first management plan was made in 1992 and was in favour of the cormorant, as the principle of prevention of damages rather than that of population management was important (Skov og Naturstyrelsen 1992). The management plan has since then been adapted to changing conditions and turned towards a more active management. Table 1 showing the inclusion of more instruments over the years illustrates this.

However, the expansion of instruments could also be a consequence of the inefficacy of the instruments actually used. In addition to the actual management options, the 2002 management plan includes recommendations of what should be investigated during the period of the management plan (Skov og Naturstyrelsen 2002).

The Ministry of the Environment, more precisely the Forest and Nature Agency, is the authority responsible for managing the cormorants, but regional State Forest Districts carry out the majority of the actual management. In Denmark, there is a strong tradition of stakeholder consultation in general and in the process of drafting management plans and over the years, stakeholders participated in an officially appointed stakeholder advisory group.

The formal set-up for managing cormorants appears to be satisfactory, but looking back at management since the first management plan, it is clear that inclusion of stakeholders or lack of such is not the only problem in the process. A major obstacle to reconciliation efforts has been that it is unclear which ministry or institutions are responsible for implementing the recommendations in the management plan. This obstacle is illustrated by the result of a recommendation for actions to be taken during the 2002 plan, i.e., the documentation of the level of damage, assessing the economic loss, and the analysis of the effects of technical mitigation instruments. The need for documentation of damages, economic losses, and effects of mitigation has been obvious for years, but the Ministry of the

**Table 1** Main instruments available in the Danish management plans

| 1980 protection | 1992 1st management plan | 1994 expansion of objectives | 2002 2nd management plan |
|---|---|---|---|
| | | | Experimental hunting |
| | | | Mitigate conflicts related to salmon and trout smolts |
| | | Culling of eggs by oiling in colonies on state owned and private land | Culling of eggs by oiling in colonies on state owned and private land |
| | | Stop for establishment of new colonies | Stop for establishment of new colonies |
| | Development of technical mitigation measures | Development of technical mitigation measures | Development of technical mitigation measures |
| Protective hunting 100 m from fishing nets all year (with permit) | Protective hunting 100 m from fishing nets all year (with permit) | Protective hunting 500 m from gear all year (with permit) | Right to protective hunting 1,000 m from gear outside breeding season |
| Permission to scare cormorants away from forestry | Permission to scare cormorants away from forestry | Permission to scare cormorants away from forestry | Permission to scare cormorants away from forestry |

Environment and the Ministry of Food, Agriculture and Fisheries have been disagreeing who should be responsible for carrying out such investigations. The result is that after more than a decade with management plans, the issue of damage inflicted by cormorants, which is the very reason for making the management plan, is still missing a proper documentation.

It can be argued that the current policy mix has evolved alongside an escalating problem. This development in the management indicates a high degree of flexibility—e.g., the decision to allow deterrence in rivers and river mouths to reduce the salmon/cormorant conflict as well as the introduction of oiling. However, this development can also be seen from a negative perspective. A possible interpretation is that the problem associated with cormorants escalated faster than the policy development. Another interpretation is that the problem was not acknowledged for too long, which has resulted in a situation that is more difficult to handle from a conflict reconciliation point of view. The different objectives of the management plans support the latter interpretation. When comparing the management plans, it seems as if the 1992 plan was based on the assumption that cormorants could not cause substantial problems whereas the 2002 plan acknowledges that problems exist.

## 3.6 Stakeholder Viewpoints

Stakeholders have strong viewpoints when it comes to cormorants. The immediate explanation could be that cormorants raise strong emotions. This is probably true as the conflict is also a clash between a rural utility-based perception of nature and an environmentalist perception of nature. For many years, the "traditional" stakeholder groups have been commercial pound net fishers and recreational (net) fishers on one side and environmental stakeholders on the other side. A simplified version of the conflict is that fishers have accused environmental stakeholders for being ignorant and environmental fundamentalists. Environmental stakeholders at the contrary have accused fishers for exaggerating the extent of the problem, making the cormorant a scapegoat for environmental degradation partially inflicted by fishers themselves and accusing fishers of wanting to eradicate the cormorants.

In the first half of the conflict duration, the Ministry of the Environment was supportive of the environmentalist nature perception and, as a result, the power relationships between stakeholders were very asymmetric. Environmental organizations had strong ties with the responsible authorities and, according to interviews, in particular the Ornithological Association believes that it had substantial influence on the Danish cormorant policy (Olesen 2005).

The fact that the pound net fishery had been in a decline for years before the cormorants became an issue, weakened the pound net fishers as a group. Therefore, pound net fishery was considered a marginal group with marginal problems compared to the fishery sector as a whole, both by the national fishery association and by authorities, and as a result, pound net fishers received little support. From the interviews carried out it is clear that fishers have not understood the debate about lack of documentation supporting their claims. To a large degree, fishers have been unaware of the difference between scientific documentation and the experience-based knowledge obtained by observing the impacts of cormorants on a daily basis, and this has weakened their position considerably when discussing the cormorant issue. Fishers have experienced this situation as not being taken seriously, which has caused a lot of frustration among them. However, according to interviews, it is the opinion of fisheries stakeholders that administration in recent years has been more sympathetic to fishers' arguments (Olesen 2005).

There has been a gradual shift in the weight of stakeholder groups. The recent research documenting a negative impact on regeneration of salmon stocks has added the anglers and fish conservationists as a new active group of stakeholders alongside the other types of fishers. The authorities have gradually shifted position in recent years towards more population management. In the particular case of salmon juveniles, authorities have recognized that cormorants can constitute a problem for depleted salmon stocks and have adjusted the management practice as a consequence.

It is interesting though that neither fisheries stakeholders nor environmental stakeholders are pleased with the 2002 management plan. It is a general concern from environmental stakeholders that the management plan is too open-ended and that the value of several of the instruments is more symbolic than actually reducing the conflict. The management plan caused a lot of debate raised primarily by environmental stakeholders but the debate calmed down when it became clear that the extended management possibilities would have an insignificant effect on the size of the cormorant population. The fishers as well complain about the management plan listing a number of mitigation instruments, which according to the fishers have proved to be of little use for not reducing the conflict (Olesen 2005).

The issue of knowledge has been crucial for the development of the conflict as well as for the relationship among stakeholders. The lack of documentation has not only been a problem in the case of damage assessments, but also when it comes to effectiveness of different mitigation instruments. The presentation of new ecological findings in the FRAP project has been the first verification of formerly undocumented claims. This also resulted in stakeholders being on terms that are more equal. Fishery stakeholders for the first time could feel at ease not being pushed in a defensive position whereas environmental stakeholders had to acknowledge that in some cases cormorants are causes of concern. The scientific documentation on cormorant predation on salmon smolts provided by the FRAP project stirred up the conflict as it was the FRAP results from the diet survey that forced stakeholders to reconsider the impacts of cormorants on fish stocks. When the ecological findings became publicly known, fisheries stakeholders successfully used this documentation to back their claims politically, resulting in a political pressure to strengthen the management of cormorants. The responsible minister avoided this by referring to ongoing evaluations of management measures, but was forced to initiate a participatory process with the objective of making a midterm evaluation of the management plan. As a result, a stakeholder-working group with an advisory role was invigorated. Both, perspectives and behavior of stakeholders did change due to the FRAP interaction. In particular for the fisheries stakeholders being in favor of increased use of population reduction measures, it has become clear that no easy solutions are in sight. The findings, which supported their claims of not being listened so far, and the experience of suddenly being listened to because of the emergence of scientific documentation has made them engage more constructively in the dialogue. Stakeholders questioning the negative effect of cormorants on the other hand have accepted that cormorant predation can endanger the survival of native salmon stocks and have consequently supported stronger management measures.

## 4 Evaluating the Management of Cormorants

In this section, the lessons learned from the past 14 years of cormorant management are discussed. We try to use this experience to make some general conclusions for the conflict reconciliation and use these to give advice for future management plans.

## 4.1 The Management Plan

The Danish management of cormorants is a policy mix of various types of instruments addressing conflict mitigation through damage reduction as well as population control, and in theory the different instruments complement each other well. The problem is that no solutions are in sight in the near future, as the current instruments in use so far have proved incapable of reconciling the conflict. It is clear from the research carried out in the FRAP project that the scope of problems necessitates a longer time horizon if population reductions is to be an element of conflict reconciliation. With the current policy in Denmark of maintaining large colonies where no intervention is allowed, research documents that population viability is secured. However, as Denmark is a core area for cormorants, it is necessary to monitor the population development carefully.

On the other hand, it is also clear that stronger efforts have to be made into developing effective deterrence instruments if damage prevention is to play a major role in conflict reconciliation. Furthermore, it is clear that the approach of the Danish management plan prescribing a range of instruments has not been sufficient to reconcile the conflict. There is a need to include a stronger element of active implementation policy including communication and information in future management plans.

Even after more than 10 years with a management plan, clear goals have not been set. Until recently, there has been no effective evaluation of existing measures, and the authorities have failed to cooperate and share responsibilities. The financial and institutional resources allocated to the tasks of management, research, the development of instruments, and implementation has not been adequate. One outcome of this is that for some of the instruments little effort has been put into their implementation. The level of information and communication regarding the conflict and implementation and effectiveness of instruments has been insufficient.

## 4.2 General Lessons Learned

One of the lessons from the cormorant case in Denmark is that prior to the protection of a species a clear set of objectives should be stated. What is the purpose of the protection and what is the exit strategy from the protection plan if a population level is reached and a protected status is no longer necessary or desirable. Approaches developed for delisting species in the Endangered Species Act of the US could serve as a model.

Another lesson is that a management plan is beneficial as long as it is adaptive and revised with reasonable intervals. However, making a management plan also needs to be coordinated between different responsible authorities. In the cormorant management plan, a range of research activities are proposed or required but it is not explicit who is supposed to finance. This generates the result that the ministry drafting the plan expects other ministries to finance which they reject without a prior agreement.

The most important lesson is that the problem of different knowledge discourses need to be addressed. There is a gap between scientific knowledge and experience-based knowledge. When arguments and perceptions originated in experience-based knowledge are occurring persistently, the decision makers and researchers need to take them seriously. Excluding certain stakeholders' knowledge undermines the legitimacy of management. If the fishers' claims of cormorant predation in pound nets had been taken seriously 15–20 years ago, the conflict would probably have evolved differently.

### 4.3 Future Management Action

Below is a list of the different conflicts encountered through this project and, based on the general lessons learned, suggestions are made for the management responses needed to reconcile those.

1. Discussions of management strategies in the public often originate in insufficient knowledge of the relationship between ineffective mitigation instruments and the population dynamics of cormorants. The level of dissemination of information regarding population dynamics and effectiveness and limitations of different mitigation instruments has to be increased.
2. The effect of existing mitigation measures must be evaluated and recommendations for improvements must be incorporated in the coming management plan.
3. The extent of damage caused by cormorants in pound nets continues to be questioned and research documenting the extent of the problem should be commenced as soon as possible.
4. If pound net fishing is to be conserved, then financial support to technical mitigation is needed. The few remaining full-time professional pound net fishers need to get full support for the costs associated with buying and installing protective means like cover nets. The European Financial Instrument for Fisheries Guidance (FIFG) funds could be a source for such support.
5. With respect to semi-commercial and recreational coastal net fishers, the conflict involves many stakeholders, and action is needed. A decrease in the numbers of cormorants feeding in the area where the fishing occurs would be of specific help. If this option is not realistic, then the relevant organizations should be directly involved in a process to help reconcile the conflict by means of cover nets, scaring, and protective shooting.
6. With respect to anglers, particularly local action is needed related to salmon and sea-trout angling. Measures to protect the native, wild populations of migratory salmonids from a high level of cormorant predation must be implemented. Regular scaring, including live shooting, should be carried out in the most sensitive areas during the few weeks of smolt migration (the time when young salmonids move back to the sea). Currently, giving permissions to anglers associations to organize scaring and shooting at the river mouths is a

test for this. The management should be ready to implement measures to deal with other species if necessary.

7. With respect to the conservation of red-listed salmon populations, conflicts have recently emerged and immediate action is needed. The measures mentioned above under item (6) need to be implemented here as well, but a significant immediate reduction of cormorant numbers in this area is also needed.

## 5 General Recommendations

The already established advisory group representing experts, stakeholders, and managers could produce a prioritized list of cormorant colonies. The colonies should be ranked from least to most conflict causing. When the list is produced, a political agreement should be made of what level of population reduction is wanted. If the most problematic colonies are effectively removed, the level of conflict will be eased without any threat to the viability of the cormorant population (Modeling results of the FRAP project shows that a considerable part of the population can be removed without endangering the population). The official political system should present the plan to the EU-environmental commissioner to be accepted as fulfilling the requirements of the Birds' Directive.

The level of documentation required for action must be stated clearly. This is important to avoid costly "over documentation" or to spend research funding on documentation that is not accepted in the end. According to this, a research program aiming at producing the required level of documentation should be initiated.

## 6 Conclusion

The population of great cormorants, residing in Denmark has reached a very high level and most areas of the country have a high density of cormorants. The population has increased by a factor 10 (1000 %) in 20 years. Through the last decades, a conflict between fisheries interests and conservationists has been evident. The coastal fishermen wish to reduce the cormorant population and the conservationists wish to leave the birds alone. The constant political pressure from the fisheries organizations has lead to a gradual change in management from total protection towards some restricted regulation. In the course of the conflict development, the participatory process of decision-making has been improved with the establishment of a permanent group of stakeholders that participates in the formulation of a revised plan. So formally, all stakeholders have influence on the national management of cormorants. However, the work done in the FRAP project indicated that the conflict was still fueled by fishers frustration of "not being heard" and that they actually had not had much influence on the management. This seemed to be related to the problems of understanding and using "knowledge and documentation" in a political process. Shortly, the fishers have (for a long period) claimed that the severe damage from cormorants on catches and fish stocks in general was a proven fact (because they all see this happening), whereas the conservationists claim that there is no real documentation of this. The political system (also in EU) needs some level of scientific documentation to acknowledge the problem and thus the fishers' claims have not resulted in a significant change in management. Through the FRAP work carried out in one Danish Fjord, some scientific evidence was found supporting the fishers claims and this clearly strengthened the influence of the fishers on the management decisions. Thus, the Danish case can be seen as a good example on how scientific documentation (and the stakeholders' perception and use of this) is crucial in conflict management. Even in a conflict where relevant governmental institutions are involved and a participatory process involving most stakeholders influence the management, the lack of scientific data can prevent reconciliation.

**Acknowledgments** This work was financed by the EU 5th Framework Program (5th FP) Project "FRAP" (Development of a procedural Framework for Action Plans to Reconcile conflicts between large vertebrate conservation and the use of biological resources: fisheries and fish-eating vertebrates as a model case), contract number EVK 2-CT-2002-00142-FRAP.

## References

Bildsøe M, Bundgård I (1998) Foraging behaviour of cormorants in pound nets in Denmark. The use of barrel nets to reduce predation. Wildl Biol 4:129–136

Bregnballe T, Olesen T (2004) Cost estimations of culling and deterrence activitities. Unpublished

Carrs DN, Marzano M (2005) Reducing the conflict between cormorants and fisheries on a pan-European scale. Summary and national overviews. EC report http://www.intercafeproject.net/pdf/REDCAFESummaryandNationalOverview.pdf. Accessed 10 Dec 2008

Carss DN (2003) Reducing the conflict between cormorants and fisheries on a pan-European scale, final report, EC report http://www.intercafeproject.net/pdf/REDCAFEFINALREPORT.pdf. Accessed 10 December 2008

Danmarks Fiskeriforening (1997) Danmarks Fiskeriforenings Bundgarnsundersoegelse, Status over bundgarnsfiskeri i Danmark

Dieperink C (1995) Depredation of commercial and recreational fisheries in a Danish fjord by cormorants, Phalacrocorax carbo sinensis. Fish Mgmt Ecol 3: 197–207

Engstrøm H (2001) Long term effects of cormorant predation on fish communities and fishery in a freshwater lake. Ecography 24:127–138

Eskildsen J (2005) Skarver 2005, Arbejdsrapport fra DMU, nr. 220. European Commission (1979) Conservation of Wild Birds. Council Directive 79/409/EEC

Jepsen N, Klenke R, Sonnesen P, Bregnballe T (2010) The use of coded wire tags to estimate cormorant predation on fish stocks in an estuary. Mar Freshw Res 61:320–329

McKay HV, Robinson KA, Carss DN, Parrott D (2003) The limitations of pellet analysis in the study of cormorant *Phalocrocorax spp.* diet. Vogelwelt 124:227–236

Miljøministeriet Skov og Naturstyrelsen (1992) Forvaltningsplan for skarven i Danmark (Management Plan for Cormorants)

Miljøministeriet Skov og Naturstyrelsen (2002) Forvaltningsplan for skarv i Danmark (Management Plan for Cormorants)

Miljøministeriet, Skov og Naturstyrelsen (1997) Lov om jagt og vildtforvaltning. LBK nr. 114 af 28/01/1997 (Hunting and Game Management Act)
Miljøministeriet, Skov og Naturstyrelsen (2003) Bekendtgørelse om vildtskader. BEK nr 869 af 10/10/2003 (Damage by Wild Life Ordinance)
Olesen T (2005) Policy instruments—practice and potential: Denmark (FRAP-project, Deliverable 14)
Pers.com Alberti P (FRAP WP8 Management and viability of target species, Dep.Ecol. Model. UFZ, Leipzig, Germany)
Pers.com Buch A (Pound net fisher)

Great cormorant (*Phalacrocorax carbo sinensis*) crossing a bay on the Baltic Sea. *Photo:* André Künzelmann

# Screening the Cormorant Conflict on the European Level

**Felix Rauschmayer and Vivien Weiss**

**Abstract** The great cormorant has fully recovered from an endangered status in the 1980s and is now present throughout Europe. Eating fish, this recovery is perceived as a danger by fishermen and anglers. The existing national or regional action or management plans are mostly ineffective due to the high mobility and numbers of the bird. At first sight astonishingly, no such plan exists at the European level. The reasons for this absence, identified through analyzing documents and interviews, are the following: value disagreement, different views on the necessity of such a plan, fear of lost autonomy, different views on species protection, and difficulties of integrating different interests. Steps are proposed how such plan could be achieved, and the final section describes the political development after the first publication of this analysis.

## 1 Introduction

This chapter differs in its structure from other chapters in this book insofar that it (a) covers the screening analysis and (b) describes in its final section the development of the conflict after the first publications on the authors' analysis (Rauschmayer and Behrens 2008; Behrens et al. 2008). The publication of the screening had an effect on the conflict development, as will be shown later.

F. Rauschmayer (✉)
Department of Environmental Politics, UFZ—Helmholtz Centre for Environmental Research, Permoserstraße 15, 04318 Leipzig, Germany
e-mail: felix.rauschmayer@ufz.de

V. Weiss
Department of Urban and Environmental Sociology, UFZ—Helmholtz Centre for Environmental Research, Permoserstraße 15, 04318 Leipzig, Germany
e-mail: vivien.weiss@ufz.de

R. A. Klenke et al. (eds.), *Human-Wildlife Conflicts in Europe*,
Environmental Science and Engineering, DOI: 10.1007/978-3-540-34789-7_9,

Great cormorants breed in colonies and hunt together in a highly unpredictable way. The great cormorant is a very mobile bird, which crosses different borders when moving in autumn several hundred kilometers from their breeding colonies in the north towards the south, and back in spring. Beside this, the species shows active dispersal behavior on a broad spatial scale. Its range of activity usually encompasses several countries, what plays an important role for discussing an international management of the European cormorant population.

Mainly through persecution, the number of cormorants dropped to a very low level all over Europe in the beginning of the 20th century and the subspecies *Phalacrocorax carbo sinensis* got extinct in several countries. In Germany, first legal protection started in 1935 (Rutschke 1998) and different conventions aiming at the protection of migrating and threatened animals were enacted in the last three decades.[1] These protection measures, also based on the increasing role of animal welfare,[2] lead to increasing cormorant numbers in the 1980s, which subsequently stabilized in the 1990s (van Eerden 2002).

Today, the number of the great cormorant is estimated to something between 500,000 and 1,600,000 birds, dependent on who is asked (cp. Bregnballe et al. 2003; cp. BirdLife International 2005; Vfg Verband für Fischerei und Gewässerschutz in Baden Württemberg 2006). Their regularly occurring large-scale movement pattern causes a strong interdependency of those countries dealing with the increasing numbers in different ways. In the last years, the great cormorant has also occupied new breeding areas probably never used before the last ice age. In these parts of Europe, it becomes an invasive species.

What is the conflict about? The different parties involved, e.g., fishers, anglers, and conservationists, of course have varying perceptions on this subject. This paper presents the identified conflicting views as well as possible reasons for these conflicts. The main line of potential conflict resolution is as follows:

Since fishers and anglers have a great interest in the protection of fish stocks for economic as well as for recreational reasons, they argue that an internationally coordinated cormorant management would provide the best solution to this problem. On the other side, bird conservationists often understand internationally coordinated management as large-scale culling and do not want it to come into being, and prefer no or only local action.

Currently, mitigation measures mainly take place on local or regional level without coordinated management of cormorants on the a large-scale. Financial compensation works only in those cases where financial losses are the problem and when enough money is available for compensation. However, there are many cases where exclusion of cormorant access to fish is impossible and where financial compensation does not appease the conflict.

---

[1] EC Birds Directive, Bern Convention, Bonn Convention and Habitats Directive.

[2] Since the early 1980s, a number of animal welfare organisations exist in several European countries. Well-being of animals is the concern of these organisations, and they started political and ethical debates about the treatment of animals.

The only mitigation, which then remains available at the local level, is scaring, working more effectively with backup by killing through shooting. Nevertheless, scaring only works on a short-term scale as the cormorant is an adaptive learner. It is rather hard to shoot them, because cormorants quickly avoid such places or stay in a good safety distance, which is increasingly the case along dams and larger rivers (Keller and Lanz 2003). However, the consequences for conflict mitigation are limited if not counter-productive: hunters and fishers usually shoot cormorants in places where it is extremely difficult to get hold of wounded animals. It is difficult to predict when cormorants appear in a given place. Consequently, their high mobility would ask for supervising huge areas simultaneously, which is expensive and difficult to manage. Additionally, in times of an increasingly sensitive society towards hunting in general, hunters are often not willing to cull cormorants, which they usually consider inedible. The result of scaring by shooting or of intended population control by shooting on a local level is of limited effect for the protection of fish (Keller and Lanz 2003), increases feeding demands of cormorants due to higher energy expenditures and has implications in terms of animal welfare (wounded birds). The only reason, why scaring by shooting becomes more and more common throughout Europe, is the growing pressure from fishers and the lack of other locally implemented mitigation measures.

It has been outlined by different scientists that culling of cormorants is not an appropriate measure to reduce the overall population but that it can be complemented by a coordinated management of breeding capacities (Frederiksen et al. 2001). As cormorants are colonial breeders, population management in the colonies could be done without the negative side effects of large-scale culling described above. Currently, such actions (egg-culling and colony destruction) do happen, but on a limited scale only and without any international coordination (see Bregnballe et al. 2005 for efforts on egg oiling).

In front of this briefly outlined specificity, dimension and severity of the conflict, we analyze the following three questions:

- Which actions had been taken until 2006 for developing a large-scale international cormorant management action plan?
- Why had such an action plan not been implemented?
- What are the current main obstacles for developing a management plan with the aim of limiting the numbers of cormorants without affecting their viability?

In order to conduct this analysis, we interviewed between 2004 and 2006 different governmental officials, scientists and representatives of stakeholder groups (see Table 1). Scientific literature, information material and various reports constitute the basis for the interviews and the analysis.

## 2 What Happened So Far?

For a couple of decades, cormorant numbers have increased in their breeding areas along the Baltic and Nordic Coast line, first mainly in Holland and Denmark, but later also in Germany, Sweden, Poland, the Baltic Countries and Finland. Outside

**Table 1** Interviewed stakeholder groups and their spatial level

| Stakeholder groups | Conducted interview | Federal state | State | EU/International |
|---|---|---|---|---|
| Conservationists | 3 | 1 | – | 2 |
| Aquaculturists/anglers | 2 | 1 | – | 1 |
| Authorities | 9 | 2 | 4 | 3 |
| Specialists/scientists | 8 | – | (8)[a] | |
| Sum | 22 | 4 | 4 | 6 + (8) |

[a] The assignment of scientists and specialists to a spatial level is ambiguous as they are often working on the international scale as well as on the national ones. However, their statements usually refer to the national context with an eye on the international situation

the breeding season, and due to growing numbers of non-breeding juvenile and adult birds partly also within this period, the increasing Cormorant numbers became also a problem for the countries in the other parts of Europe. In order to resolve the growing conflicts, some countries asked for coordination on a pan-European level because they considered it impossible to solve the problem of this migrating bird within their own boundaries.

The Convention on Migratory Species (CMS, also known as Bonn Convention) was the first international framework for dealing with this issue. The aim of CMS is to protect migratory species and furthermore to establish cross-boundary cooperation with regard to their conservation. The CMS debated the cormorant topic in its 1994 meeting and gave a recommendation, which aimed at the conservation and a special management of the cormorant population (CMS 1994). The CMS invited Denmark and the Netherlands, as the countries with the largest breeding colonies to draft a pan-European action plan.

The next step was an expert meeting in 1996 in the Netherlands. The outcome was a report, which gathered scientific information about the continuing increase of cormorants, the expanding geographical range, and management options (CMS 1997a). At the fifth CMS meeting in 1997, Denmark and the Netherlands were reminded to take the necessary steps and to involve the affected countries for creating an action plan and finalize it before the end of 1997 (CMS 1997b). Another expert meeting followed in the same year, where representatives of the concerned countries as well as non-governmental organizations acting at different scales and representing different interests discussed the drafted action plan.

The plan was mainly a compilation of different measures, which can be used by the countries to lower the damages caused by cormorants (CMS 1997a). Measures split up in four main categories, including both, lethal and non-lethal measures:

- Technical/physical measures taken outside the breeding colonies
- Technical/physical measures within the breeding colonies
- Financial compensation measures
- Monitoring

The plan mentions that measures in breeding colonies have to be in accordance with EU and international legislations. In its preface, the plan clearly points out that a substantial reduction of cormorant numbers will not necessarily lead to a reduction of the extent of conflicts in problem areas due to the fact that killed cormorants will be replaced by other cormorants (CMS 1997a). The parties did not agree to a clear aim (such as a predefined damage or population reduction).

The proposed measures are currently applied in an uncoordinated manner in several European countries, but seem not to mitigate the conflict especially for the most concerned countries (for the Danish and Italian case compare Wilson et al. 2004, cp. also Wilson (2012) in this book). In fact, this set of measures proposed possibilities, which the relevant authorities have to select and combine carefully to achieve comprehensive conflict mitigation. Since an uncoordinated management might endanger the species, it is necessary to coordinate the measures taken (Frederiksen et al. 2001). The main actors Denmark and the Netherlands did not intend to achieve a European-wide population management through this plan (personal communication by a cormorant expert), and any such listing could not achieve this aim.

This loose set of management options was not the expected outcome for some of the affected countries in Central Europe as they pointed out that the problem would have to be solved through a consequent population control as solution for the problem (CMS 1997b). It became quickly obvious that many different interests were (and still are) present in Europe, which cannot be easily combined in a common policy.

In this second expert meeting, the German Federal Government and the representatives of the European Inland Fisheries Advisory Commission (EIFAC) did not support the adoption of the action plan. They stated that it was not the expected outcome and that they did not consider it sufficient to the specific conflicts (CMS 1997b). Nevertheless, Denmark and the Netherlands sent a finalized plan to the CMS in 1998. The members of the CMS did not ratify this action plan and so it went into a deadlock. This is puzzling because the majority of the concerned countries agreed to the plan, as stated by a CMS-official.

In 2002, the European Anglers Alliance (EAA) initiated a meeting in Strasburg to replace the cormorant issue on the political agenda. As a result, a recommendation, which suggested resuming the work on an action plan, was sent to the European Commission (EC) (EAA—European Anglers Alliance 2002). On the European level, the ORNIS Committee, which is responsible for the implementation of the EC Birds Directive, could be an adequate body to deal with the topic. However, on their meeting in 2003, the ORNIS Committee stated that they could not find any agreement for the need of an international management. Therefore, they did not fulfill the demand of the EAA.

Currently, almost all European states have their own national or provincial regulations to mitigate the conflicts (Carss 2003). These regulations contain different measures which range from strictly protecting the species, e.g., in the Netherlands, to shooting large numbers of birds, e.g., culling with departmental quotas in France. These broadly varying approaches to handle the cormorant problem show that the interests of the respective countries are very different and hard to combine in a management plan.

In comparison, the United States have launched a federal management plan on the double-crested cormorant (*Phalacrocorax auritus*) in 2003 (USDI Fish and Wildlife Service 2003). The relevant authority defined six different management alternatives including the results of a participatory decision-making approach. The Fish and Wildlife Service favored one of these proposals, the "Public Resource Depredation Order". This option relates to more flexibility of the agencies dealing with this problem and allows 24 States to cull cormorants if deemed necessary.

A difference to the European efforts lies in the fact that a clear responsibility for the mentioned agency is given. This agency can take up the viewpoints of different stakeholders, employ professionals, and therefore increase the chance of implementing the action plan. Another difference observed is the participatory decision-making including citizens. In Europe, scientists, NGOs as well as governmental representatives are involved in the discussion about the topic [e.g., in the EU projects REDCAFE and INTERCAFE (see www.intercafeproject.net for both)], but no legitimate and inclusive forum exists, which could also include statements of non-organized citizens.

To summarize the European case: some actors made efforts in order to find a solution to this cross-national problem, but did not succeed to generate a general solution. On the European level, the ORNIS Committee apparently remained the only institution to take up the issue again.

## 2.1 Different Perceptions of Damage

Due to bird migration, the specific damages to fish stock occur in most regions only at a specific time of the year and on specific parts of the fish production or resource. Usually, it is not clear to which extent cormorants contribute to the overall damage on fish, as compared to egrets or other fish-eating birds, as well as to other fish-eating animals or to fish diseases. This lack of clarity is mainly due to a general ignorance about the amount of fish present in a water body, as well as to the high mobility of the cormorant.

In aquaculture, damage is—taking into account the lack of clarity mentioned above—more easily measurable than on the coast, in lakes and rivers. The single and annual loss in aquaculture can be substantial, e.g., if a large flock of cormorants forages at a pond during several days. Damages are not only caused by direct consumption of fish by the cormorant but also due to secondary damage which occurs through injuries, spread of diseases (Carss 2003), and disturbance/disruption of the production cycle (heavy losses in only one age class). In aquaculture and contrary to angling or fishing in coastal waters, lakes and rivers, the monetarization of identified losses of marketable fish does not pose a major problem.

Since anglers have an interest in the maintenance of different fish species for economic and recreational reasons, the damage for this stakeholder group is often less of financial character, but anglers rather complain about the impact on specific, sometimes threatened fish populations (Keller and Vordermeier 1994; Carss 2003).

The same non-financial aspect applies as well to conservationists or ornithologists. These stakeholder groups (often appearing as one group) want to keep the population and the colonies to enjoy or research on them and make use of them in terms of birdwatchers. Therefore, they reject the interference in breeding colonies and deny large-scale shooting of cormorants.

Damages may have social character additionally to their material and financial aspects (Wilson et al. 2004). In many cases, fisheries do not have high economic relevance in terms of employees or profit; they rather contribute to the social identity of a region as for example in the German FRAP model region Upper Lusatia. If one aims at successful conflict mitigation comprising ecological, economic and social aspects, different mitigation measures as well as policy instruments have to be embedded in the institutional and legal frame on the European level as well as in the respective countries.

## 2.2 Institutional and Legal Frame

The Birds Directive as well as fishery and agri-environmental schemes constitute the European legal frame for the resolution of the conflict. The European and member states' policies usually distinguish between nature conservation and resource-use administrations. Their specific distinction, authority, and political level vary between the member states.

The protection status of the cormorant has changed at the European level. Until 1997, the Birds Directive listed it on the Annex 1, which meant that also the habitats of this species should be protected. However, due the ongoing increase of cormorant populations the ORNIS committee removed the cormorant from Annex 1. Some stakeholder groups called for its conclusion in Annex 2, the listing of huntable species, but without success. As the ORNIS committee did not transfer the cormorant to Annex 2, a general protection status remains, and culling is only allowed with derogation from the Birds Directive. Derogations are possible, referring to article 9 of the Birds Directive. This allows countries to give permissions for hunting a species to prevent serious damages to crops, fisheries, livestock, forests and water. This article applies, if no other satisfactory solution to prevent damages is available. The article, though, does neither define "damage" in terms of economic losses, nor "satisfactory solution". However, even with permission, cullers have to keep the rules of the Birds Directive, e.g., not hunting during the breeding season or only with certain methods. If one country does not ensure sufficient protection of wild birds, the European Commission can take legal actions against this country. This applied to a Swedish case where authorities permitted shooting of cormorants and destroying eggs during the breeding season (European Commission 2005).

### 2.3 *Different Stakeholders on Different Levels: a Complex Arena*

The local level is the place where damage to fisheries occurs, and where mitigation measures are implemented. The specificity of local sites often asks for locally specific technical mitigation measures. According to the REDCAFE project, this level is the most important to mitigate the conflicts (Carss 2003). Especially if one considers a general cormorant population reduction impossible or ineffective, as some cormorant experts do (Carss 2003), then the local level is the only place where to implement conflict mitigation measures.

The regional or state level (according to the institutional setting in the respective country) can determine the political style of conflict handling (for the differences between the neighboring provinces of Ferrara and Ravenna, Italy cp. Wilson 2004), or even design specific economic or institutionalized mitigation measures (such as the cormorant enactments interpreting the Birds Directive derogation possibilities in German states, compare Thum 2004, 2005).

The national level represents the countries' interests in international bodies, such as the ORNIS Committee on the European level or the Convention on Migratory Species, and, in some countries, determines the national policy concerning the cormorant (such as in France). National governments tend to represent a balance of positions of "their" stakeholders: The strength of nature conservation NGOs is in line with the Dutch government's position, as it is the relative strength of natural resource users such as fishers and hunters with the French government's position.

The international level is characterized on the one hand by the EU and its relevant bodies and on the other hand by non-governmental organizations like the CMS, the African-Eurasian Waterbird Agreement (AEWA) , and the Cormorant Research Group (CRG) of Wetlands International. The latter organizations are also open for states from outside the EU: 95 countries from all over the world cooperate within the CMS with the aim to conserve migrating species throughout their range. Within the CRG, scientists and experts from different countries exchange information on cormorants and on conflicts between cormorants and human fisheries' interests. On the European level, the ORNIS committee, which assists the European Commission in implementing the Birds Directive and which is composed of representatives of each member state, can be the appropriate body to reconcile the problem.

It seems as if the European level is relatively isolated from local conflicts. At this level, it is more common to discuss in rather abstract terms. Calls for action at the European level are prevalent (e.g. EAA—European Anglers Alliance 2002; Bundesrat 2003), but got stuck due to the institutional and consensus-oriented design of European nature conservation policy. The need for an agreement with different stakeholders especially with The Netherlands and Denmark rendered action towards an European management plan difficult.

One possible solution would be that an international institution would take up the problem. At the same time, some stakeholder groups do not want the issue to become overly bureaucratic what they expect if the EU is dealing with this problem.

## 3 What was Impeding a Common Solution?

Based on the efforts undertaken to create a European management plan, it becomes clear that some stakeholder groups aim at a common solution, but some reasons prevented the development of an effective action plan. In the following, we will discuss the reasons for this failure, as extracted from the discourse (Fig. 1).

Stakeholders have different views on whether and how to manage a population—differences mainly caused by different views on species protection and ecosystem management in general. Whereas some stakeholders are calling for large-scale culling and an overall reduction of the population, others state that the population regulates itself by access to food supply and that there is no reason for human management.

Behind the data disagreement, though, seems to lay a value disagreement about the appearance and rights of animals and humans respectively. Animal right activists are strong in the Netherlands, and their point of view influences national hunting legislation. Until now, the Netherlands refused to take action concerning population control. If someone judges "natural" population size the "right" population size, then human population management is unacceptable.

Some states are not interested in a coordinated management, as they are afraid of loosing autonomy. They associate transferring some kind of coordinating authority to the European Commission to the worry about a high degree of EU-bureaucracy. According to the perception of some stakeholders, they consider transferring responsibility to the European Commission not as good as to the Cormorant Research Group of Wetlands International coordinating actions and exchanging scientific data. This group mainly consists of experts regarding the ecological and biological features of cormorants.

It is unclear, how to integrate different interests in the development of a management plan. A characteristic of pan-European issues is the extremely complex composition of involved parties what makes it difficult to communicate and balance power relations. Bird conservationists of the CMS strongly influenced the 1997 action plan, as a fisheries organization stated. Fishers, anglers, and states with high rates of wintering birds were and still are doubtful whether conflict mitigation would be possible without reducing the population numbers to a high extent. Therefore, they did not consider the drafted plan sufficient.

Conflict-mitigating measures have been adapted to local conditions of the conflict and are thus site-specific. Statements in the conducted interviews point out that, due to the uniqueness of each conflict situation, there would be no need for a pan-European management. Twenty-five countries with differences in their cultural, social, and economic conditions are involved and it is hard to combine their

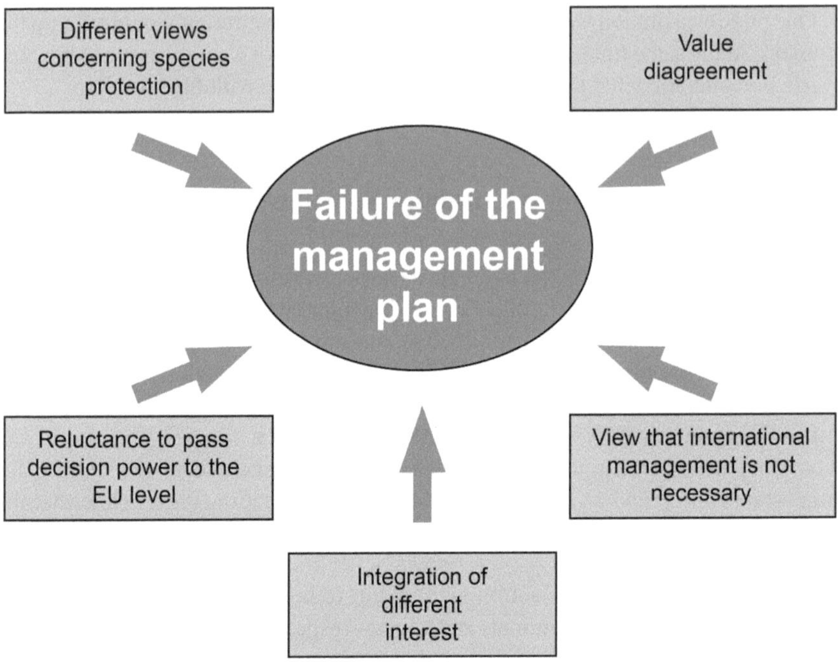


**Fig. 1** Reasons for the failure of a pan-European management plan

different needs with regard to a management plan. As conflicts mainly occur at the local level, it is difficult to develop a wider picture of the conflict, which includes all specific factors.

Therefore, an international management would only be necessary if the local and national measures are ineffective. However, even if they were ineffective, it still would have to be proven that a regulation on the European level would bring more benefit than locally applied measures.

Given the difficulties, which led to the failure of an European-wide coordination of cormorant management, the question is now, whether the advantages of such coordination would be high enough to surmount the difficulties.

## *3.1 On a Way to an International Action Plan*

From the statements in the interviews, it seemed as if the conflict about the cormorant had gone into a stalemate by 2006. Attempts to find a pan-European solution were undertaken but failed. As the concerned countries have not changed their viewpoints dramatically, but only incrementally, any new attempt to bring this issue up again on an international level, would have to consider the reasons analyzed above.

In Denmark, e.g., a management plan was set up in the early 1990s and has been modified over the years with regard to the complaints of the fishers. The measures shifted from protecting the species to a kind of population management. The change of the existing management measures shows that the chosen techniques were not appropriate to calm down the conflict, which made an adjustment necessary. We observed the same shift in most European countries, the Netherlands being a notable exception from the rule.

The value difference between the concerned parties and countries, e.g., expressed by the attitude of the Netherlands, is a crucial point, which cannot easily be resolved. Treating value differences at the international level and agreeing on how to translate value differences into specific policy schemes or plans seems to be rather difficult. According to Rijsberman (1999), it is impossible to negotiate about values because they are deeply rooted beliefs which are gained through a long period of cultural and personal experiences. Besides values, interests and aims play an important role for the characterization of conflicts. These can be partly accommodated for in negotiations to achieve an agreement on a solution to a conflict. In the specific setting concerning population management of European birds, each party concerned remains autonomous. So, if power relations are such that any management needs the agreement of a specific actor, and if this actor will not agree to change its policy, then the parties would not agree to a management scheme. Treating value differences at this level and agreeing on how to translate value differences into specific policy schemes or plans seems to be rather difficult.

Due to the still growing numbers of cormorants, the conflict is sharpening. On the sub-national level, solutions were found fitting to the needs of the local fishers. However, complaints by fishers concerning the high and still growing damage caused by cormorants have been increasing.

Results of population modeling show that measures taken in one country may influence the population in other countries and therefore the uncoordinated handling of conflicts via culling may endanger the complete European population. For that reason, the concerned countries should consider the establishment of an international cooperation once more.

## 4 Range of Potential New Attempts

The benefit of a pan-European coordination will depend on the design. Two extremes of such a management plan determine the range of potential plans: The mere listing of mitigation measures without any common aim or schedule contrasts with a detailed plan obliging parties to take actions under conditions specified in the plan itself. If the plan is merely a loose set of techniques, which authorities apply more or less anyway, then the benefit will not be high. This was one major problem of the drafted CMS action plan. Some parties did not consider it as appropriate due to the lack of binding duties and aims. The other extreme is unrealistic and therefore inappropriate as well:

States will not oblige themselves to undertake specific actions to lower the cormorant population.

In every new attempt of drafting an action plan, the reasons, which already led or very probably will lead to failure, have to be considered carefully narrowing down the range of options. A clear objective with respect to the different stakeholders' interests may enhance their cooperation in implementing such action plan. Any effective and efficient management plan would need a better understanding of the social, ecological and economic costs and benefits linked to the cormorant and to the policy options to be developed. Policy options would cover sharing of knowledge, financial measures, such as compensation payments, and the definition of duties and rights.

As the mere listing of different mitigation measures in the CMS plan was not successful and as an obligatory large-scale culling is not enforceable, a solution between these poles could be possible.

## 5 Possible Design and Steps of an Action Plan

The thorough understanding of the current European structure of institutions and policies dealing with the cormorant at the various levels is a necessary basis for designing a management plan.

1. As the analysis of reasons showed data disagreement, additional scientific information could solve the problem. Spreading knowledge on effective local mitigation measures, as initiated by REDCAFE, would be a first step towards less damage and therefore more efficient conflict mitigation. The exchange of scientific data between the countries advising for the application of different measures seems to be another necessary component of any management plan.
2. Modeling the European population would be a further step towards a scientific basis for deciding about management options. A regional component of this modeling would then indicate the regional distribution of the population and herewith of damages.[3]
3. Ecological, economic and social damages still have to be assessed unanimously, because a picture of these damages in the different member states is still missing.
4. Integrating different management scenarios in the modeling would be a step towards defining different management options. The regional adjustment of this modeling could refine and specify the options.
5. The next challenge would consist of linking local, regional, national and European policy processes together in an appropriate manner, which is far from

[3] Population modeling of the cormorant has been done within FRAP—the regional aspect would still have to be integrated.

evident seeing the different competencies and ways of handling the problem in the different parts of Europe. This might be the largest challenge because it means reducing autonomy of the respective levels if a supra-national institution would coordinate the policies.

6. The highest level of coordination would be a common management plan detailing who would do what and when, potentially including a common budget for implementing the measures. However, as the interests and values differ in the countries and in between them, it will be difficult for the parties to agree to a binding multi-level agreement. Therefore, any solution would have to be composed of binding and voluntary parts. Every country needs to have the possibility to take action if deemed necessary. This would leave autonomy to actors on the national level to balance the internal interests.

Undertaking the first four steps does not lead to actions automatically, but would constitute a necessary basis for a substantial evaluation of different management options. The mentioned components provide a database of scientific knowledge, which includes the specific views of different countries. If all stakeholder groups provide data to build up a comprehensive database, then any plan based on such comprehensive knowledge will be less contested. If one wants to consider the social, ecological, economic and political costs and benefits of a coordinated management of the population, then the steps one to five are necessary components of the way towards such consideration. However, if one is sure that steps five and six are politically unrealistic, ecologically ineffective or economically inefficient, then the necessity of any of the steps one to four might be challenged. Until now, participating countries took steps one to three to some extent without any clear agreement on how to continue to deal with the European-wide conflict about the great cormorant.

## 6 Evolving Situation

One of the authors had the possibility to present the results of the screening analysis twice: first, to the Scientific Advisory Body to the ORNIS committee in November 2006 and, second, in November 2008, to a meeting on the European management of cormorants, organized by the rapporteur of the European Parliament on this issue. While the first meeting was rather "bird-friendly", the people attending the second (and most of the speakers) were definitely "fisher-friendly". Reactions in both meetings showed that the proposed steps towards an action plan were neither fully appreciated by the "bird lovers", nor by the "fishermen".

The publication of the steps and the presentation at the second meeting (Behrens et al. 2008) gave scientific backing to the institutionally weaker fisher and angler associations: The cormorant issue has been dealt with under the authority of nature and environmental administrations whose call for scientific evidence as basis for environmental policies did not match the comparatively weak scientific backing of anglers' and fishermen's arguments.

The political pressure for action was such that the European Commission announced at this second meeting in November 2008 (still before a resolution, in December 2008, of the European Parliament urging the Commission to take action towards managing the cormorant on a European level) to organize meetings in 2009 on this issue. Two meetings held in January and March 2009 served to gather information on the interactions between cormorants and fisheries in the EU member states, and to discuss these among member states and stakeholder representatives.

The participants gathered information about the situation in the member states (parts of step 1), discussed the possibilities to use more widely the derogation clause of the Birds Directive, and, particularly amongst the stakeholders, argued about the actual numbers of cormorants and the appropriate counting schemes. It was made clear by the European Commission that no European management plan was envisaged, but that the endeavours would go into a guidance document allowing a more autonomous management of cormorants at the national (or state) level. Fisher and angler associations doubt whether a more permissive use of the derogation would reduce the damage, let alone solve the conflict.

Evaluating these discussions, one might say that still a lot of energy is spent on discussing the exact number of cormorants instead of first agreeing on a common aim. A profound institutional change, e.g., the creation of a new authority dealing with management of "problem species", is—understandably (cp. Rauschmayer et al. 2008)—out of question for the European Commission. The Commission is not yet decided fully on how to proceed further, but apparently tries to take a mediating role between member states that receive more autonomy in this regard. It remains to be seen whether this inter-level play is a step towards resolving the conflict, or just muddling through in multi-level governance.

# 7 Conclusion

Resolving the conflict between fisheries/anglers and the cormorant without seriously taking at least some of the steps mentioned above does not seem realistic. Considering the institutional setting, an organization, which might be willing and considered legitimate to deal with the topic, is not in view. As resource-using stakeholders perceive the existing decision bodies as biased by conservationists' view, such as the European Commission's Directorate General Environment, an appropriate institution might be necessary.

Stakeholders would have to recognize clearly the interests and values of the others EU member states, represented by their officials, play a special role here, due to their legal competency. Any process addressing the possibility of a management plan would also have to address the institutional possibilities of any management, even before going into the contents of the plan. It is necessary to sound out the possibilities of institutional change due to the challenges of this novel European-wide species management.

Restarting the discussion concerning a general management plan should also comprise discussing an appropriate institutional design, which can coordinate both, the process of the discussion itself as well as the following exchange of data and scientific knowledge as the very basic option of coordination. Some stakeholders do not want this topic to shift to EU competency because of the high extent of bureaucracy which they relate to the European Commission. Therefore, it has to be considered whether any other existing agency could deal with this problem and the variety of stakeholders, or whether the costs induced by a conflict solution are sufficient to legitimize the creation of a new institutional body to deal with such multi-level problems.

As shown above, only an internationally coordinated effort has the potential to mitigate the conflict sustainably. Mere data exchange aiming more at building up trust between the stakeholders than at conflict mitigation itself could be a starting point. Which steps could then follow, would turn out with the relations between stakeholders developing further.

It is evident that any further and more substantial coordination would require a rather important effort if the aim is an effective and efficient management. It can be questioned whether the gains obtained through such a management would outweigh the costs related to it, but this would have to be analyzed.

**Acknowledgments** This work was financed by the EU 5th Framework Program (5th FP) Project "FRAP" (Development of a procedural Framework for Action Plans to Reconcile conflicts between large vertebrate conservation and the use of biological resources: fisheries and fish-eating vertebrates as a model case), contract number EVK 2-CT-2002-00142-FRAP.

## References

Behrens V, Rauschmayer F, Wittmer H (2008) Why pan-European cormorant management is so difficult: a contribution to managing international "problem" species. Environ Conserv 35(1):55–63. doi:10.1017/S037689290800444X

BirdLife International (2005) Species factsheet. Phalacrocorax carbo. http://www.birdlife.org. Accessed 27 Feb 2006

Bregnballe T, Engström H, Knief W, Van Eerden M, Van Rijn S, Kieckbusch JJ, Eskildsen J (2003) Development of the breeding population of great cormorants *Phalacrocorax carbo sinensis* in The Netherlands, Germany, Denmark and Sweden during the 1990s. Vogelwelt 124:15–26

Bregnballe T, Sterup J, Eskildsen J (2005) Oiling of great cormorant eggs in Denmark and behavioural responses to oiling. In: 7th International Conference on Cormorants, Villeneuve, Switzerland. Proceedings under preparation. http://web.tiscali.it/sv2001/index.htm. Accessed 7 July 2010

Bundesrat (2003) Entschließung des Bundesrates zur Kormoranproblematik. Bundesrat. Drucksache 716/03. http://www.naturschutzrecht.net/Materialien/BR-DRS_716_03.pdf. Accessed 21 Aug 2009

Carss D (2003) Reducing the conflict between cormorants and fisheries on a pan-European scale. REDCAFE—Final Report, Banchory, Centre for Ecology & Hydrology. http://www.intercafeproject.net/pdf/REDCAFEFINALREPORT.pdf. Accessed 10 Dec 2008

CMS (1994) Recommendation 4.1—Conservation and management of cormorants in the African-Eurasian Region. Conference of Parties—Fourth Meeting, Nairobi. http://www.cms.int/bodies/COP/cop4/cop4_documents_overview.htm. Accessed the 21 Aug 2009

CMS (1997a) Action plan for the management of the great cormorant—minutes of the meeting. Expert meeting. Copenhagen, Mimeo

CMS (1997b) Development of an action plan for the great cormorant in the African-Eurasian region. Conference of Parties—Fifth Meeting. Geneva. http://www.cms.int/bodies/COP/cop5/cop5_documents_overview.htm. Accessed 21 Aug 2009

EAA—European Anglers Alliance (2002) EAA attends cormorant meeting arranged by european parliamentary intergroup "fieldsports, fishing and conservation". http://www.eaa-europe.org/2003/PFCasework/Cormorants-Strasbourg-15-05-2002-EN.htm. Accessed 2 July 2005

European commission (2005) Environment: commission takes legal action over breaches of environmental law in Finland, Sweden and Denmark. http://europa.eu.int/rapid/pressReleasesAction.do?reference=IP/05/34&format=HTML&aged=0&language=EN&guiLanguage=en. Accessed 10 June 2006

Frederiksen M, Lebreton JD, Bregnballe T (2001) The interplay between culling and density-dependence in the great cormorant: a modelling approach. J Appl Ecol 38(3):617–627

Keller T, Lanz U (2003) Great cormorant *Phalacrocorax carbo sinensis* management in Bavaria, southern Germany—what can we learn from seven winters with intensive shooting? Vogelwelt 124:339–348

Keller T, Vordermeier T (1994) Abschlussbericht zum Forschungsvorhaben Einfluss des Kormorans (*Phalacrocorax carbo sinensis*) auf die Fischbestände ausgewählter bayerischer Gewässer unter Berücksichtigung fischökologischer und fischereiökonomischer Aspekte. Bayrische Landesanstalt für Fischerei. Starnberg. p 442

Rauschmayer F, Behrens V (2008) Legitimacy of species management: the great cormorant in the EU. In: Keulartz J, Leistra G (eds) Legitimacy in European nature conservation policy—case studies in multi-level governance. Springer, Heidelberg, pp 55–74

Rauschmayer F, Wittmer H, Berghöfer A (2008) Institutional challenges for resolving conflicts between fisheries and endangered species conservation. Marine Policy 32:178–188. doi:10.1016/j.marpol.2007.09.008

Rijsberman F (1999) Conflict management and consensus building for integrated coastal management in Latin America and the Caribbean. Inter-American Development Bank. Delft, The Netherlands, Inter-American Development Bank: 60. http://idbdocs.iadb.org/wsdocs/getdocument.aspx?docnum=359595

Rutschke E (1998) Der Kormoran: Biologie, Ökologie Schadenabwehr. Parey Buch, Berlin, p 161

Thum R (2004) Rechtliche Instrumente zur Lösung von Konflikten zwischen Artenschutz und wirtschaftlicher Nutzung natürlicher Ressourcen durch den Menschen am Beispiel Kormoranschutz and Teichwirtschaft. Nat Recht 9:580–587

Thum R (2005) Zur Rechtmäßigkeit so genannter Kormoranverordnungen. Agrar- und Umweltrecht 5:148–152

USDI Fish and Wildlife Service (2003) Final environmental impact statement—double-crested cormorant management in the United States. U.S. Department of interior. http://library.fws.gov/Bird_Publications/CormorantFEIS.pdf. Accessed 21 Aug 2009

van Eerden M (2002) Managing cormorants in Western Europe: mission impossible? In: Sächsische Landesstiftung Natur und Umwelt (ed) Der Kormoran (*Phalacrocorax carbo*) im Spannungsfeld zwischen Naturschutz und Teichbewirtschaftung. Dresden. www.lanu.de

Vfg Verband für Fischerei und Gewässerschutz in Baden Württemberg e.V. (2006) Kormorane und Fische. http://www.vfg-bw.org/seite295.htm. Accessed 10 Dec 2008

Wilson DC (2004) Discourse analysis, UFZ-Centre for environmental research Leipzig-Halle, unpublished document

Wilson DC (2012) Module 6—stakeholder analysis. In: Klenke R, Ring I, Kranz A, Jepsen N, Rauschmayer F, Henle K (eds) Human-wildlife conflicts in Europe—fisheries and fish-eating verte-brates as a model case. Springer, Heidelberg

Wilson, DC Antunes P et al. (2004) The social impact assessment report, UFZ-Centre for environmental Research Leipzig-Halle, unpublished document

Golden eagle (*Aquila chrysaetos*). *Photo:* Dana Krimmling - Fotolia.com

# The Golden Eagle Compensation Scheme in Finland as an Example of Incentive Measures

## Potential for Conflict Management?

**Leila Suvantola**

**Abstract** This article is an introduction to an incentive compensation scheme and its comparison to the traditional compensation for confirmed damages. The scheme is for losses caused to the reindeer (*Rangifer taranus*) husbandry by golden eagles (*Aquila chrysaetos*). In the scheme payments are based on nesting and reproduction of the species. The legal research framework is widened with regulation theory to evaluate the benefits of the different compensation schemes. The incentive scheme turns out to be more capable of achieving the targets of providing a continuing incentive, effectiveness and dependability, financial efficiency, administrative feasibility, equity and acceptability than the confirmed damage compensation scheme. The incentive scheme builds upon deliberative follow-up and shared information and has proved its ability for conflict management.

## 1 Introduction

The Convention on Biological Diversity (1992) encourages member countries to use incentive measures to promote conservation of biodiversity. Can incentive measures combined with participatory management be used as conflict management tools in nature conservation?

This paper introduces the compensation scheme for losses caused by golden eagles (*Aquila chrysaetos*) to the reindeer (*Rangifer tarandus*) husbandry. Compensation is a nature conservation instrument used in Finland and it can be described as a dynamic incentive. The paper compares the traditional compensation

L. Suvantola (✉)
Department of Law, Faculty of Social Studies and Business Studies,
University of Eastern Finland, Joensuu campus, 111 80101 Joensuu, Finland
e-mail: leila.suvantola@suursaimaa.com

R. A. Klenke et al. (eds.), *Human-Wildlife Conflicts in Europe*,
Environmental Science and Engineering, DOI: 10.1007/978-3-540-34789-7_10,

for confirmed damages and the new incentive scheme. According to the new scheme, payments are based on nesting and reproduction of the species. The aim of the incentive scheme is to promote the conservation status of the species. On the one hand the scheme discourages disturbance of the eagles during nesting and on the other it encourages creation of nesting sites rather than their destruction. Those concerned can participate in the scheme. They have access to information on nesting sites. There is also an ongoing deliberative follow-up of the scheme. This feature may increase trust between the authorities and those who are subject to the negative impacts of the recovery of the species.

The research framework of this article is legal. Yet, the aim is to widen the scope by utilization of regulation theory in order to evaluate the benefits of the different compensation schemes. The article uses the findings of recent sociological and ecological research on the topic.

## 2 The Golden Eagle and Damages to Reindeer Husbandry in Finland

The golden eagle was protected in Finland in 1962 after the population had declined to 20–50 pairs. In the 1960ies there were about 70 known nesting territories of the species. Since then, conservation effort has led to a significant recovery of the species. In 1999 there were 175 known nesting territories. In 2004 the population had risen to 400 pairs (Ollila 2004; Forest and Park Services 2005). 90 % of the Finnish population lives in the reindeer herding area. Recovery of the species has led to a conflict between species conservation and reindeer husbandry because golden eagles prey on reindeer calves (Fig. 1).

Reindeer husbandry is a traditional livelihood of the Sami people. The Sami are an indigenous people in Finland, Sweden, Norway, and the North-West of the Russian Federation. In Sweden, the right to carry out reindeer herding is restricted to the Sami people. In Finland there are no such restrictions.

Reindeer owners in Finland are organized in 56 cooperatives each one with its own grazing areas. The land is not owned by the cooperatives. However, reindeer owners have a legal right to carry out reindeer herding undisturbed on all land outside private house yards and gardens. The reindeer are owned by individual families or persons and they are identified by earmarks.

The number of reindeer killed by golden eagles in Finland was recorded until the end of 1998. Between 1996 and 1998 the annual damage was 700 reindeer on average, mostly calves. Losses were recorded mainly in the 14 mountainous cooperatives where almost 80 % of all damage occurred. In 16 cooperatives there were almost no losses (Ministry of the Environment 1999). Economic impacts of the losses are significant in the heavily impacted cooperatives.

Recent research in six different reindeer cooperatives showed that the total loss of reindeer calves (irrespectively of cause) was between 6 and 14 % of born calves on average. Golden eagles were the cause in 31 % of identified cases (Fig. 2).

**Fig. 1** Reindeer herd. *Photo:* Andreas Gradin - Fotolia.com

**Fig. 2** Golden eagle (*Aquila chrysaetos*) on reindeer carcass in Utajärvi, Finland. *Photo:* Allessandro Carboni

However, in 38 % of the cases the cause remained unidentified (Norberg et al. 2005, pp. 34–39). In one of the six cooperatives there were no known golden eagle territories, no observed killings by eagle, thus the golden eagle prey was proportionally higher in the other five.

# 3 Two Approaches to Cover Losses Caused by Wildlife

## *3.1 The Development of Compensation Schemes in Finland*

Until the end of 1998 all losses caused by large predators to the reindeer husbandry in Finland were covered by a general compensation system. All the other large predators (brown bear *Ursus arctos*, lynx *Lynx lynx*, wolf *Canis lupus* and wolverine *Gulo gulo*) are game species, whereas golden eagle is a protected species. The Ministry of Agriculture and Forestry is responsible for compensation of damages as hunting falls into its responsibilities.

Compensation is based on the number of found reindeer carcasses which can be identified to have been killed by one of the large predators. In this paper I call this compensation system as the "confirmed damage compensation scheme".

In 1998 the Parliament transferred the responsibility for compensation of losses caused by golden eagles to the Ministry of the Environment as conservation of protected species belongs to its responsibilities. The Parliament requested that the subsequent preparation of the compensation scheme would be based on the Swedish model where compensation for losses caused by wolverine is now based on the estimated killings and not on damage caused in single cases (Reply of the Parliament of Finland 228/1997). In Sweden the wolverine compensation is supposed to reflect loss of income plus additional damage to the reindeer owners. Compensation is based on the headcount of the wolverines. In this paper this approach is called the "incentive scheme".

The duty of the state to pay compensation for losses caused by wildlife is not legally clear, rather debatable. On the one hand, the state is not the owner of the wild animal species. Instead, they are regarded *res nullius*, nobody's property (de Klemm 1996, p. 34). Wildlife is part of nature, and the related risks are natural risks, which are carried by private actors in principle. Thus, monetary payments for "natural" losses are by definition rather *subsidies* based on the voluntariness of the state than compensation based on legal duty. Accordingly, for example in the common law countries such as the United States and Canada, the courts have refused to grant damages against the state on the basis of natural losses (de Klemm 1996, p. 35). On the other hand, it has been argued that by protecting the damage causing species the legislator has removed the "natural" right of the people to defend themselves and their property against this natural risk (de Klemm 1996, pp. 35, 37). Thus, the state has the responsibility to cover the losses caused by protected species.

Reindeer in Finland are semi-domesticated and identifiable private property. Considerable financial losses are caused by golden eagle predation to reindeer owners annually. These losses have been compensated in Finland by the state since 1950s at least in part. However, the Finnish Supreme Administrative Court found in its decision to an appeal on the new compensation scheme that even a consistent compensation practice based on state budget financing does not create a permanent right to full compensation (KHO 12.9.2002 file 2154).

Whether there is a legal duty to compensate the losses or not, it is certainly rational to cover them. Refusal to compensate would give those suffering losses an incentive to kill damage causing animals (de Klemm 1996, p. 53)—in this case the reindeer owners an incentive to kill eagles threatening their calves. Over the years there had been recurrent golden eagle killings as well as obvious disturbance of nesting, despite a strict prohibition and criminal sanctions imposed in the nature conservation legislation and despite the confirmed damage compensation scheme. Thus, there was a conservation interest to provide a disincentive to activities threatening the species. The new system was also to act as an incentive to reindeer owners to inform the authorities of known nests and maybe even to search them. Nests unknown to the authorities would mean loss of income to the reindeer cooperative.

Issues relating to fairness and economy also supported changes to the scheme. Reindeer owners were known to have different capabilities to find reindeer carcasses. Some of the cooperatives carry out shepherding whereas others tend to the herd at intervals. Those reindeer owners with another employment had less opportunities and time to search their reindeer on ground. Economically, the new scheme made it unnecessary to look for the carcasses, which meant saved costs in fuel, time, and verification. Nests of eagles are checked and counted as part of the conservation work and research, thus the new scheme did not cause additional need for field work. Furthermore, administrative simplicity of the scheme was to reduce costs, even though the need to find out the golden eagle nesting in the border zone required information exchange among neighboring countries.

Both systems are explained in detail below. The confirmed damage compensation scheme applicable to large predators will be explained in its current form as amended since 1998.

## 3.2 The Confirmed Damage Compensation Scheme

The losses to reindeer husbandry caused by large game predators are compensated according to the Council of State Decree on Compensation of Damage Caused by Predators (277/2000, amended 821/2001). The current legislation is quite similar to the provisions in force since 1956 (Act No. 574/1956 on Compensation of Reindeer Killed by Predators) as amended in 1973. Payments are based on the number of reindeer carcasses that can be confirmed to have been killed by brown bear, lynx, wolf, or wolverine, and before 1999 by golden eagle. Since 2000 the payment is twice the value of a killed reindeer. The value of an individual reindeer varies according to its use. The reindeer used for meat production are less valuable, whereas reproduction and draught animals are more valuable individuals. Multiple payments also cover the loss of individuals that cannot be found. It is assumed that only 50 % of the killed individuals are found and identified. A reindeer carcass must be found, marked on ground, ears must be removed for confirmation and identification, damage must be reported and a compensation request form submitted to the responsible authority. The responsible authority may require that an authorized controller confirms the damage and its cause. The cost of the control visit is paid by the reindeer cooperative up to a maximum of 100 €.

In the beginning of 2002 a personal deductible was adopted whereby the state pays compensation only for damages exceeding 250 € per reindeer owner annually (amendment 821/2001). The purpose of the deductible is to encourage reindeer owners to protect their calves from predators.

## 3.3 The Newly Adopted Incentive Scheme

The losses caused by golden eagles to the reindeer husbandry are compensated according the Council of State Decree on Compensation of the Losses Caused to Reindeer Husbandry by the Golden Eagle 8/2002, which replaced the 373/1999 Council of State Decision on the same issue. The content of the decree is similar to the previous decision and the revision was merely a formality due to the adoption of the new Constitution in 1999.

The incentive scheme is not related to individual losses. The compensation rate aims to reflect the average costs of conservation to the reindeer husbandry: the estimated killing—and thus the value—of reindeer killed by a golden eagle on its territory annually.

The incentive scheme is based on information of known nesting territories of the golden eagle. These are controlled twice annually: decorated nests (occupied territories) are counted in May, and the offspring are counted on occupied territories in June. A territory is defined as occupied even if a nest cannot be found, if in the two previous years decorated nests or offspring have been found (amendment 839/2005). The amendment is based on the scientific fact that golden eagle has several nests in a single territory, all of which may not be known, and a pair may swap its nesting site from year to year.

The incentive scheme relies on shared information. The Forest and Parks Services representative informs the representatives of each reindeer cooperative annually of the known nests. In turn, the reindeer owners inform the Services representative of potential or suspected nests. These sites are controlled by the Forest and Parks Services.

There is an annually revised compensation rate for an occupied territory. In 2005 the rate was 516 €. In mountain areas the payment for an occupied territory is twofold. This is based on research, which indicates that golden eagles prey more on reindeer in mountainous areas (Norberg at al. 2005, pp. 33–42). If a golden eagle pair produces offspring on its territory, the compensation is threefold in forest areas and fivefold in mountain areas. The purpose is to cover the additional prey of the species needed for feeding the offspring. However, it also rewards cooperatives where the species succeeds in producing offspring.

In case a territory extends to several reindeer cooperatives (a nest is located close to the boundary of the cooperatives), it is regarded as a joint territory, and the compensation is split in equal shares.

In the national border areas compensation is paid also for nests outside the borders of Finland. Thus the damage caused by golden eagles nesting in Sweden, Norway or Russia is also covered. The compensation for those territories is always twofold in mountain areas, elsewhere the standard rate. If a territory ranges partly south of the reindeer herding area within Finland the compensation is partial.

The financial payment is made to individual reindeer cooperatives. This is a significant difference compared to the Swedish model, where the payments are made to the political Sametinget, the Sami Parliament (Sellethin and Skogh 2004, p. 242). If it can be proved that a golden eagle has killed a reindeer belonging to an individual reindeer owner, the cooperative has to compensate the value of the reindeer to the owner. Otherwise, the cooperative decides how to use the annual payments. For example, the funds can be divided between the reindeer owners who have suffered losses or used for collectively beneficial projects, such as product development or meat production facilities.

The Supreme Administrative Court of Finland had to review the incentive scheme as one of the cooperatives made an appeal against the compensation decision in 1999 because the compensations did not cover the confirmed losses of the cooperative. The cooperative consisted mainly of Sami people and it argued that the scheme was contrary to the Article 27 of the United Nations International Covenant on Civil and Political Rights (1966) stating that in those States in which ethnic minorities exist, persons belonging to such minorities shall not be denied the right, in community with the other members of their group, to enjoy their own culture. The Supreme Administrative Court found in its decision (KHO 12.9.2002 file 2154) that the incentive scheme did not as such deny the right of the Sami people to enjoy their culture.

## 3.4 Deliberative Design and Development

The incentive scheme was designed by the Ministry of the Environment in collaboration with the Forest and Parks Services. The proposed design was consulted with the representatives of the Sami people, the Association of the Reindeer Cooperatives, the Regional Environmental Centre of Lapland and a representative of individual reindeer owners. Consultations led to amendments of the proposal and the final scheme. For example, the golden eagles nesting in neighboring countries were taken into account.

In conjunction to the adoption of the new scheme, a negotiating group was set up. The group consists of the representatives of the Forests and Parks Services, the Regional Environmental Centre of Lapland, the Association of the Reindeer Cooperatives, the Sami Parliament, the Fish and Game Research Institute, and the Ministry of the Environment. The task of the negotiating group is to follow ongoing research on the effects of large predators on reindeer husbandry. This research examines the loss of calves in general and in particular the losses caused by large predators, such as the golden eagle. This research makes it possible to assess the accuracy of compensation. The group can and has made proposals to refine the scheme. Some of the proposals have led to amendment of the scheme, some have not. One of the proposals has been to combine the incentive scheme with compensation of confirmed damages. It was rejected by the legislator because it would have partly led to double compensation.

# 4 Evaluation and Comparison of the Alternative

## 4.1 Evaluation Criteria

As the aim of this article is to evaluate the different approaches of how to cover losses caused by a protected species, it is necessary to establish a framework within which to make comparisons. There are a number of different criteria in regulation theory for the evaluation of environmental policy instruments (for an illustration of differences and variety of assessment criteria for policy instruments see Sullivan 2005, pp. 6–13). This article uses a criteria consisting of seven factors (Young et al. 1996, p. 106; Gunningham and Young 1997, pp. 252–253; OECD 1996, p. 84; OECD 1999, p. 68):

1. *Environmental effectiveness*: the desired biodiversity target is achieved;
2. *precaution*: serious or irreversible consequences are avoided even in case of scientific uncertainty;
3. *dynamic and continuing incentive*: improvement of biodiversity is encouraged beyond the official policy target and adapted to changing conditions;
4. *economic efficiency*: the target is achieved at least cost and the biodiversity objectives are improved without making someone worse off;
5. *administrative feasibility and transaction cost*: the information and monitoring costs are minimal, enforcement is cost-effective, and decision-making processes are simple and transparent;
6. *equity*: no group (including future generations) is unfairly advantaged or disadvantaged; and
7. *community and political acceptability*: community regards the conservation objectives as legitimately formulated and delivered and the instrument is consistent with government commitments and politically supported.

These factors are interrelated. For example an instrument cannot be economically efficient if it is not also effective. The political acceptability of an environmental policy instrument rests on the other criteria and is a good reminder of the significance of each of the other. It may be impossible to achieve all the above mentioned factors (OECD 1999, p. 68). However, failure to address in particular the effectiveness, economic efficiency, or equity factors would render an environmental policy instrument vulnerable to changes in the public opinion and thus without long term political acceptability.

## 4.2 Effectiveness, Precaution, and Continuing Incentive

The *effectiveness* of the compensation scheme is evaluated with respect to nesting territories of the golden eagle. Although there was a significant increase of nesting territories from the 1960s (70 known nesting territories) towards the end of the 1990s (175 nesting territories) due to increased protection activities and the confirmed damage compensation scheme, the new incentive scheme seems to be much more effective in a shorter time frame. The effectiveness of the incentive scheme has been good as golden eagle nesting has increased from 175 nesting territories in 1999 to 288 nesting territories in 2004. This increase may partly be due to increased control and better information of nesting sites. The latter is the result of information exchange between the authorities and reindeer owners. The number of chicks has correspondingly risen from the average of 94 chicks annually in the 1990s to the average of 147 chicks annually in 2000–2005 (despite a crash in the year 2005 due to weather conditions). The number of failed nesting has decreased after the adoption of the incentive scheme by 2 % compared with the 1990s (Ollila 2004). This is not a significant change, yet it does indicate that the change of the compensation approach has definitely not increased disturbance of nesting.

The incentive scheme works on top of the comprehensive protection of the species, thus the precaution requirement is met. The instrument provides a *continuing incentive* as the amount of compensation increases with increased nesting and reproduction of the species.

## 4.3 Economic Efficiency and Administrative Feasibility

When the incentive scheme is examined from the *economic efficiency* point of view, it appears that the expected economic benefits have been achieved. The administration is more efficient and the costs borne by the reindeer owners have decreased. Payments are based on the nesting reports of the Forest and Parks Services. The costs of time and fuel for search and verification of carcasses are saved. There is no need for filling and handling of compensation request forms or costs of the control visits of the authorities.

In contrast, the search, verification, and compensation request costs related to the confirmed damage compensation scheme of other large predators are quite significant. The annual costs in these cooperatives in 2002–2003 were on average close to 25,000 € per cooperative (Sippola et al. 2005, p. 125). The economic inefficiency of the confirmed damage compensation scheme relates also to what is called a *moral hazard*. This refers to a situation where those suffering the losses receive full compensation and thus have no incentive to carry out damage abatement measures (Rollins and Briggs 1996, p. 369). Full compensation may even attract activities that otherwise would be avoided such as deliberately letting one's animals into risk areas (Nyhus et al. 2005). The confirmed damage compensation scheme attempts to solve the problem by the adoption of the above mentioned annual deductible of 250 €. Within a reindeer cooperative the deductible may cumulate to several thousand euros as often every family member within the cooperative owns at least one reindeer. However, as the deductible is annual for each reindeer owner, there is no incentive to abate damages after the first annual damage. It must, however, be noted, that it is difficult to abate wildlife damages to reindeer as most of them range free. It would at least require that during and after birth of calves the females were restricted to guarded sites.

## 4.4 Equity

*Equity* must be examined both as an *intergenerational* (between generations) and *intragenerational* (within a generation) question. The intergenerational equity is achieved if the biodiversity target of favorable conservation status of the species is achieved i.e. if the instrument is *dependable*. This appears to be the case in relation to the incentive scheme (see above). Within a generation the key question of equity is the fair distribution of costs (Doremus 2003, p. 223; Suvantola 2004). The instrument is not equitable if the costs of conservation are carried by the reindeer owners alone. Thus the question can be simplified as the correctness of compensation of the damages. Recent research has shown that loss of reindeer calves according to the incentive scheme is mainly (in 4 of the 5 cooperatives examined) compensated financially accurately. However, losses caused by other large predators are practically never met in the confirmed damage compensation scheme (Sippola et al. 2005, pp. 140–142).

In the Finnish incentive scheme the payments are made to the reindeer cooperatives where the golden eagle territory is located. Thus the compensation is likely to meet the occurring costs of conservation. The Swedish model, where the payments are made to the Sami Parliament, is regarded unfair, as the compensation does not reach the villages or reindeer owners suffering the loss (Sellethin and Skogh 2004, p. 242). Sellethin and Skogh also regard the Swedish model unable to provide a disincentive to poaching.

## 4.5 Community Acceptability

The *community acceptability* of the incentive scheme has not been examined generally in the Finnish society. However, recently a research group made a survey of the attitudes of the reindeer owners towards both the golden eagle incentive scheme and the confirmed damage compensation scheme. The research conclusions shed light on the acceptability of the incentive scheme among those suffering the implications of the recovery of the species. The survey showed that the reindeer owners were roughly split into those who were fully or mostly satisfied (48 %), and those not (41 %), even though a larger share had a positive sentiment (Table 1). This result is somewhat similar to the attitudes towards the confirmed damage compensation scheme where the majority were totally or mostly satisfied, whereas one third were totally unsatisfied and willing to change the scheme (Sippola et al. 2005).

The attitudes, however, changed when the reindeer owners were asked about their willingness to maintain or change the existing incentive scheme. One fifth of the reindeer owners wanted to maintain the golden eagle incentive scheme as such and half of them wanted to maintain it, yet with some amendments. Only 3 % were willing to restore the confirmed damage compensation scheme, and 6 % regarded some other schemes better (Table 2). It appears that the overall principle of the incentive scheme is strongly accepted among those concerned even though the reindeer owners do not regard the scheme design to be satisfactory.

In turn, a large number of the reindeer owners were willing to have the incentive scheme applied also to other large predators. For brown bear and lynx the majority even preferred the incentive scheme, whereas for wolf and wolverine the incentive scheme was slightly less preferred (see Table 3). More than a quarter of the respondents were unable or unwilling to state their opinion on the issue.

In the attitude survey the reindeer owners were asked about their views on the need to amend the incentive scheme. The most popular proposal was that eagles visiting from other countries or other reindeer cooperatives should be taken into account (14.4 %). As the eagles nesting outside the borders of Finland and within hunting range are taken into account, and the territories ranging to several reindeer cooperatives are regarded as joint territories, the researchers suspected unawareness of the details of the scheme (Sippola et al. 2005, p. 71). Another strongly supported proposal was to pay compensation for uninhabited nests (13.1 %). This is partly already the case according to the amendment made in 2005 explained above. Most other proposals were also in line with the principles of the incentive scheme. It was proposed that compensation rate for each territory were bigger (13.1 %), there should be better search for nests and clarification of the size of a territory (10.8 %), and better cooperation with authorities (5 %). The last point mentioned was regarded by researchers as a sign of distrust towards the authorities carrying out nest controls (Sippola et al. 2005, p. 74). However, it may also indicate information problems within the cooperative as there are regular meetings of the Forest and Parks Services representatives and the head of each cooperative. A popular proposal was to pay compensation also for young and not nesting eagles (12.9 %). This would be in line with the principles of the scheme, even though it does not provide a disincentive to disturbance of nesting.

Some quite popular proposals, however, would have combined the incentive scheme with some of the features of the confirmed damage compensation scheme.

**Table 1** Satisfaction of the reindeer owners with the confirmed damage scheme and the incentive scheme (Sippola et al. 2005)

| | Confirmed damage scheme (%) | Incentive scheme (%) |
|---|---|---|
| Fully satisfied | 8 | 6 |
| Mostly satisfied | 43 | 42 |
| Do not know | 14 | 11 |
| Mostly unsatisfied | – | 19 |
| Fully unsatisfied | 35 | 22 |

**Table 2** The willingness of the reindeer owners to maintain the incentive scheme (Sippola et al. 2005)

| Attitude | Proportion (%) |
|---|---|
| Willing to maintain the scheme as it is | 22 |
| Willing to maintain the scheme refined | 52 |
| Do not know | 17 |
| Rather the confirmed damage scheme | 3 |
| Rather some other compensation system | 6 |

**Table 3** The willingness of the reindeer owners to have the incentive scheme applied to other large predators (Sippola et al. 2005)

| Willingness | Wolverine (%) | Wolf (%) | Lynx (%) | Brown bear (%) |
|---|---|---|---|---|
| Yes | 33 | 32 | 41 | 45 |
| Do not know | 28 | 28 | 27 | 25 |
| No | 39 | 40 | 32 | 30 |

It was proposed that compensation should be based on loss of calves and the share of reindeer in eagle feed (10 %) and direct compensation for found reindeer carcass to the owner from separate funds (12.9 %). Some also proposed control of the population growth of the golden eagle (5.5 %).

## 5 Conclusions

The incentive scheme provides a continuing incentive and appears effective and dependable, financially more efficient and administratively more feasible, more equitable, and more acceptable than the confirmed damage compensation scheme for the compensation of the losses to reindeer husbandry caused by golden eagles. The total number of reindeer in Lapland is constantly very high, almost too high compared to the grazing resources available. In southern and mid-Lapland reindeer cooperatives, the main winter food, lichen has diminished about or over 80 % since 1970s (Mattila 2006, p. 52). The above explained moral hazard theory would combine the high number of reindeer with the confirmed damage compensation scheme. In theory, it is sensible to have high numbers of calves, because there is double compensation of their value, if they fall prey to large predators. The detachment of the incentive scheme from the occurred damage also detaches it from the head-number of reindeer and reduces the moral hazard.

As the incentive scheme builds upon deliberative follow-up and shared information, and the participants express trust in research, the scheme has proved its potential for conflict management. The increase of the eagle population after the adoption of the incentive scheme has not lead to increased conflict. The experience encourages examination of the potential of expanding the approach to other similar human-wildlife conflicts, e.g., in Finland to the other large predators with sufficiently identifiable territories.

**Acknowledgments** This article has been written as part of the Legitimacy project of the Environment and Law Research Program 2005–2008 financed by the Academy of Finland. The writer participated in the development of the incentive scheme as senior adviser in the Ministry of the Environment.

## References

de Klemm C (1996) Compensation for damage caused by wild animals. Nat Environ No 84, Council of Europe, Strasbourg

Doremus H (2003) A policy portfolio approach to biodiversity protection on private lands. Environ Sci Policy 6:217–232

Forest and Park Services (2005) Maakotkaraportti (Annual Report of the Golden Eagle, in Finnish). http://www.metsa.fi/binary.aspx?Section=2106&Item=5069. Accessed 10 Dec 2008

Gunningham N, Young MD (1997) Toward optimal environmental policy: The case of biodivers conserv Ecol Law Q XXXIX:243–298

Mattila E (2006) Porojen talvilaitumien kunto poronhoitoalueen etelä- ja keskiosien merkkipiireissä 2002–2004 ja kehitys 1970-luvun puolivälistä alkaen. METLAn työraportteja 27/2006. http://www.metla.fi/julkaisut/workingpapers/2006/mwp027.htm. Accessed 10 Dec 2008

Ministry of the Environment (1999) Esittelymuistio 18.3.1999: Ehdotus valtio-neuvoston päätökseksi maakotkien porotaloudelle aiheuttamien vahinkojen korvaaamisesta. (Memorandum 18.3.1999: proposal for the council of state decree on compensation of the losses caused to reindeer husbandry. In Finnish). Unpublished

Norberg H, Nieminen M, Kumpula J, Kojola I, Maijala V (2005) Poronvasojen kuolleisuus ja kuolinsyyt: yhteenveto telemetriatutkimuksista Suomen poronhoitoalueella vuosina 1997–2004. Kala- ja riistaraportteja 355, Kaamanen

Nyhus PJ, Osofsky SA, Ferraro P, Madden F, Fischer H (2005) Bearing the costs of human-wildlife conflict: the challenges of compensation schemes. In: Woodroffe R, Thirgood S, Rabinowitz A (eds) People and wildlife: conflict or coexistence. Cambridge University Press, Cambridge, pp 107–121

OECD (1996) Saving biological diversity—Economic Incentives. OECD, Paris

OECD (1999) Handbook of incentive measures for biodiversity—design and implementation. OECD, Paris

Ollila T (2004) Maakotkaraportti 2004. Unpublished golden eagle nesting report 2004. The forest and parks services

Rollins K, Briggs HC (1996) Moral hazard, externalities and compensation for crop damages from wildlife. J Env Econ Manage 31:368–386

Sellethin MO, Skogh G (2004) Property rights in endangered species: the wolverine case. Eur J Law Econ 18:239–247

Sippola AL, Norberg H, Renko M, Suopajärvi K, Sutinen K (2005) Petovahinkojen sosioekonominen merkitys porotaloudelle Suomessa—loppuraportti. Arktisen keskuksen tiedotteita 44, Rovaniemi

Sullivan R (2005) Rethinking voluntary approaches in environmental policy. Edward Elgar, Cheltenham

Suvantola L (2004) A legal perspective on cost sharing in biodiversity conservation. Proceedings of the conference on policy instruments for safeguarding forest biodiversity—legal and economic viewpoints. The 5th international BIOECON conference 15th–16th January 2004, Helsinki, 69–82. http://www.metla.fi/julkaisut/workingpapers/2004/mwp001-03.pdf. Accessed 10 Dec 2008

Young MD, Gunningham N, Elix J, Lambert J, Howard B, Grabosky P, McCrone E (1996) Reimbursing the future. An evaluation of motivational, voluntary, price-based, property right, and regulatory incentives for the conservation of biodiversity. Biodiversity series, Paper no. 9. Department of the environment, sport and territories, commonwealth of Australia

# Part II
# A Generic Framework for the Development of Conflict Reconciliation Action Plans

**Klaus Henle, Irene Ring, Reinhard A. Klenke, Niels Jepsen, Andreas Kranz, Carsten Neßhöver and Felix Rauschmayer**

**Abstract** Attempts to reconcile human-wildlife conflicts are usually developed on a case-by-case approach. A generic framework can offer help in the reconciliation of such conflicts by building on common elements and the experience gained in case studies. Here we introduce the concept and structure of such a generic framework that is based on the experience from the joint work of natural and social science researchers. It is a structured collection of good advice presented as

K. Henle (✉) · R. A. Klenke · C. Neßhöver
Department of Conservation Biology, UFZ—Helmholtz Centre for Environmental Research, Permoserstraße 15, 04318, Leipzig, Germany
e-mail: klaus.henle@ufz.de

I. Ring
Department of Economics, UFZ—Helmholtz Centre for Environmental Research, Permoserstraße 15, 04318, Leipzig, Germany
e-mail: irene.ring@ufz.de

R. A. Klenke
e-mail: reinhard.klenke@ufz.de

N. Jepsen
National Institute of Aquatic Resources, Section of Freshwater Fisheries Ecology, Technical University of Denmark, Vejlsøvej 39, 8600, Silkeborg, Denmark
e-mail: nj@aqua.dtu.dk

A. Kranz
Institute of Wildlife Biology and Game Management, University of Natural Resources and Life Sciences, Gregor Mendel Str. 33, 1180, Vienna, Austria
e-mail: andreas.kranz@aon.at

C. Neßhöver
e-mail: carsten.nesshoever@ufz.de

F. Rauschmayer
Department of Environmental Politics, UFZ—Helmholtz Centre for Environmental Research, Permoserstraße 15, 04318, Leipzig, Germany
e-mail: felix.rauschmayer@ufz.de

guidelines for people who have to deal with human-wildlife conflicts. Its purpose is to both facilitate the identification of gaps in existing reconciliation approaches that may hamper their success and to help design reconciliation action plans for newly arising or previously ignored conflicts. The generic framework is designed in a modular way, each module addressing a different main topic in the assessment and reconciliation of conflicts. The following chapters in this book explain these modules in detail.

**Purpose of the Generic Framework**

Competition between humans and wildlife for biological resources has led to the persecution of wildlife for centuries. With changes in civil society's attitudes towards wildlife and success of conservation management wildlife species increase again in number and return to areas from where they had disappeared. Consequently, conflicts between humans about wildlife and biological resources are on the rise in many parts of Europe and throughout the world.

The need to reconcile such conflicts is the logical response to this change in fundamental paradigms in human-wildlife relationships. Reconciliation means managing the competition between people and wildlife as well as the conflict among different interest groups that are quoted in these changing attitudes.

Attempts to reconcile conflicts are usually developed on a case-by-case approach. As a consequence, attempts differ greatly among conflicts and among countries, both in approaches followed and in their successes and failures. Nevertheless, there are many common elements to human-wildlife conflicts around the world. A generic framework can offer help in the reconciliation of human-wildlife conflicts by building on such common elements and the experience gained in case studies. Increasing demands for harmonizing management, especially across the EU but also elsewhere in the world, also calls for the development of a generic framework for human-wildlife conflict reconciliation.

The generic framework presented here is based on the experience from the joint work of natural and social science researchers who studied model conflicts over a 3-year period within the EU project FRAP. It is a structured collection of good advice presented as guidelines for people who have to deal with human-wildlife conflicts. Its purpose is to both facilitate the identification of gaps in existing reconciliation approaches that may hamper their success and to help design reconciliation action plans (RAPs) for new or previously ignored human-wildlife conflicts. It suggests approaches to characterize the conflicts and to identify and assess key elements that drive the conflicts, as well as measures that may help to reconcile them and thus ultimately supports the sustainable use of biological resources.

The target groups are mainly managers and decision makers in such conflicts, representatives of the affected stakeholder groups that are involved in negotiations for reconciliation action plans, and scientists that contribute to conflict analysis and the development of reconciliation strategies.

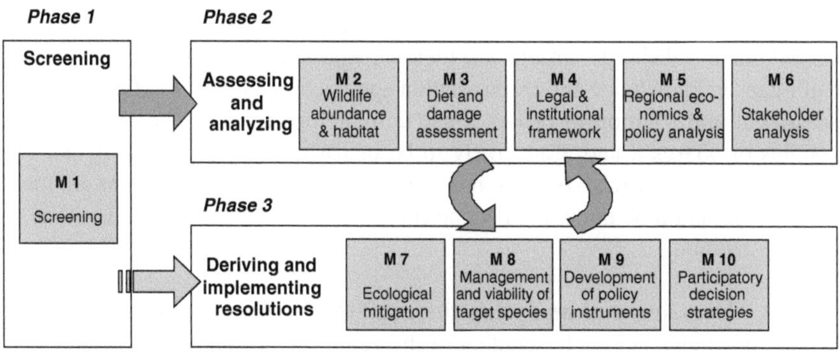


**Fig. 1** Generic framework for the development and implementation of reconciliation strategies or action plans, consisting of 3 phases and 10 modules (M)

## Content and Use of the Generic Framework

We designed the generic framework in a modular way, each module addressing a different main topic in the assessment and reconciliation of conflicts (Fig. 1). The modules are allocated to three essential phases in the development of reconciliation action plans: the screening phase, the assessment phase, and the phase for deriving and implementing resolutions.

The development of reconciliation strategies is an iterative process. At the beginning (screening phase) one should develop an impression regarding what approaches may be particularly relevant for the derivation of resolutions, i.e., which modules in phase 3 (deriving and implementing resolutions) will need to get which level of attention. This first impression will then be refined through the information collected in phase 2 (assessment phase), since the data might be challenged by some stakeholders, if the method to create them has not been agreed upon beforehand. This notwithstanding, the information gained through the modules in phase 2 may necessitate a change in the initial impression about potentially successful ways of mitigating the conflict. Therefore, it is necessary to cover all modules of phase 2, as each of them represents different though essential aspects of conflict management. This may just result in a change of focus for phase 3, but it may also require re-addressing one or several modules in greater depth. Conflict resolution may need to proceed repeatedly through phases 2 and 3 in an iterative way to find optimal solutions (Fig. 1).

**Phase 1 "Screening the conflict"**, contains one module. Its purpose is to raise awareness of different aspects to consider in the development of a reconciliation action plan by a preliminary evaluation of the history, intensity, and dimension of the conflict. It facilitates identification of the relevant area/region for the assessment and reconciliation. Thus, the screening phase helps to identify relevant scales: ecological (activity range and distribution of wildlife, type, and magnitude of damage), political and administrative (administrative units, relevant decision bodies, laws, and policy instruments), economic (economic activities and sectors affected, including relevance of economic damage), and social (intensity of conflict, main stakeholder groups and their perceptions). It further helps to identify major gaps in knowledge and reconciliation approaches and thus facilitates conflict assessment and solution in phases 2 and 3 by helping to evaluate, which of the modules of phases 2 and 3 are likely to be of particular relevance (Fig. 1). The screening phase is crucial, because it is in the beginning, where decisions are made about where and how much money and time will be spent and whether it is in an effective method.

**Phase 2 "Assessing and analyzing the conflict"** has the purpose of improving factual knowledge and understanding of the conflict by evaluating the ecological, legal, and socio–economic basis of the conflict. This understanding is essential to identify opportunities and limitations for conflict reconciliation. The modules of phase 2 are designed to help identify major ecological, institutional, economic, and social drivers of the conflict. Phase 2 comprises two ecological and three socio-economic modules. Module 2 provides guidelines to assess wildlife abundance as well as key habitat and landscape factors that determine a spatially variable exposure of the biological resource to wildlife. Module 3 provides guidelines for an assessment of direct and indirect damage caused by wildlife. Module 4 addresses the legal and institutional framework for conflict reconciliation at various governmental levels, with a special emphasis on European regulations. Module 5 provides guidelines to assess the regional relevance of the conflict in economic terms and relevant policy instruments. Module 6 is dedicated to a stakeholder analysis to assess their perceptions and viewpoints. The ecological and socio-economic key factors of the conflict as well as opportunities and limitations identified by the modules of phase 2 point to promising resolutions and thus help to identify which of the modules of the resolution phase are of particular relevance (Fig. 1). This process may confirm the assumptions of the screening phase or lead to an adaptation and change of focus.

**Phase 3 "Deriving and implementing resolutions"** is dedicated to supporting the development and implementation of reconciliation strategies or action plans (RAPs). Note that a single approach rarely works for conflict resolution. Therefore, the modules of phase 3 facilitate the identification of combinations of effective approaches, their opportunities, limitations, and risks, and of relevant persons and institutions for the implementation process. The results of this phase will be a conflict management strategy or a RAP, which includes actions on the ground, work with stakeholders, and recommendations for policy changes, but may also include a return to phase 2 for more in-depth scientific assessment of specific components. Phase 3 contains four modules. Module 7 helps to assess ecological mitigation methods. Module 8 provides guidelines for monitoring wildlife abundance and discusses the advantages and limitations of various modeling approaches to assess wildlife population viability under different management scenarios. Module 9 evaluates policy instruments and provides advice for their improvement and the development of new instruments. Module 10 delivers guidelines for effective stakeholder interaction and the design of participatory decision strategies.

Except for the screening phase and module 4, we identify for each module minimum requirements and make recommendations for standard and advanced approaches. In general, we recommend following standard approaches. In conflicts that are difficult to reconcile or in which stakes are particularly high, we suggest using advanced approaches at least for the key elements for assessing and reconciling the conflict. Background and technical information regarding each module are provided in boxes.

To use our generic framework efficiently, we recommend starting with module 1, the screening phase. Then continue to the modules indicated by the results of your screening process, unless you are well familiar with the multidisciplinary and temporal facets of human-wildlife conflict reconciliation and are looking only for recommendations and support material in relation to specific components of conflict assessment and reconciliation.

**Acknowledgments** This work was financed by the EU 5th Framework Program (5th FP) Project "FRAP" (Development of a procedural Framework for Action Plans to Reconcile conflicts between large vertebrate conservation and the use of biological resources: fisheries and fish-eating vertebrates as a model case), contract number EVK 2-CT-2002-00142-FRAP.

# Module 1: Screening of the Conflict

**Andreas Kranz, Felix Rauschmayer and Irene Ring**

**Abstract** The basis and first step of an efficient conflict reconciliation and management is a proper screening study. Screening aims at yielding the best picture achievable through collecting existing information on key aspects without going deeply into own investigations. The outcome of the screening phase provides information on how to proceed with, where to focus on, and how much effort to spend on which aspect. This chapter provides checklists of different questions concerning the history, the current scale, ecology, damage, legal, social and economic aspects of the issue, as well as possible ways out of the conflict. The screening analysis indicates whether further steps should be taken along a minimum, a standard or an advanced approach. Already this first phase of a conflict management should be carried out by a knowledgeable person (conflict manager), who is accepted by all stakeholders who might have a crucial impact on conflict reconciliation.

A. Kranz (✉)
Institute of Wildlife Biology and Game Management, University of Natural Resources and Life Sciences, Gregor Mendel Street 33, 1180 Vienna, Austria
e-mail: andreas.kranz@aon.at

F. Rauschmayer
Department of Environmental Politics, UFZ—Helmholtz Centre for Environmental Research, Permoserstraße 15, 04318 Leipzig, Germany
e-mail: felix.rauschmayer@ufz.de

I. Ring
Department of Economics, UFZ—Helmholtz Centre for Environmental Research, Permoserstraße 15, 04318 Leipzig, Germany
e-mail: irene.ring@ufz.de

R. A. Klenke et al. (eds.), *Human-Wildlife Conflicts in Europe*,
Environmental Science and Engineering, DOI: 10.1007/978-3-540-34789-7_11,

## 1 Rationale and Objectives

The aim of screening is to get a first impression of the factors and circumstances which may be crucial in understanding and reconciling the conflict. The screening should provide a basis for deciding how to approach the conflict: The intensity and type of conflict and the lack of information and understanding of aspects will indicate whether a minimum, a standard, or an advanced approach is most appropriate. It may well be that in some aspects there is no need to go into details, whereas in others it is necessary to do so. The screening phase is crucial, as it will aid in making decisions on how and where to allocate resources:

- Which information is still needed?
- Which stakeholders are crucial?
- How efficient could technical measures, economic incentives, regulations, social processes, or information and education campaigns reconcile the conflict?

After the screening process (phase 1), one should be able to decide, whether to use the minimum, standard, or advanced approach indicated in each of the following modules, and how to link the modules and their results in the reconciliation process. Although certain flexibility will be required due to new knowledge gained, any substantial rearrangement later on will require more money and time, and retardation may have a significant effect on the success of conflict resolution. On the one hand, screening for potential ways of reconciling the conflict in phase 3 gives a first indication on the depth of information needed in the assessment (phase 2). On the other hand, the deeper understanding of the issue gained in the assessment phase, refines the ideas on potential ways of conflict reconciliation. The screening should include persons with both ecological and socio-economic analytical skills.

## 2 Methods and Approaches

The first task in this screening process is to gather information about:

- the history and scale of the conflict,
- the basic ecological features of the targeted wildlife species,
- the basic legal, institutional, and economic context of the conflict,
- the identity of the stakeholders and their attitudes to this species, and
- the amount of caused damage and related costs.

This should be done by a succinct review of journals and scientific literature, and by interviewing few important persons from administrations, resource users, NGOs, and science. These interviews are conducted with a guideline consisting of a set of relevant topics, but in a rather passive way: Let the interviewees talk while giving as few input as possible in order to influence them as little as possible. It may be that not all stakeholders accept and trust the conflict manager. In these cases, it is of utmost importance to select a neutral person to conduct interviews. It is not the aim to state who is wrong and who is right but to understand the different points of view.

Concerning a first appreciation of the history and scale of the conflict, the questions in Box 1 might guide the conflict manager. Usually, they will have a profound understanding of the situation. However, they should be invited to crosscheck their understanding with other stakeholders in the conflict, complemented by literature.

**Box 1** General Questions on History and Scale of Conflict

- When did the conflict occur first? Recently or in times long ago?
- Has there been a change in conflict intensity, and why/why not? (change in human use of resource or other human behavior/change in population or behavior of wildlife)
- Has there been a change in the way stakeholders relate to each other?
- Are there existing working groups or stakeholder fora dealing with the conflict at a regular basis?
- Is traditional ecological knowledge on the species and damage reductions available?
- How would the conflict likely evolve without any intervention?
- Is the species mobile on a local, regional, national, or international level?
- Is the social conflict on the same spatial scale as the species mobility?

More specific questions with regard to assessing and analyzing the conflict (Box 1–Box 5) then help to identify major gaps in knowledge. Answers to these questions, together with answers to questions focusing on reconciliation strategies (Box 6–Box 9), will give an idea about the intensity of work to be done in each module of the framework, i.e. whether to use a minimum, standard, or advanced approach.

Regarding conflict analysis and assessment, the first questions deal with the abundance of the wildlife species, their dependence on specific land- or seascape features, and damage caused by them (Box 2).

**Box 2** Questions on Wildlife Abundance and Habitat (Module 2)

- What are the ecological key features, e.g., is it a gregarious or solitary animal (hunter)?
- What data on abundance and on population structure are available?
- What are its basic habitat needs?
- Are there local effects triggered by conditions or processes at another site or on another scale?

Due to the tradition in European natural resource and biodiversity management, some ecological knowledge on the wildlife will be available. In stakeholder interviews or analysis of their documents, their understanding of these facts should be checked in order to reveal disparities. Disparities in the second block of questions (Box 3 may need a more thorough interaction than a mere information campaign, potentially sufficient for disparities in the first block.

**Box 3** Questions on Damage Caused by Wildlife (Module 3)

- Which types of damage occur: food, habitat alteration, secondary losses, surplus killing, damage to gear, spreading of diseases, etc.?
- Is the damage caused by the animal documented or is it difficult to prove?
- Is damage exactly or reasonably quantifiable? Is quantification difficult? Is a minimum estimate possible?
- Are technical mitigation measures in place and are they believed to help reduce damage? Are specific factors provoking damage known or unknown (type of resource production, landscape factors, etc.)?
- Does damage occur all year round, seasonally, sporadically, or is it unpredictable? Is it a long-term or short-term damage?
- Is damage immediately visible for resource holders? Later? Is it never clear?

Questions concerning the importance of damage (Box 4) also cut across different modules, as they concern the diet of the conflict species (module 3), institutional and economic aspects (modules 4 and 5), as well as the perception of the stakeholders (module 6):

**Box 4** Questions on the Importance of Damage (Modules 3, as well as 4, 5, and 6)

- Is damage measurable (monetary or otherwise)?
- How significant is the single or annual amount of damage for individual stakeholders and a region?
- Is damage evenly distributed?
- Who has property rights to the resource (farmed or wild resource)?
- Are there any policies compensating damage?
- How is damage perceived by the main stakeholders?

The legal and socio-economic context of the conflict (modules 4 and 5) constitutes the next issue in the screening process (Box 5).

The first contact with stakeholders (module 6) will also indicate whether the conflict is rather dormant, slowly evolving, or explosive (Box 6). It may also cast

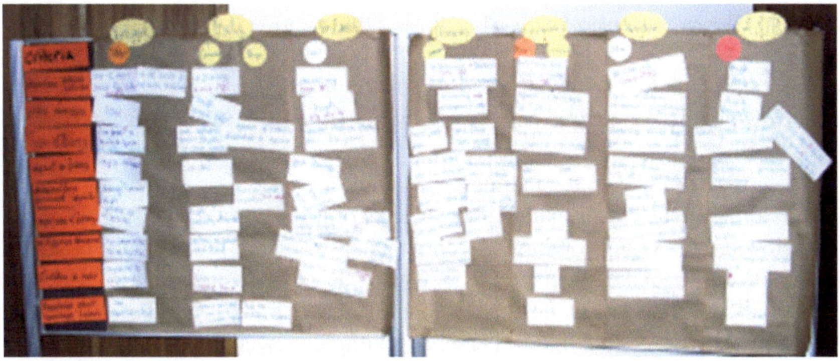

**Fig. 1** First results of a screening process. *Photo:* Oliver Zwirner

light upon the importance of the conflict for society, aspects all together of significant influence on the subsequent intensity of conflict resolution and stakeholder participation. Stakeholder opinions will indicate whether the conflict is mainly due to different interpretation of facts, different interests, or different values. All three elements are usually intermingled to varying degrees. Fact-related conflicts come into being, when stakeholders do not agree upon data due to lack of knowledge or to a different framing of the issue. In contrast, interest dominated conflicts arise when stakeholders want to use the same resource differently, more or less excluding the others. Value conflicts, finally, develop from different convictions on the rights and duties of stakeholders, or of animals, species, or nature as a whole (Fig. 1).

**Box 5** Questions on Legal, Institutional, and Socio-Economic Background (Modules 4 and 5)

- Which laws, regulations, and institutions may be most relevant to the conflict at hand?
- On which level (local, regional, national, and international) are the institutions responsible for conflict management and the actual implementation of policies?
- What is the economic and social relevance of the resource and the conflict?
- What is the economic and social relevance of the economic sector or of the stakeholder groups affected?
- Which are the existing policy instruments dealing directly or interfering with the conflict, such as species management plans, hunting regulations, compensation schemes, subsidies, or communicative instruments, and are there de facto institutional practices?
- What laws and policies are implemented in similar conflicts in the same or other areas?

**Box 6** Questions on Stakeholder Situation and Attitudes (Module 6)

- Who is affected by damage and by the conflict?
- How do affected people perceive costs and benefits of the conflict species' presence?
- Who is involved in regulating the conflict?
- Who is interested in maintaining or changing the *status-quo*?
- How are the groups organized—on which level (local, regional, national)?
- How do the stakeholder groups relate to each other?
- What do they perceive as the core of the conflict? Conflict species, resources, interests, values, personal relations, power?
- The attitudes of the stakeholders towards the involved wildlife and resource are an important momentum in the conflict and are decisive for its reconciliation:
- Is the conflict species perceived as a "natural" part of the environment with its own rights or as an intruder?
- What are the emotion towards the conflict species compared to the resource competed for?

Through the analysis of the views and perceptions of the stakeholders (module 6), a basic understanding can be achieved on the appropriateness of the actual decision process, and on where to put the focus of an intervention, if such is deemed necessary. Different views on data, interests and values are all reasons for the same conflict, but may be of different importance, also in the different stages of the conflict.

Conflicts that are mainly based on data need another type of intervention than conflicts primarily emerging out of interest or value differences. Thus, it is important to notice whether all knowledge holders have the impression to be heard in the process, and whether they are trusted by the other stakeholders. The legitimacy of interests may be evaluated quite differently by the stakeholders—the weight given to fulfilling these interests will vary accordingly. The same is true for values, but those are often not outspoken. Focusing on social dynamics of the process and on ways to improve its legitimacy is appropriate for interest and value conflicts.

Institutional and policy analysis (modules 4 and 5) help to understand the institutional history of the conflict, which is relevant for the range of policy options easily achievable. Each institutional redesign demands an extra effort, which has to be accounted for. Enriching the range of policy options or technical mitigation measures with lessons from similar cases can lead out of a deadlock. A well-informed economic analysis is relevant for the costs of the methods proposed and for having a basis for deciding who should bear the costs (which sometimes may not be too high for direct stakeholders, e.g., with the support of EU agri-environmental schemes). More thorough socio-economic analyses are usually appropriate in interest conflicts.

Through interviews, the conflict manager can get an impression on the type of process the main stakeholders might be willing to undertake. Sometimes, they may not be interested in the interference of a third party external to the conflict, sometimes they judge that such intervention from a neutral side gives a new legitimacy to a process and enables solutions that would have been impossible otherwise. The stakeholders know of, and can sometimes create, windows of opportunities for new processes.

Thus, the screening process will not only indicate which information is needed in greater detail, i.e. which module should be done in a standard or even advanced way, but will also give a first indication on how to reconcile the conflict. Ideas on how the conflict might be reconciled, in turn, influence the need for information on conflict details.

Consequently, the second task of the screening process is to develop a clearer picture of the strategies which one wishes to use to reconcile the conflict. Four main types of strategies can be distinguished: dealing with the management of the species population itself (module 8), mitigating the conflict through technical measures (module 7), policy instruments (module 9), as well as, finally, addressing the conflict resolution in a participatory process (module 10). In most cases, a combination of several, perhaps of all four strategies will be useful. Answers to the following questions make it clearer from the outset, which strategy could be the predominant. This first picture will then be refined during the assessment phase.

Possible approaches for conflict reconciliation from an ecological perspective are dealt with in module 7 and 8. The screening has to cover a set of questions, which are outlined in Box 7.

**Box 7** Questions on Population Management and Ecological Mitigation (Modules 7 and 8)

- Are technical mitigation measures currently available, uncertain or unknown?
- Is a combination of mitigation measures used or are all activities concentrated on one approach?
- What is known about change in abundance of the concerned population?
- How easy will it be to monitor change in abundance?
- Do we need absolute or relative abundance for choosing management options?

The list of known mitigation measures and those already applied in the conflict area will give a reasonable clear picture of the depth the approach in module 7 may need to take. Since technical mitigation was the classical approach and often still is the only one in place in damage reduction, some methods will be known. However, in the screening phase it is crucial to get a feeling of the probability of reconciling the conflict with technical mitigation. The existence of measures alone does not tell much about their power for conflict resolution. Spatial characteristics of the problem species as well as predator prey relationships may modify their suitability considerably. The screening process should provide an indication whether such differences are known or should be addressed in the assessment and the conflict resolution phases. This is also true for estimating how difficult modeling, population management, and monitoring of managed populations may be and how difficult it may be to get relevant data for population parameters, e.g., growth rates of the wildlife population. If stakeholders accept that the population should be maintained on the current level, but not allowed to grow further, monitoring of relative abundance on managed sites, a minimum approach for management and monitoring, may be justified. It is advisable to consult field specialists as well as modelers in this part of the screening in order to decide, whether population management may be an option or not (is it technically feasible for the species and what would be the magnitude of costs to achieve the goal?). However, even when it is quite clear from the beginning that population management is an unlikely option in practice, modeling may have its value in the process of conflict management to assess the potential damage that may be caused by current or expected future abundance of the wildlife and the viability of the wildlife species at levels below the current population size.

Policy instruments (module 9) in the form of conservation regulations already exist in all European wildlife conflicts, but most conflicts are also addressed by specific conflict mitigation policies that play an important role in their reconciliation. In the screening phase, the conflict manager can get a first understanding of the factual and potential importance of human-wildlife conflict policies (see questions in Box 1–Box 7). If basic and sufficiently working conflict management is already in place, the focus could be on improving single policy instruments, e.g., by increasing their ecological effectiveness or cost-effectiveness. In all other cases, a policy-mix consisting of regulative, economic, communicative, and educational instruments should be investigated and assessed for combined impacts. If no instrument yet exists (apart from species conservation regulation) or the challenge of coordinating several instruments in a species-specific management plan is on the agenda, an advanced approach is required for module 9. The latter would also mean following an advanced approach for the policy analysis of implemented instruments (module 5). If the introduction of certain policies will only succeed with broad public support, an advanced approach in the form of a survey on attitudes of different stakeholders towards policies is also recommended for module 6.

**Box 8** Questions on Policy Instruments (Module 9)

- What is the perceived effectiveness and efficiency of existing policy instruments?
- Are there conflict issues which are not covered by instruments?
- Are European or national funds available for the provision of ecosystem services or for mitigating or compensating damage, but not used?
- Is a species-specific management plan in use, to be revised or newly developed?

Processes of stakeholder participation (module 10) are increasingly used when reconciling biodiversity-related conflicts as they usually need the cooperation of all parties concerned. Characterizing conflicts is a first step to decide whether to undertake a participatory process and for the selection of the appropriate method. In this respect, it is the aim of the screening phase to assess whether there is a window of opportunity for such a process, and if so, how large it might be. The existence or creation of such window relates to the existence of an institutional frame for participatory processes, the contentment of the stakeholders with the processes and the general dynamics of the situation. The design of appropriate processes will be facilitated by a deepened understanding of the stakeholder perceptions of the conflict (module 6). Often, one aim of participatory processes is the creation or adaptation of policy instruments—therefore, conducting modules 5 and 9 in some depth will be productive in most cases of participation.

**Box 9** Questions on Participatory Processes (Module 10)

- Is there an institutional setting for existing or new participatory processes?
- Are stakeholders satisfied with the existing reconciliation processes and their role within them?
- Do stakeholders expect a change in the conflict situation due to social, organizational, economic, or ecological changes?

The results of the whole checklist consisting of all questions should be presented in a concise conflict description, which in its conclusion identifies the intensity of the conflict and the importance of different tasks. The description should explicitly include all deficiencies in data, information, etc. The decision of how to proceed with the reconciliation process should ideally be taken in a participatory way, bringing together the interests and the knowledge of the most important different stakeholders. This might be done in a workshop, if the situation is appropriate (e.g., existing and not explosive conflict). The decision on how to proceed, however, will be the privilege of those willing to pay for the actions and analyses needed within the reconciliation process.

## References

Conover M (2002) Resolving human-wildlife conflicts: the science of wildlife damage management. Lewis Publishers (CRC Press LLC), Boca Raton

Moll P, Zander U (2006) Managing the interface: from knowledge to action in global change and sustainability science. Oekom Verlag, München

Woodroffe R, Thirgood S, Rabinowitz A (2005) People and wildlife. Conflict or coexistence? Cambridge University Press, Cambridge

## Journals

Journal of Wildlife Biology http://www.wildlifebiology.com
Journal of Wildlife Management http://www.wildlifejournals.org
Journal of International Wildlife Law and Policy http://www.jiwlp.com
Ecology and Society http://www.ecologyandsociety.org
Human Dimensions of Wildlife http://www.informaworld.com/smpp/title ~ db=all ~ content=t713666717 ~ tab=issueslist
Wildlife Monographs http://www.wildlifejournals.org

## Webpages

Large Carnivore Initiative for Europe http://www.lcie.org
People and Wildlife Initiative http://www.peopleandwildlife.org.uk
The Wildlife Society (TWS) http://www.wildlife.org
COEX: Sharing the Land with Wildlife http://www.coex-wildlife.org

# Module 2: Wildlife Abundance and Habitat

**Margarida Santos-Reis, Reinhard A. Klenke and Klaus Henle**

**Abstract** The shared use of natural resources by humans and wildlife is the basis of long-lasting conflicts whose reconciliation is an urgent need. To fully understand the degree of damage and the conflict intensity, detailed knowledge is required on the distribution and abundance of animals and thus their visiting rate and contact with vulnerable resources, which in turn vary spatially depending on habitat characteristics and regional resolution instruments. The purpose of this module is to assess how landscape factors and resource management factors influence wildlife abundance and the exposure and vulnerability of the resource to wildlife. Abundance estimates and species-habitat associations are the two key factors in this context. Three approaches are presented (minimum, standard and advanced), varying in time and funding needs, and these range from literature-based educated guesses to powerful predictors using field-survey datasets, as described. We suggest as a first step the standard approach and only if evidences support a strong influence of landscape factors select an advanced approach in a later step of the reconciliation process.

M. Santos-Reis (✉)
Centro de Biologia Ambiental/Departamento de Biologia Animal, Faculdade de Ciências, Universidade de Lisboa, Campo Grande, Bloco C2, 5° Piso, 1749-016 Lisbon, Portugal
e-mail: mmreis@fc.ul.pt

R. A. Klenke · K. Henle
Department of Conservation Biology, UFZ—Helmholtz Centre for Environmental Research, Permoserstraße 15, 04318 Leipzig, Germany
e-mail: reinhard.klenke@ufz.de

K. Henle
e-mail: klaus.henle@ufz.de

R. A. Klenke et al. (eds.), *Human-Wildlife Conflicts in Europe*,
Environmental Science and Engineering, DOI: 10.1007/978-3-540-34789-7_12,

# 1 Rationale and Objectives

Landscape factors influence the distribution and abundance of animals and thus their visiting rate and contact with vulnerable resources. Therefore, conflicts may strongly differ from landscape to landscape and from location to location within a landscape. The purpose of this module is to assess how landscape factors and resource management factors influence wildlife abundance and the exposure and vulnerability of the resource to wildlife. Together with modules 3 (Damage assessment), 5 (Economic and policy analysis), and 6 (Stakeholder analysis) it will provide spatially differentiated information about the degree of damage and the conflict intensity.

The module "Wildlife abundance and habitat" is important whenever the screening process indicates strong spatial differentiation of the conflict and locally or regionally differentiated resolution instruments are available or a feasible option. It helps addressing the following questions:

- How abundant is the wildlife species and how is their distribution within the conflict area?
- How frequent is its contact with the resource?
- Are landscape factors influencing the local status of wildlife and, therefore, influencing the conflict, e.g., by promoting or preventing losses of commercially important fish or damage to fishing gear?
- Do we have local problems triggered by conditions and processes at another site or scale?

Different landscapes are not equally attractive and do not have the same carrying capacity, e.g., rivers, streams, or small and large ponds in the case of fish predators. Characteristics of water bodies (e.g., water surface of ponds, percentage of bank cover, fish density) and of the surrounding area (e.g., land cover, density of water bodies) explain spatial differences in wildlife occupancy with implications for the intensity of the conflict. In the marine area the concept of landscape factors is not yet as well developed as for terrestrial habitats, but like in these, structure of the sea floor, currents, ice cover, distance to breeding sites, etc. are corresponding factors that may influence the distribution of the wildlife and/or the resource and cause spatial differences in conflicts. The same applies to estuarine areas where additional factors, such as the salinity gradient and the distance to freshwater sources, are spatially differentiated and may influence both the fish-eaters and the wild fish species that may act as important alternative prey to the former.

# 2 Methods and Approaches

Abundance estimates and species-habitat associations are the two key factors that must be assessed in order to understand how landscape factors and resource management factors influence the exposure and vulnerability of the resource to wildlife.

**Table 1** List of potential methods to be applied for a minimum, standard, and advanced approach

| Parameters | Approaches | | |
|---|---|---|---|
| | Minimum | Standard | Advanced |
| Distribution | Literature | Wildlife survey | Wildlife survey |
| Abundance | Expert guess | Presence/absence data or relative abundance | Population size |
| Visiting rate | Traditional knowledge | Standard field surveys (direct observations or search of signs of presence) | Advanced field surveys (direct observations or video surveillance) |
| Habitat availability | Literature | Quantification from map (e.g., CORINE land cover map) | Remote sensing techniques (e.g., aerial surveys, satellite imagery), mapping of special habitat structures |
| Key landscape factors | Expert guess | Analysis of landscape composition and use of descriptive statistical methods for analysis of maps and site distributions (habitat models based on field surveys—Habitat Suitability Indexes) | Analysis of landscape composition and configuration as well as spatial distribution and neighbourhood relation of maps and site distributions; analysis of population development in time and space; use of inferential statistics for hypothesis testing (habitat models based on capture/recapture, telemetry—GIS based models) |

**Fig. 1** Well trained professional or voluntary fieldworkers are mandatory for estimation and monitoring of abundance even with simple methods using sight counts or signs of presence. *Photo:* André Künzelmann

## 2.1 Minimum Approach

A minimum requirement should be a review of readily accessible literature on distribution, habitat requirements, and habitat preferences of the conflict species (see Table 1). In addition, expert knowledge may be consulted with the advantage that experts are usually familiar with the relevant literature, but having the disadvantage that they may belong to specific interest groups within the conflict. If available data prove to be insufficient or inadequate to assess key factors, fieldwork, as outlined in the standard approach, is mandatory.

## 2.2 Standard Approach

A standard approach, which will suffice for most purposes, will be to collect data on the presence and relative abundance of the wildlife species and on landscape and human resource factors (e.g., fish stocking densities) assumed to be important for the attraction of the wildlife to the resource (see Table 1). The extent of damages depends on a mixed set of factors (e.g., availability of alternative prey, multi-species-interactions), more than just the distribution and abundance of the wildlife species. However, the contact with vulnerable resources (visiting rate) may contribute to the understanding of the species occupancy pattern and may explain perceived damage.

Measurements done by direct observation have the advantage of providing the number of animals visiting the resource, but may be time consuming or impossible (e.g., for carnivores), and the mere presence of the observer may influence the visiting rate (Fig. 1). The number of signs or frequency with which they are observed in repeated surveys, provides a relative index of visiting rates, not absolute numbers of animals visiting the resource. In most cases this should be sufficient. When using signs of presence, it is mandatory to have trained field-workers. For methodological recommendations to assess visiting rates see Box 1.

**Box 1** Visiting Rates and Population Size

Visiting rates are a measurement of the number of times a species encounters the resource during a certain time span (e.g., day, week). This may be done through direct observation of the animals (e.g., birds) or indirectly by searching specific signs of presence (e.g., carnivores) or by assessing damage caused to fishing gear (e.g., marine mammals). Measurements should be repeated systematically in order to have a reliable value. It might also be important to replicate this measurement in time (e.g., seasonally) if the biology of the species indicates that its presence varies significantly with time (e.g., migratory species, species with changes in habitat use during breeding). Replicates will allow a better assessment of wildlife impacts on contested resources.

The search of tracks and/or faeces is a common field method when studying cryptic animals, such as the otter. In the frame of the FRAP project, otter visiting rates to selected fish ponds were assessed based on repetitive monitoring of old and fresh spraints. Ponds were checked weekly and the presence (or absence) of fresh (from previous night) and old (from the other six nights) spraints was recorded. A new maximum likelihood approach was developed to estimate the probability of the species visiting the resource per day. Results from a simulation experiment show that the estimator provides an unbiased and precise estimate if the sampling effort is strong (5–20 weeks) and the weekly visiting rates lie between 0.1 and 0.6.

In a more advanced approach, field surveys may assess visiting rates with video surveillance and supply the number and behavior of animals involved in those visits. Important limitations are that video-surveillance is expensive, time-consuming in the tape analysis, and cannot be applied to a large scale, besides needing to be optimized case by case (e.g., number and position of video cameras). It may be, however, the only solution to understand the reason why in specific situations high visiting rates, assessed through scat and footprint counting, do not correlate with the damage extent in terms of diet (e.g., otters in Sado Estuary, Portugal—Sales-Luís et al. 2009) and *vice versa*.

Visiting rates may be used as a surrogate of population abundance but do not necessarily correlate to population size, a parameter required in advanced approaches. For some species population size can be calculated precisely through direct observation (e.g., birds at breeding colonies) and regular monitoring provides time series data on trends, most useful in advanced approaches. For others, however, to obtain even a relative index of abundance (standard approach) is a rather difficult task, as is the case of most carnivores that are mainly solitary, nocturnal, extremely elusive, and live at low densities. Absolute numbers, necessary for advanced approaches, can be estimated only under special environmental conditions (e.g., new snow cover) or taking profit of recent advances in molecular tools using scats (molecular scatology). Molecular scat analysis is a costly and time-consuming technique that still has a relatively low rate of success in samples collected in adverse environmental conditions (e.g., extremely high temperatures or level of humidity) and/or incorrectly preserved until analysis, two factors that contribute to DNA degradation. As a methodological guideline for estimating relative and absolute numbers of wildlife species, Caughley (1980) is still the best reference for non-specialist users. For users with more advanced statistical skills, Williams et al. (2002) is recommended. For species specific survey methods Cooperrider et al. (1986) is a very useful resource. The estimation of visiting rates, as developed and used in the FRAP-project, is described in detail by Gruber et al. (2008).

Based on distribution pattern and habitat requirements, simple species-habitat association models can be developed. The advantage of habitat modelling procedures is that the probability of presence or relative abundance of the wildlife species can be extrapolated to larger areas and distribution maps can be produced quickly and efficiently, especially if inexpensive environmental data that cover large regions are available. See Box 2 "Habitat analysis and habitat modeling" for technical and statistical recommendations. Limitations do exist, however, and these are related to the scale and quality of the datasets (e.g., whereas data may be very concrete and easily available for some resources, such as fish in small wintering ponds, they may be extremely difficult to collect in some cases, such as in marine fish resources) as well as to the time and funding available. Also, when extrapolating across large spatial scales, one needs to keep in mind that the importance of landscape factors may change in different types of landscapes.

**Box 2** Habitat Analysis and Habitat Modeling

Presence/absence data, either geo-referenced or mapped over a grid system, can be associated to a series of environmental descriptors. They are widely used to develop habitat models that allow extrapolation to non-sampled areas. Various statistical methods exist to derive habitat models from field data, such as simple habitat suitability models, which group environmental data into classes based on species distribution data, or multivariate statistical models (logistic regression, discriminant analysis, Generalized Linear Models—GLM, Generalized Additive Models—GAM) to predict the location and extent of habitat for wildlife species. Recent developments in the field of computational ecology allow the use of mixed ecological datasets, including individual capture/recapture and radio-tracking data. Geographic Information Systems (GIS) facilitate an optimal use of existing spatial data, e.g., on land cover, soils, geology, or topography. By combining mathematical and computer models of natural systems with geographically explicit details of the biotic and abiotic components of the environment, the understanding of natural systems and the ability to properly manage it and to predict the effects of alternative management options is improved.

Most methods require species presence and absence data, with presences attesting a suitable habitat and absences an unsuitable habitat. An overview of relevant statistical methods can be found in Corsi et al. (2000). There is, however, an important constraint, which is the fact that blank areas are ambiguous as to whether the species is really absent or no records were available. Recent developments in this area include modeling species habitat suitability with presence only data, a new area that deserves to be further explored (e.g., Hirzel et al. 2002; Pearce and Boyce 2006). Model validation techniques are however needed to ensure that model predictions are not statistic artifacts.

The sampling points selected for the development of habitat models in the first step should cover the whole range of variation of the landscape factors and should be spaced as far apart as feasible to avoid autocorrelation and pseudo-replication.

The advantage of habitat modeling procedures is that, if inexpensive environmental data covering larger regions are available, the probability of presence or relative abundance of the wildlife species can be estimated and distribution maps can be produced quickly and efficiently.

## 2.3 Advanced Approach

If the screening process indicates that the conflict may have very strong spatially differentiated components and spatially differentiated mitigation or management instruments may be feasible, one should consider the advanced approach, which consists of estimating the population size (density) of the wildlife species (Box 1), which then can be related to landscape factors using more sophisticated statistical methods (Box 2). The calculation of real loss, based on consumption rates of the resource by wildlife, also requires knowledge of absolute numbers of wildlife visits to the resource (modules 3 and 5) obtainable only with advanced approaches. Finally, knowledge of the population size of the wildlife species or the carrying capacity of the habitat is usually needed for the modeling of the viability of the wildlife species under different management scenarios. Such more in-depth analyses are usually called for if the adequacy of financial compensations for loss of resources to wildlife may be a key driver of the conflict.

# 3 Recommendations and Conclusions

Each wildlife species is associated with key landscape factors that are undoubtedly driving its distribution and density at a certain geographic area. These associations may change through time (e.g., seasonally) or space (within its full range) and in turn have a differentiated impact upon the extent of damage the wildlife species may cause. Responses of wildlife to changing landscapes or human resource use furthermore influence population viability (module 8) and determine which technical mitigation measures are adequate, and where and when they have to be used (module 7). The use of spatially differentiated approaches and the coordination of these with mitigation measures (module 7) and policy instruments (module 9) are crucial to mitigate conflicts (Reimoser 2002).

Therefore, it is mandatory to have updated information on the distribution and status of the wildlife species involved in the conflict. Table 1 summarizes potential approaches. This may be comparably easy with terrestrial species that are easily seen and identified (e.g., cormorants) or, although elusive, do leave very conspicuous and characteristic signs of presence (e.g., otters), but may be more difficult with wide-ranging marine species (e.g., seals) that need appropriate technical means, leaving as an option the carrying out of oral inquiries to relevant interest groups. However, the presence of a predator at a given conflict area alone does not provide sufficient information about the extent of damage the species may cause. To accurately predict the predatory impact, an abundance estimator is highly recommended. Varying with the species body size, life-style (solitary *vs.* gregarious), activity pattern (diurnal *vs.* nocturnal), and habitat requirements (generalist *vs.* specialist), this is usually difficult, time-consuming, and needs highly qualified personnel, being however the only option in some situations.

Complementarily, to identify key landscape features that are influencing the occupancy pattern and abundance of the wildlife species, the use of the freely available CORINE land cover map is recommended as the best possible approach with limited resources. More expensive (high resolution maps) or time-consuming (field surveys) methods should be restricted to situations where conflict reconciliation clearly requires a spatially differentiated approach.

Identification of landscape factors promoting damage to fish farms requires good datasets. As outlined in module 3 (Damage assessment), wildlife may have a large local impact, either directly (fish eaten, damaged or disturbed) or indirectly (damage to catch and gear) (Freitas et al. 2007; Santos-Reis et al. 2013). Landscape factors that contribute to the carrying capacity of an area may be useful to assess the risk of impact to fisheries. These, however, act at different scales according to the species and the area. For migratory species (e.g., birds) and for wide ranging marine species (e.g., seals) the landscape scale seems to be the appropriate one, although the relevant factors may be difficult to manage directly and indirect measures may be needed. For small sized terrestrial species, however, local scale factors better explain damages, and relevant factors may be modified to reduce predation.

In summary, the minimum approach only provides an educated guess about the key landscape factors that may be influencing wildlife numbers and distribution, and hence about the potential extent of damage (low, medium, or high). With the standard approach the reliability of the results provides a much better insight into the conflict. Advanced approaches result in more powerful predictors but are more expensive, more time-consuming, and require specialized manpower. Specifically targeted research projects may be essential for an adjustment of methods to the specific problem at hand. The obvious way to handle these trade-offs is to use as a first step the standard approach and then, if the results reveal that landscape factors are crucial in conflict reconciliation, select an advanced approach in a later step of the reconciliation process or try to get the support of a research organization that has the expertise and scientific interest to pursue an advanced approach.

## References

Caughley G (1980) Analysis of vertebrate populations. Wiley, New York

Cooperrider AY, Boyd RJ, Stuart HR (1986) Inventory and monitoring of wildlife habitat. United States Department of the Interior, Bureau of Land Management, Service Centre, Denver

Corsi F, de Leeuw J, Skidmore A (2000) Modelling species distribution with GIS. In: Boitani L, Fuller T (eds) Research techniques in animal ecology: controversies and consequences. Columbia University Press, New York, pp 389–434

Freitas D, Gomes J, Sales Luís T, Madruga L, Marques C, Baptista G, Rosalino LM, Antunes P, Santos R, Santos-Reis M (2007) Otters and fish farms in the Sado Estuary: ecological and socio-economic basis of a conflict. Hydrobiologia 587:51–62

Gruber B, Reineking B, Calabrese JM, Kranz A, Poledníková K, Poledník L, Klenke R, Valentin A, Henle K (2008) A new method for estimating visitation rates of cryptic animals via repeated surveys of indirect signs. J Appl Ecol 45:728–735

Hirzel AH, Hausser J, Chessel D, Perrin N (2002) Ecological-niche factor analysis. How to compute habitat-suitability maps without absence data. Ecology 83:2027–2036

Pearce JL, Boyce MS (2006) Modelling distribution and abundance with presence-only data. J Appl Ecol 43:405–418

Reimoser F (2002) Wildlife ecological spatial planning (WESP): an instrument for integrating wildlife into comprehensive land management. In: Proceedings of the XXIVth congress of the international union of game biologists (Thessaloniki, 1999), pp 176–185

Santos-Reis M, Santos R, Antunes P, Sales-Luís T, Gomes J, Freitas D, Madruga L (2013) Reconciliation of the conflict between otters and fish farmers: lessons from Sado estuary in Portugal. In: Klenke R, Ring I, Kranz A, Jepsen N, Rauschmayer R, Henle K (eds) Human-wildlife conflicts in Europe—fisheries and fish-eating vertebrates as a model case. Springer, Heidelberg, doi:10.1007/978-3-540-34789-7_4

Santos-Reis M, Santos R, Antunes P, Sales-Luís T, Gomes J, Freitas D, Madruga L (2013) Reconciliation of the conflict between otters and fish farmers: lessons from Sado estuary in

Portugal. In: Klenke R, Ring I, Kranz A, Jepsen N, Rauschmayer R, Henle K (eds) Human-wildlife conflicts in Europe—fisheries and fish-eating vertebrates as a model case. Springer, Heidelberg, pp 30–30
Williams BK, Nichols JD, Conroy MJ (2002) Analysis and management of animal populations. Academic Press, San Diego

## Free and Commercial Software for Statistical Analysis and GIS

ArcGIS—ESR. http://www.esri.com. Accessed 08 July 2009
BioMapper. http://www2.unil.ch/biomapper/. Accessed 08 July 2009
GRASS GIS—The world leading free software GIS. http://grass.fbk.eu. Accessed 08 July 2009
R Development Core Team (2006) R: a language and environment for statistical computing. R foundation for statistical computing, Vienna, Austria. http://www.r-project.org. Accessed 28 Aug 2006
SAGA GIS—A system for an automated geographical analysis. http://www.saga-gis.org/en/index.html. Accessed 27 June 2012

## Web Pages

CRAN Task View: Analysis of ecological and environmental data. http://cran.r-project.org/web/views/Environmetrics.html. Accessed 08 July 2009
European Environmental Agency (EEA) dataservice. http://dataservice.eea.eu.int/dataservice. Accessed 08 July 2009
Free GIS software and data. http://www.freegis.org. Accessed 08 July 2009
Habitat Evaluation Procedures Handbook. http://www.fws.gov/policy/ESMindex.html. Accessed 08 July 2009
Patuxent Software Archive. http://www.mbr-pwrc.usgs.gov/software.html. Accessed 08 July 2009
phidot.org home page. http://www.phidot.org. Accessed 08 July 2009
USGS National Wetlands Research Center: Habitat Suitability Index Models Introduction. http://www.nwrc.usgs.gov/wdb/pub/hsi/hsiintro.html. Accessed 08 July 2009

# Module 3: Diet and Damage Assessment

**Håkan Westerberg and Klaus Henle**

**Abstract** Data on the diet of an aquatic top predator are essential to assess the impact of predation on the fisheries resources. Quantitative data are sparse however. Field data is usually collected by analysis of undigested parts in the food remains of scats, pellets, or in stomach and intestine of dead animals. Advanced techniques as fatty acid or stable-isotope analysis makes it possible to assess diet of components not leaving identifiable remains and to look at diet over larger timescales. A general observation is that aquatic predators tend to be opportunistic in their choice of prey. In the absence of field data the assumption can be made that diet reflects local availability of prey of appropriate size for the predator. To analyse real or perceived resource competition calculation of the amount consumed by the predator should be based on current and methodologically sound data. A simple comparison of the mass or number of a resource species taken by the predator and the fishery respectively can be highly misleading. An analysis should always be made in an ecosystem context, taking size distribution and density dependence into account.

H. Westerberg (✉)
Department of Research and Development, Swedish Board of Fisheries,
Box 423 40126, Gothenburg, Sweden
e-mail: hakan.westerberg@slu.se

K. Henle
Department of Conservation Biology,
UFZ—Helmholtz Centre for Environmental Research,
Permoserstraße 15, 04318, Leipzig, Germany
e-mail: klaus.henle@ufz.de

R. A. Klenke et al. (eds.), *Human-Wildlife Conflicts in Europe*,
Environmental Science and Engineering, DOI: 10.1007/978-3-540-34789-7_13,

# 1 Rationale and Objectives

The conflict between wildlife and humans is primarily one of competition for the same wild or cultivated food resource. Secondarily, structural damage or competition for space can be part of the conflict. In many cases it is unclear whether food competition is real or perceived. The conditions for reconciliation of the conflict are fundamentally different in these two cases, and in order to resolve this question it is necessary to have quantitative data on how large a share of the contested resource is removed by the wildlife species causing conflict. The objectives of this module are to help answer the following questions arising from the screening phase (module 1):

- What is the quantitative impact on the resources caused by the wildlife species through consumption and surplus killing?
- What other ecosystem level effects are caused by the wildlife species?
- What indirect impact is exerted on human structures and activities?

The first question requires diet studies. Here, recommendations are primarily given for aquatic carnivores. For herbivores and specialized terrestrial predators, we refer the reader to the general references given below (Useful literature). The second and third questions are usually complex and require approaches that are adapted to the specific circumstances.

Knowledge of the diet composition has a direct relevance for module 2 (Wildlife abundance and habitat). In turn, information of abundance and the habitat are important in the deeper analysis of diet data. The spatial distribution of ecosystem productivity in terms of amount and availability of a resource is very often one of the main factors determining not only the distribution and abundance of a species feeding on this resource, but also its behavior and social organization. The advanced approach, whereby ecosystem effects are integrated into diet analysis, has close ties to module 8 (Management and viability of target species). Quantification of resource competition is essential for establishing the nature of the conflict, and, thus, is important for all modules 5 through 10.

# 2 Methods and Approaches

Examination of diet composition is central in the assessment of the interaction between species through predation, consumption, or competition. Direct observations of food choice are limited to species that occupy open habitat or to observations of food brought to juveniles at nests or dens and can only be used in special situations. A widespread approach to study diet composition is analysis of food remains in scats, pellets, or the stomach and intestine of dead animals (Hyslop 1980; Pierce and Boyle 1991; Litvaitis 2000). Assessment of undigested food remains can be applied to both carnivores (skeletal bones, feathers, arthropod exoskeleton, scales etc.) and herbivores (seeds, fruits etc.) (Litvaitis 2000). In the case of large herbivores, studies of bite marks are useful.

Advantages with scats and pellets are the possibility to collect large samples throughout the year and the fact that collection is a non-destructive and, ideally, non-intrusive method. This advantage is partially offset by the very labor-intensive analysis and the fact that it is impossible to differentiate the individual predator by age or sex, unless DNA-based methods are used (Fedriani and Kohn 2001). Whenever possible, the hard part remains should be identified as to species and should—using a reference collection of animal or plant remains—be measured to estimate the size of the prey item.

With a sufficient sample size the average prey species distribution and size composition of the diet can be established (Jacobsen and Hansen 1996). Using the consumption rate—measured in physiological experiments or estimated from the feeding of captive animals—and the total population in the area of interest (module 2), the amount of different prey species consumed can be calculated. An alternative method for estimating the impact of predation is to tag a number of prey species and then calculate predation rate from recovery of tags from scats or regurgitates. This method is independent of knowledge of prey population size.

Box 1 provides summarized information from the diet studies in the FRAP project for Eurasian otter (*Lutra lutra*), grey seal (*Halichoerus grypus*), and great cormorant (*Phalacrocorax carbo sinensis*) and generalizes it to other aquatic top predators. Please see Freitas et al. (2007); Lundström et al. (2007), Jepsen and Olesen (2012) for more details.

**Box 1** Diet of Otters, Seals, Cormorants, and Other Aquatic Top Predators

Cormorants, grey seals, and, with some reservation, otters, are fundamentally opportunistic in their choice of food, both in terms of fish species and size. The diet reflects the availability of resources—in terms of abundance and ease to capture—rather than a special preference. This makes evolutionary sense for aquatic predators, characterized by few competitors and a highly variable and unpredictable food base. This may be generalized to other top predators targeting fish. Among terrestrial predators the larger biodiversity and competition makes specialization more common.

Even if the marine top predators are opportunists or generalists regarding diet seen as a whole they often specialize on a shorter time scale. This means that they can optimize the time used for searching for food by targeting the species which gives the highest caloric intake per unit time. Thus, prey species vary depending on the situation, but there is an advantage in foraging by following a specific search image.

In addition to what is actually consumed by the predator, there may be surplus killing and indirect damage of the prey, which increases the impact on the resource. Healed wounds from unsuccessful attacks by cormorants and seals are commonly seen on fish in areas where there is a conflict between fishery and those species. Evidently, a proportion of the wounded fish will die due to trauma and infections. This effect is usually difficult to quantify but can be significant.

In many situations there is a direct interaction between a wildlife species and human activities. For aquatic predators the most obvious case is damage to fishing gear and catch or to aquaculture facilities. Traditionally, the damage to the catch is assessed by counting the remains of fish in the gear. For cormorants the number or proportion of fishes showing bite marks is commonly used as damage parameter. Loss from aquaculture pens can be massive, especially if a net cage is damaged, and estimation is a matter of measuring the amount of fish remaining after a damage incidence.

Counting fish remains will underestimate the actual loss if the whole fish is taken out of the fishing gear or when damage to the gear leads to escape of captured fish. Such "hidden loss" can be measured in different ways (Fjälling 2005). Comparing the mean catch during days when damage to catch is observed to the mean during adjacent days without evident damage is one approach. Another method, which works well both for bird and mammal caused damage in the gillnet fishery, is to redeploy the net with the captured fish still entangled in the net after they have been counted and marked. The loss of marked fish can then be measured after a known time of exposure to the wildlife species.

## *2.1 Minimum, Standard, and Advanced Approaches*

Diet studies are a large subject in ecology and the methodology varies both with trophic level, at which the wildlife species operates, and with the kind of specialization involved. This requires specific approaches to diet and damage assessment for different types of wildlife, making it difficult to provide general recommendations that apply to all wildlife. Our recommendations are based on our experience with fish-eating vertebrates and mainly illustrated with such cases.

### 2.1.1 Minimum Approaches

In the absence of relevant diet data a minimum approach, which can be used for opportunistic predators, is to assume that the diet reflects the relative abundance of the prey species. Exception should be made for species that are too big or small to be likely prey for the predator in question. Combined with data on population size and overall consumption rate of the predator this will give a rough estimate of the consumption of different commercial species of interest in a wildlife conflict.

In many wildlife conflicts involving indirect damage, claims from stakeholders about the extent of damage can strongly vary between the nature conservation side and the users of a resource. Reliable data about the frequency and extent of damage are thus important and should be collected by interviews and field

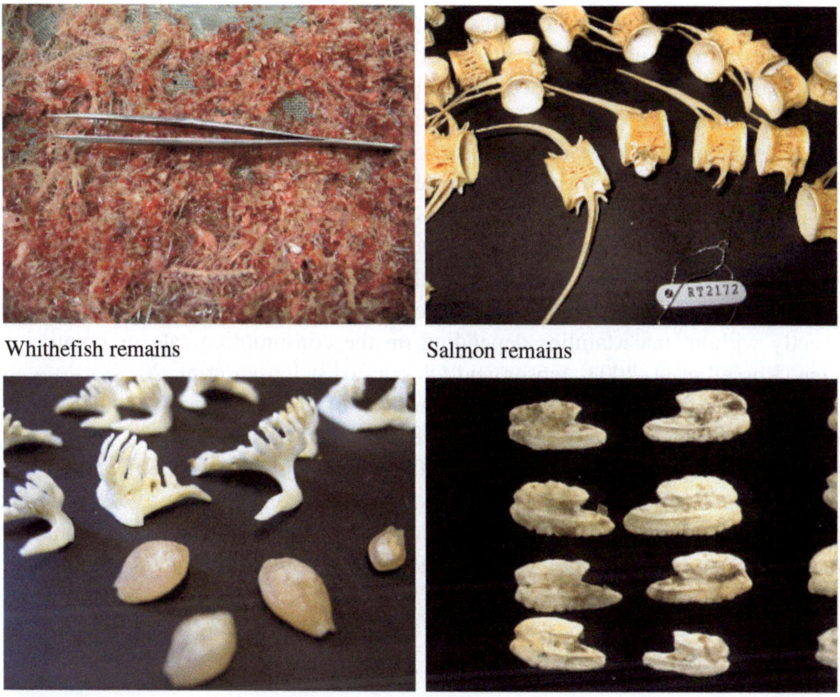

Whithefish remains Salmon remains

Cyprinid pharyngeal teeth and chewing pads Herring otoliths

**Fig. 1** Hard parts from food remains of different fish species from grey seal stomach after different stages of preparation. *Photos:* Karl Lundström

observations. Statistics that are accepted by all sides in the conflict are a prerequisite for successful conflict reconciliation.

### 2.1.2 Standard Approaches

The determination of diet composition from analysis of hard parts in food remains can be regarded as the standard approach (Fig. 1). If possible, the data should be examined for dietary patterns in space and time. For the otter, for example, dietary data can be used together with an estimated visiting rate (module 2) to calculate the loss of fish in an aquaculture pond. The same applies for cormorants and aquaculture. Data on the consumption is more problematic and prone to misinterpretation in an open system (see "Advanced approaches" below).

Assessment of diet composition provides qualitative and quantitative information on which species are affected by the predator, but biases in the dietary analysis can lead to over- or under-representation of particular prey species. Several sources of error can influence conclusions about diet composition from an analysis of hard parts (Jobling and Breiby 1986; Pierce and Boyle 1991). Due to the decomposition of bony structures in the stomach, prey items with small otoliths and bones become underrepresented. For large specimens on the other hand, just the soft tissues may be consumed, without leaving identifiable remains. In between those extremes there is an unknown degree of erosion that makes back-calculation to size of prey uncertain. Correction factors should be used to account for bias due to digestion (e.g., Tollit et al. 1997; Brzezinski and Marzec 2003).

A completely different method is to assess the impact of the predator by tagging a known proportion of the prey with indestructible tags and then search the scats or regurgitated pellets for tags. This method has been used for cormorants in Denmark and has the advantage that the effect on the prey population is measured directly without uncertainties depending on the consumption rate or population sizes (Russell et al. 2003, Jepsen and Olesen 2013; Jepsen et al. 2010).

To study indirect damage, such as the loss of catch and gear damage in the fishery, a system of representative persons keeping a detailed journal of their damage can be used (Lunneryd et al. 2005). If the journal keepers constitute a random sample of the fishery, then an extrapolation can be made to calculate the total structural damage. Where hidden losses can be expected, those should be quantified using the methods discussed above and added to the total.

### 2.1.3 Advanced Approaches

Both scat and digestive tract analysis provide rather short-term pictures of the diet and it is seldom possible to get a sample large enough to cover spatial and temporal variations in prey choice. If possible, several different methods should be combined to get a picture of the diet in a particular case, as it has been done in Denmark for cormorants (see Jepsen and Olesen 2013).

Studies using new approaches in digestive tract and scat analysis have been included in the FRAP project. Fatty acid analysis may give a more representative picture of the average consumption (Iverson et al. 2004). Stable-isotope analysis can also be used to monitor the dietary intake over a longer time scale (Szepanski et al. 1999). Both fatty acid analysis and stable isotope analysis are important in cases in which no identifiable hard parts can be retrieved, but do not provide as detailed information on specific prey species as does the identification of hard part prey remains. Another method that may be used as a supplementary tool with hard part analysis, in particular to identify prey that provide no or few identifiable remains, is DNA-based identification of prey species from pellets, scats, and digestive tracts, which has proven useful for a variety of predator species (Symondson 2002). A method to improve the assessment of herbivore diet composition is the use of indigestible species-specific plant wax components (Dove and Mayes 1996).

A simple comparison mass-by-mass or number-by-number between predator consumption and human catch is not a relevant measure of impact and can easily be misleading (Trites 2003). A large consumption by the predator may be targeted on small, non-commercial size classes and if there is density dependant mortality or size dependent growth limitation of the prey, the predator may actually improve the production of the resource of commercial size. On the other hand a relatively small quantity, which is consumed at a sensitive stage in the life cycle of the prey, can have significant effect on the stock. An example is the cormorant predation on salmon or trout smolts at the stage when they migrate out from a river. Even if smolts constitute a small part of the yearly consumption of the cormorants, the effect on a wild salmon population can be disastrous (Dieperink et al. 2002).

Studies of diet composition should aim for a more complete picture of the resource use and how it varies by specific segments of the actual population. An advanced approach to the analysis of diet data thus has to take an ecosystem approach, i.e. to assess the role of the predator in the ecosystem.

Particularly in inland waters there are scale effects. Both cormorants and otters can locally have a large impact, especially in small ponds irrespective of whether these are stocked or not, even if the average over a larger area is small. Locally, seals can be a severe problem for the fishery due to damage to the catch and the gear, even if the actual consumption associated with this is an insignificant part of the diet and the overall ecological effect may be limited.

At an advanced level, indirect effects other than structural damage should be included. An effect, which can have very important economic consequences for human use of a resource, is the spreading of parasites and diseases by wildlife. This is a subject beyond the scope of this module.

## 3 Recommendations and Conclusions

Basic diet information is a prerequisite for the reconciliation of all conflicts between human use of natural resources and the real or perceived competition with wildlife. Information about the amount consumed is an important argument in the reconciliation. To make such a case convincing, it is necessary that the diet studies fulfill the following requirements:

- Current, so that they reflect the actual availability of resources;
- Comprehensive, so that the whole wildlife population is included, geographically and during all seasons;
- Methodologically sound, so that all parts of the diet are included quantitatively in the analysis.

Those requirements are always important, but a larger degree of uncertainty can be accepted in a situation where the data show a considerable degree of competition. In that case the main difficulty lies in interpreting and communicating the result to avoid misuse in the debate, as discussed above in the section "Advanced approaches".

The Danish cormorant RAP (Jepsen and Olesen 2013) and the Swedish-Finnish grey seal RAP (Bruckmeier et al. 2013) both demonstrate the difficulties and necessity of fulfilling those requirements. With opportunistic predators, diet data are transitory. The diet of the grey seal has changed radically during recent decades. However, the methods available for diet analyses are labor-intensive. Therefore, it is expensive and difficult to cover spatial and seasonal variations fully. The standard method is based on the analysis of hard part remains, which is prone to bias both in size and species composition. In addition to this, there are other aspects of the overall analysis of competition that should be stressed. Those are the need to know the size of the wildlife population (module 2) and the consumption rate of free-ranging animals. For many marine predators these data can be more difficult to obtain than a rough diet estimate. If the predator is judged to be an opportunist, it may be a more cost-efficient approach to estimate the diet indirectly from an assessment of the fish species balance in the area and to concentrate the effort and resources on a population census or on experiments to determine the consumption rate.

The main research need is in developing alternative, cheaper, and unbiased methods to determine diet composition. The fatty acid signature method seems promising but there is still much research on fundamental methodological questions that has to be finalized before it can be used routinely. A comprehensive database on the fatty acid signature of the major fish stocks in Europe is lacking and would be necessary to apply the method widely for fish-eating predators.

## References


Bruckmeier K, Westerberg H, Varjopuro R (2013) Reconciliation in practice: the seal conflict and its mitigation in Sweden and Finland. In: Klenke R, Ring I, Kranz A, Jepsen N, Rauschmayer R, Henle K (eds) Human-wildlife conflicts in Europe–fisheries and fish-eating vertebrates as a model case. Springer, Heidelberg

Brzezinski M, Marzec M (2003) Correction factors used for estimating prey biomass in the diet of American mink (*Mustela vison*). Acta Theriol 48:247–254

Dieperink C, Bak BD, Pedersen LF, Pedersen MI, Pedersen S (2002) Predation on Atlantic salmon and sea trout during their first days as postsmolts. J Fish Biol 61:848–852

Dove H, Mayes RW (1996) Plant wax components. A new approach to estimating intake and diet composition in herbivores. J Nutr 126:13–26

Ferdiani JM, Kohn MH (2001) Genotyping faeces links individuals to their diet. Ecol Lett 4:477–483

Fjälling A (2005) The estimation of hidden seal-inflicted losses in the Baltic set-trap salmon fisheries. ICES J Mar Sci 62:1630–1635

Freitas D, Gomes J, Sales Luís T, Madruga L, Marques C, Baptista G, Rosalino LM, Antunes P, Santos M, Santos-Reis R (2007) Otters and fish farms in the Sado Estuary: ecological and socio-economic basis of a conflict. Hydrobiologia 587:51–62

Hyslop EJ (1980) Stomach content analysis—a review of methods and their application. J Fish Biol 17:411–429

Iverson SJ, Field C, Bowen WD, Blanchard W (2004) Quantitative fatty acid signature analysis, a new method of estimating predator diets. Ecol Monogr 74:211–253

Jacobsen L, Hansen HM (1996) Analysis of otter (*Lutra lutra*) spraints. Part 1, comparison of methods to estimate prey proportions; part 2, Estimation of the size of prey fish. J Zool 238:167–180

Jepsen N, Klenke R, Sonnesen P, Bregnballe T (2010) The use of coded wire tags to estimate cormorant predation on fish stocks in an estuary. Mar Freshw Res 61:320–329

Jepsen N, Olesen T (2013) Cormorants in Denmark—re-enforced management and scientific evidence. In: Klenke R, Ring I, Kranz A, Jepsen N, Rauschmayer F, Henle K (eds) Human-wildlife conflicts in Europe—fisheries and fish-eating vertebrates as a model case. Springe, Heidelberg

Jobling M, Breiby A (1986) The use and abuse of fish otoliths in studies of feeding habits of marine piscivores. Sarsia 71:265–274

Litvaitis JA (2000) Investigating food habits of terrestrial vertebrates. In: Boitani L, Kuller TK (eds) Research techniques in animal ecology. Controversies and consequences. Columbia University Press, New York, pp 165–190

Lundström K, Hjerne O, Alexandersson A, Karlsson O (2007) Estimation of grey seal (*Halichoerus grypus*) diet composition in the Baltic Sea. In Haug T, Hammill M, Olafsdóttir D (eds) NAMMCO Sci Publ 6. NAMMCO Tromsø, pp 177–196

Lunneryd SG, Hemmingsson M, Tärnlund S, Fjälling A (2005) A voluntary logbook scheme as a method of monitoring the by-catch of seals in Swedish coastal fishery. ICES CM2005/X:04, 10

Pierce GJ, Boyle PR (1991) A review of methods for diet analysis in piscivorous marine mammals. In Barnes M, Barnes H (eds). Oceanogr Mar Biol Annu Rev 29:409–486

Russell I, Kinsman DA, Ives MJ, Finney J, Murrell M (2003) The use of coded-wire microtags to assess prey selectivity and foraging behaviour in Great Cormorants. Vogelwelt 124:245–253

Symondson WOC (2002) Molecular identification of prey in predator diets. Mol Ecol 11:327–641

Szepanski MM, Ben-David M, Van Ballenberghe V (1999) Assessment of anadromous salmon resources in the diet of the Alexander Archipelago wolf using stable isotope analysis. Oecologica 120:327–335

Tollit DJ, Steward MJ, Thompson PM, Pierce GJ, Santos MB, Hughes S (1997) Species and size differences in the digestion of otoliths and beaks, implications for estimates of pinniped diet composition. Can J Fish Aquat Sci 54:105–119

Trites AW (2003) Food webs in the ocean, who eats whom and how much? In: Sinclair M, Valdimarsson G (eds) Responsible fisheries in the marine ecosystem. FAO. Rome and CABI Publishing, Wallingford, pp 125–141

# Module 4: Legal and Institutional Framework

**Jukka Similä, Riku Varjopuro, Randi Habighorst and Irene Ring**

**Abstract** This chapter focuses on the legal and institutional framework that is relevant for the reconciliation of human-wildlife conflicts. The identification of the framework is important, because laws, regulations and norms may impose severe restrictions for reconciliation policies, but also because they open avenues of action. The legal and institutional framework involves governmental levels from local to international as well as different policy sectors from nature conservation and hunting to trade and competition. Relevant governmental levels and policy sectors vary greatly, depending on the conflict and national politico-administrative traditions. Presentation of the framework is organized by using three categories of possible policy instrument types that progress from stringent to least binding measures: "nature conservation regulation", "economic instruments" and "information and training".

J. Similä (✉) · R. Varjopuro
Environmental Policy Centre, Finnish Environment Institute (SYKE),
Box 140 00251, Helsinki, Finland
e-mail: jukka.simila@ymparisto.fi

R. Habighorst
Department of Environmental and Planning Law,
UFZ—Helmholtz Centre for Environmental Research,
Permoserstraße 15, 04318, Leipzig, Germany
e-mail: randi.habighorst@gmx.de

I. Ring
Department of Economics, UFZ—Helmholtz Centre for Environmental Research,
Permoserstraße 15, 04318, Leipzig, Germany
e-mail: irene.ring@ufz.de

R. A. Klenke et al. (eds.), *Human-Wildlife Conflicts in Europe*,
Environmental Science and Engineering, DOI: 10.1007/978-3-540-34789-7_14,

# 1 Rationale and Objectives

This module aims to identify the legal and institutional framework that is relevant for the reconciliation of human-wildlife conflicts. Here "institution" is mostly understood in a narrow sense as referring to organized institutions with specialized personnel, acting bodies and budget. In a broader sense, it can also refer to the laws, regulations and norms relevant to the conflict. The identification of the legal and institutional framework is important, because these may impose severe restrictions for reconciliation policies, but also because it opens avenues of action. These frameworks do not only set conditions for reconciliation; preparation of a reconciliation action plan may also aim to influence the legal and institutional setting by recognizing needs and opportunities for adjustment. Thus, the objective of this module is to identify the existing framework as well as the needs and opportunities for developing it.

This module sets an indispensable basis for modules 5 (Regional economic and policy analysis) and 9 (Development of policy instruments). A basic knowledge of the legal and institutional framework is very important for all aspects of conflict reconciliation and should be gained as early as possible. Conflict managers need a thorough overview of the basic legal framework and relevant institutions, for example to identify further relevant stakeholders to be interviewed as part of module 6 and to understand and categorize their answers relating to different policy instruments. Results are also relevant to inform modules 7 (Ecological mitigation) and 8 (Management and viability of target species) regarding legal or illegal technical mitigation or population management measures.

# 2 Methods and Approaches

In contrast to the other modules, module 4 does not include a minimum, standard and advanced approach. This is due to the fact that a basic identification of the relevant legal and institutional framework is an indispensable step for any conflict reconciliation approach and further differentiation only takes place in following modules. Therefore, the objective of this module is to provide a brief overview of potentially relevant legal and institutional frameworks.

The legal and institutional framework involves governmental levels from local to international as well as different policy sectors from nature conservation and hunting to trade and competition. The number of governmental levels and policy sectors relevant for conflict reconciliation may vary greatly, depending on the conflict and national politico-administrative traditions. Usually, upper or centralized governmental levels set the basic framework regulation through, e.g., international treaties, European Union (EU), or national regulation. It depends on the administrative structure of a country, which administrative level is responsible for implementing these regulations respectively designing reconciliation policies.

In centralized states, national and local administrative bodies may be most relevant; in federalist countries, an intermediate state level, such as the Bundesländer in Germany and Austria or the Provinces in Italy, additionally comes into play for authorized law-setting. Identification of the relevant legal and institutional framework can normally be done on the basis of documents and, if needed, interviews. It is a task that many people can do, though previous knowledge about the legal-administrative traditions is of immediate help.

In non-EU-countries, the structure and content of regulation may differ, though most of the legal and institutional frameworks have similar functions in one or another form. The integration of EU-countries means that many areas of regulation are much more developed at EU level than is generally the case at the international level. All EU Directives have to be implemented by the member states and non-compliance can result in conviction by the European Court of Justice. Therefore, the relevance of the Habitats and Birds Directives for reconciliation policy in the EU member states tends to be greater than relevance of other international conventions and treaties.

It is useful to approach the legal and institutional framework from the perspective of policy instruments to understand which instruments reconciliation policy could employ. Typically, a reconciliation policy consists of three categories of policy instruments involving a broad variety of measures that mostly aim at minimizing damage either to economic activities or the wildlife species:

1. "*Command and control*" are regulatory instruments that oblige regulated persons or groups to act in a certain way required by authoritative decision. Most important measures in terms of conflict reconciliation relate to the management and control of protected species (module 8) that are usually implemented by way of nature conservation regulation.
2. *Economic instruments*, which aim at reaching a certain goal through an economic or financial incentive, while allowing individuals more flexibility in the choice of action courses. Financial support for technical measures (module 7) and compensation payments are the most important measures (modules 5 and 9).
3. *Action directed at civil society*: These measures are based on information that "uses the transfer of knowledge, communication of reasoned argument, and persuasion as means of influence" (Vedung 1997). Training, education, spreading information or the creation of a stakeholder forum are measures belonging to this category (module 10).

The use of each category of instruments is linked to legal regulation in two ways: legal regulation at various governmental levels provides options for or restricts the use of different kinds of measures and often the use of measures is linked to the adoption and implementation of new legal regulation. A species management or a reconciliation action plan itself can be seen as a meta-instrument, i.e. an instrument of instruments that defines how other instruments and measures are used or might be used.

The involvement of stakeholders is also defined in the legal framework. In most states governed by law, those directly affected by administrative decisions (e.g., single citizens, associations, NGOs, or other administrative bodies) have a right to be heard during the process of decision-making. There are also permanent participatory arrangements for management issues, e.g., management or advisory boards in parks. In addition, during preparation of policy decisions and in designing policy instruments there are arrangements for participation, for example in the form of public hearings and different kinds of committees or councils. The issue of participation and the role of non-governmental actors are discussed in more detail in modules 6 and 10.

**Box 1** Property rights and the management of natural resources

A relevant institutional feature regarding the use and management of natural resources is property rights. For instance, for many fisheries operations one needs a permit. This stems from user rights of water areas. Often it is the authorities that give permission for fish farms or setting fishing gears in certain locations, but not always. In some countries water areas are in private ownership even in coastal areas or the task of fisheries management in watercourses is delegated to private associations. From this may follow that authorities do not have a direct decision-making power in these areas. For the making of reconciliation policies or action plans, property and user rights of the land and waters constitute an institutional setting to be taken into account.

In the following, the most important policy instruments relevant to conflict reconciliation will be presented in more detail. Concerning regulatory instruments, nature conservation regulation is usually most important for setting the frame. With respect to the economic activity in conflict with species conservation, property rights regimes must not be neglected (see Box 1). Regarding the category of economic instruments, subsidies for technical mitigation measures and compensation payments will be dealt with. Concerning action directed to civil society, the focus will be on information and training.

## 2.1 Nature Conservation Regulation

Nature conservation regulation has a dual relationship to human-wildlife conflicts. On the one hand effective nature conservation regulation may contribute to conflicts; on the other hand it regulates reconciliation activities. Though international treaties and institutional arrangements still have many deficiencies with regard to conflict management, they may play some role in the management of conflicts. An example is the HELCOM agreement for the Baltic Sea. HELCOM is in a special position regarding the conflict between grey seal conservation and coastal fisheries in the Baltic Sea, since its recommendation 9/1 on seal protection (adopted in 1988) sets a common framework for all HELCOM parties. Concrete actions recommended are to ban hunting and to establish protection areas for seals. The HELCOM recommended ban on hunting has been eased in 1996 and it allows hunting for scientific purposes and to minimize damage that seals may cause.

The protection under nature conservation law is rarely absolute. Most of these laws contain mechanisms to balance different interests by way of derogation clauses. For reconciliation policy, these mechanisms are of crucial relevance because they make possible, if need be, to allow and regulate the management and control of the target species. Both basic categories of instruments of nature conservation law, i.e. protection of species and protection of areas, may turn out to be relevant.

With regard to the EU, the Habitats Directive and the Birds Directive are clearly the most important regulations for species protection. According to these Directives mitigation measures, which involve deliberate capture or killing, deliberate disturbance, deliberate destruction or taking of eggs from the wild, deterioration or destruction of breeding sites or resting places are prohibited with regard to strictly protected species (listed in Annex IV). However, a member state may allow mitigation measures, if certain conditions are met. According to the Habitats Directive, mitigation measures must fulfill two general conditions: there is no satisfactory alternative and mitigation is not detrimental to the maintenance of the populations of the species at a favorable conservation status in its natural range. Secondly, its aim must be one of the legitimate purposes listed in Article 16. Among them is the prevention of serious damage (e.g., to fisheries) and "interest of public health and safety or other imperative reasons of overriding public interest, including those of a social and economic nature" Alternatively, it is possible to allow, under strictly supervised conditions, on a selective basis and to a limited extent, the taking or keeping of certain specimens of the strictly protected species in numbers specified by the competent national authorities. The Birds Directive has slightly different derogation rules. Firstly, there is no reference in the Birds Directive to "other imperative reasons of overriding public interest, including those of a social and economic nature". Secondly, there is no direct reference to hunting, although it is possible to permit capture, keeping, or other judicious use. It is doubtful, whether protective hunting could be considered as "judicious use". If so, the only option for allowing protective hunting for reconciliation purposes would relate to the prevention of serious damage to crops, livestock, forests, fisheries, and water. The seriousness of the damage apparently should be interpreted as referring to the economic sector (such as fisheries) as a whole, not to an individual farmer or fisherman. Due to the fact that damage is often serious "only" in relation to an individual resource user, this derogation cannot always be used. Examples for derogation measures are given in Box 2. Member states must report on a regular basis on the derogations from the Habitats and Birds Directives.

**Box 2** Protective hunting, killing, and disturbance of animals in Europe

Protective hunting and other means of local damage control are used as mitigation measures in many member states. Finland and Sweden, for example, use protective hunting to mitigate conflicts between grey seal (*Halichoerus grypus*) and fisheries (see Bruckmeier et al. 2013) and Denmark, Sweden, and Italy to mitigate cormorant-related conflicts. German states either adopted cormorant regulations or use single permits for allowing the killing of cormorants (*Phlacrocorax carbo*) to avoid damage to fisheries (Thum 2004, 2005). Denmark is also oiling cormorant eggs to try to control the cormorant population at certain sites (see Jepsen and Olesen 2013). The use of such measures is not limited to these two species. Despite established practices in member states, there is an ongoing discussion of what the exact legal limitations for protective hunting or other means of disturbing or harming animals for protective purposes are. In case (C-342/05), the European Court of Justice came into conclusion that the member state is required to establish, while allowing protective hunting, that the hunting is such as to prevent serious damage within the meaning of Article 16(1)(b) of the Habitats Directive.

## 2.2 *Economic Instruments*

### 2.2.1 Compensation Payments

In many countries, various types of compensation schemes address wildlife conservation or damage caused by wildlife. Some schemes focus on the ecological service provided by the land user for wildlife conservation. This is the favored type of compensation payments, both in academic literature and also increasingly in practice (see Box 3 and Bulte and Rondeau 2005; Suvantola 2013), but requires that there is still a focus on the conservation of the target species. This is difficult, when species abundance has become very high and conflicts have erupted due to considerable damage to land users. Therefore, other schemes focus on compensating damage caused by wildlife. However, some countries have adopted a policy of generally not compensating damage caused by wildlife.

Compensation may be granted either after damage has occurred (*ex post facto*) or before. In theory, ex post facto compensation enables to adjust compensation to actual loss, although in practice this is often a complicated matter (module 3). Uncertainty stresses the importance of procedural aspects of assessment (involvement of stakeholders—module 6), which is relevant for the acceptability of the outcomes of the assessment. Future damage can be compensated through specific instruments solely designed for this purpose, or as a part of other support mechanisms for fisheries or other economic sectors concerned (Schwerdtner and Gruber 2007).

Partial compensation is a common approach and advisable for economic reasons. Otherwise, there is no incentive for the resource user to engage in damage mitigation activities. Therefore, a common condition to be eligible for compensation is that all possibilities to undertake protective measures (hunting, technical measures etc.—module 7) have been used. Most schemes only cover primary losses, i.e. the resources eaten by the wildlife species. Few schemes also cover secondary loss, such as injured fish.

In many cases it is very difficult to know exactly how much of a resource is eaten or damaged by wildlife (module 3). One reaction to this problem could be to pay a lump sum for each resource user in a certain category, although this approach does not seem to be common. If the presentation of full evidence of occurrence and amount of loss is required, it will lead to substantial practical problems and frustration. In that case the economic instrument does not necessarily contribute to reconciliation of a conflict. In practice, full evidence is seldom required.

**Box 3** Compensation payments in Saxony, Germany

The Saxon Support Program for Environmentally Sound Agriculture is based on the European Agricultural Guidance and Guarantee Fund (EAGGF). The program includes funding of agri-environmental measures to protect threatened, historically valuable fish ponds. The Saxon aquaculture has been supported by this program with 2.3–2.6 million € between 2000 and 2002, covering more than 95 % of the Saxon pond area used for carp (*Cyprinus carpio*) farming. One measure specifically aims at reconciling the conflict between fish farming and otter (*Lutra lutra*) conservation. Pond farmers are paid a fixed amount per hectare for surplus stocking of ponds to compensate for otter-induced mortality, a measure unofficially named "otter bonus". In this way, farmers provide feeding habitat for otter. This is regarded as an "ecological service" to society. The state of Saxony also pays for damage caused by wild animals (including cormorants and otters). The damage compensation scheme is funded from the state budget without any European support. Payments require significant damage. In agriculture and fisheries, the annual eligible damage per applicant has to exceed 100 € per hectare and 1,000 € in total. In forestry, a minimum of 50 € per hectare and 50 € in total have to be reached for eligibility (Thum et al. 2003).

In the EU, competition law regulates national compensation schemes. This means that EU state aid rules must be considered when designing compensation schemes (see Box 4). In addition, there are special rules for fisheries production and trade in fisheries products. The Commission has adopted guidelines for the examination of state aid to fisheries and aquaculture (OJ 2008/C 84/06). State aid has to be notified to the Commission and then the Commission will consider whether it can be allowed or not. Article 87 of EC treaty provides an exhaustive list of grounds on which state aid is either always compatible or may be considered to be compatible with the common market.

**Box 4** Compensation of loss of catch in Finland: a conflict with state aid

A compensation scheme for seal-induced loss of the catch of professional fishermen was first planned as a permanent compensation scheme to be funded purely from national sources. However, on the basis of the European state aid law, the European Commission could accept only a temporary compensation scheme for 2001 and 2002. The Commission stressed in its decision that the compensation scheme was allowed due to its temporary nature and due to the fact that the occurrence of damage was considered an exceptional event (unexpected increase of seal population). The Commission noted that according to legal practice, compensation paid for damage caused by public bodies to private persons is not state aid. However, it claimed that in this case the nature of compensation was different. Taking into consideration that it had trade effects, it was to be considered state aid.

One must note that EU state aid regulation refers to funding from national or sub-national sources. Funding mitigation measures from European structural funds is not considered as state aid (see e.g., Box 4). It is also worth noticing that aid to research, when it complies with the community framework for state aids for research and development, and aid to experimental fishing projects provided that its aim is the conservation of fisheries resources and they implement more selective techniques, may be deemed to be compatible. An additional condition is that the amount does not go beyond what is strictly necessary in order to achieve the purpose of aid and the aid is of limited duration (for a detailed analysis see Similä et al. 2006).

### 2.2.2 Subsidies for Technical Development

Subsidies for technical measures aim to foster technical development and management practices, which prevent or reduce damage. The same source of funding may be used for promotion of technical development and diffusion of innovations, but these two functions should conceptually be kept apart. The developers and those adopting innovations are not necessary the same people.

In many countries subsidy schemes are used to support investments in technical mitigation measures (Box 5). These may be administered by the authorities responsible for the economic activity in question or, as in some countries, by the nature conservation or hunting authorities.

In addition to national funding in many EU member states also European structural funds are used for conflict mitigation purposes, for instance to develop and spread technical measures.

**Box 5** Damage compensation and support to technical measures in Sweden

Under its nature conservation legislation Sweden has an ordinance concerning damage caused by wild animals. It outlines the possibilities to grant economic compensation to persons affected by damage to gear, stock, catch, or person caused by wildlife. The compensation can be given as means to prevent further damage or as actual economic compensation for values already lost. As a consequence, Sweden has compensated losses, but also supported research and development of technical measures to prevent seal damage on fishing. In addition, Sweden has also subsidized fishermen's investments on technical measures, especially trap-nets that are not vulnerable to seal damage (Bruckmeier et al. 2013).

## *2.3 Information and Training*

Information and training instruments are the least coercive of policy instruments. It can mean general distribution of information or it can be more targeted training of relevant individuals or groups of people. Information about technical solutions and good practices to mitigate human-wildlife conflicts effectively enhance the diffusion of them. Also basic information about the species in question may help in conflict mitigation. Training of hunters is one example of information-based instruments. If hunting is used as a measure to mitigate the conflict, hunters can be trained to hunt in a more efficient manner. The hunting of the species concerned may be difficult and a certain method of hunting may give better results than others. The need for training may also stem from previous hunting prohibitions, which have interrupted hunting traditions. Similar kinds of training can aim at improving the resource users' know-how about the use of technical mitigation measures.

# References


Bruckmeier K, Westerberg H, Varjopuro R (2013) Reconciliation in practice: the seal conflict and its mitigation in Sweden and Finland. In: Klenke R, Ring I, Kranz A, Jepsen N, Rauschmayer F, Henle K (eds) Human-wildlife conflicts in Europe—fisheries and fish-eating vertebrates as a model case. Springer, Heidelberg, doi:10.1007/978-3-540-34789-7_3

Bulte EH, Rondeau D (2005) Why compensating wildlife damages may be bad for conservation. J Wildl Manag 69:14–19

Jepsen N, Olesen T (2013) Cormorants in Denmark—re-enforced management and scientific evidence. In: Klenke R, Ring I, Kranz A, Jepsen N, Rauschmayer F, Henle K (eds) Human-wildlife conflicts in Europe—fisheries and fish-eating vertebrates as a model case. Springer, Heidelberg, doi:10.1007/978-3-540-34789-7_8

Schwerdtner K, Gruber B (2007) A conceptual framework for damage compensation schemes. Biol Conserv 134:354–360

Similä J, Thum R, Varjopuro R, Ring I (2006) Protected species in conflict with fisheries: the interplay between European and national regulation. J Eur Environ Plan Law 3(5):432–445

Suvantola L (2013) The Golden eagle compensation scheme in Finland as an example of incentive measures—potential for conflict management? In: Klenke R, Ring I, Kranz A, Jepsen N, Rauschmayer F, Henle K (eds) Human-wildlife conflicts in Europe—fisheries and fish-eating vertebrates as a model case. Springer, Heidelberg, doi:10.1007/978-3-540-34789-7_10

Thum R (2004) Rechtliche Instrumente zur Lösung von Konflikten zwischen Artenschutz und wirtschaftlicher Nutzung natürlicher Ressourcen durch den Menschen am Beispiel Kormoranschutz und Teichwirtschaft. Nat Recht 9:580–587

Thum R (2005) Zur Rechtmäßigkeit so genannter Kormoranverordnungen. Agrar- und Umweltrecht 5:148–152

Thum R, Schwerdtner K, Ring I (2003) Artenschutz und Teichwirtschaft. Rechtliche und institutionelle Rahmenbedingungen in Deutschland unter besonderer Berücksichtigung des Freistaates Sachsen. UFZ Discussion Papers 11/2003, UFZ Centre for Environmental Research, Leipzig pp 1–48

Vedung E (1997) Public policy and programme evaluation. Transaction Publishers, New Brunswick

# Module 5: Regional Economics and Policy Analysis

**Rui Santos, Paula Antunes and Irene Ring**

**Abstract** This chapter describes the module Regional Economics and Policy Analysis of the FRAP framework. This analysis aims to derive an understanding of the regional socio-economic context underlying a particular conflict between biodiversity conservation and economic activities and to study the role of policy instruments, which are in place (or were used before) to deal with the conflict. This information is critical to understand the conflict in the study area, as well as the reasons for the success or failure of the adopted policies. Lessons derived at this stage are fundamental for the development of new policy instruments, or for the improvement of existing instruments. The chapter identifies the main issues that should be addressed in the regional economics and policy analysis, presenting guidelines and methods that may be used for the different tasks. The approaches to be used at three different levels of analysis—minimum, standard and advance—are summarized in the end of the chapter.

R. Santos (✉) · P. Antunes
CENSE—Centre for Environmental and Sustainability Research,
Faculdade de Ciências e Tecnologia, Universidade Nova de Lisboa,
Campus de Caparica, 2829-516 Caparica, Portugal
e-mail: rfs@fct.unl.pt

P. Antunes
e-mail: mpa@fct.unl.pt

I. Ring
Department of Economics, UFZ—Helmholtz Centre for Environmental Research,
Permoserstraße 15, 04318 Leipzig, Germany
e-mail: irene.ring@ufz.de

R. A. Klenke et al. (eds.), *Human-Wildlife Conflicts in Europe*,
Environmental Science and Engineering, DOI: 10.1007/978-3-540-34789-7_15,

# 1 Rationale and Objectives

"Regional economics and policy analysis" is the tool to describe and analyze the regional socio-economic context, with a particular focus on the direct and indirect economic and social importance of the activities related to the conflict (Santos et al. 2005). It also serves to identify the role of relevant past and present policy instruments, which have been used to deal with the conflict. Such information is critical to understand the importance and evolution of the conflict in the study area, focusing mainly at the regional level, as well as the reasons of success or failures of the adopted policies. Two types of analysis should be undertaken to accomplish these objectives (Barbier 1994; Development Bank of Southern Africa 2001; Karanth and Gopal 2005):

- a "regional economic analysis" describing the study regions and assessing the role that economic activities and biodiversity conservation play within the regional context;
- a "policy analysis" to describe the past use and present implementation of instruments to address the conflict between economic activities and protected species.

The results of these assessments are of special relevance for the development of improved or new instrument mixes (module 9), as well as to support the definition of participatory strategies and the development of reconciliation action plans. It aims, in particular, to present policy recommendations on how to design these tools for increased local effectiveness (Gutman 2003; McNeely 1988; Nyhus et al. 2005; Western and Waithaka 2005).

# 2 Methods and Approaches

## *2.1 Regional Socio-Economic Analysis*

The socio-economic analysis aims to achieve the following objectives:

1. to characterize the region in economic and social terms, such as the general economic structure, population, income situation, and labor market characteristics;
2. to evaluate the economic and social importance of the activities related to the conflict (e.g., fisheries) in a regional context;
3. to evaluate the economic and social impacts of potential conflicts, such as potential production losses, and to clarify whether the conflict is marginal or of prime interest for the region.

A minimum approach should be primarily based on the collection of published and available statistical data. A set of indicators for socio-economic characterization of the study region has to be established, including, when appropriate, country/case specific indicators. It is important to select indicators that focus on the general economic structure (e.g., gross domestic product—GDP), the economic sector(s) related to the conflict (e.g., for aquaculture: number and size of fish farms regarding area and employees, legal status, production, productivity, relevant market, competitiveness, business constraints and opportunities), human population (e.g., population density, growth pattern, age structure), income situation (e.g., sources of income), and labor market characteristics (e.g., employment).

In a standard approach, this data analysis may be complemented with literature reviews and expert interviews, mainly when data availability and reliability are not assured. Expert knowledge about specific local socio-economic contexts, uncertainty, and complex ecological frames may be incorporated. It is important to broadly cover the past development, the actual situation, and, when possible, future perspectives of the conflicting sectors (e.g., fisheries, biodiversity, and, where applicable, conservation based recreation and tourism). In a more advanced approach, this analysis may be complemented by the realization of fieldwork, namely interviews, for additional data collection.

An adequate time frame must be adopted. When possible, at least the past 10 years should be covered. The selection of the spatial scale of analysis is also an important issue. The administrative structure, as well as the level at which statistical data are produced in each country should be considered when deciding on area size. In the EU, the collection of data for the NUTS (Nomenclature of Units for Territorial Statistics) levels, from the national to the local level, allows identification of possible particularities of the focus region. The ecological dimension should also be considered in the selection of the spatial scale (it depends, for example, on the species involved and ecosystem characteristics—see module 2), since the integration of both is a key issue in the development of reconciliation action plans.

The integration of the socio-economic analysis with information from ecology is very important at this stage, in order to evaluate the importance of the conflict, namely the economic damage (e.g., fishing—stock size, population size of wildlife, diet and daily consumption, consumption from fishing gear, impact on fish stocks; compare modules 2 and 3), to help reducing misunderstandings and to support the design of more effective and efficient policies. This information may not always be available as early as desired and with the necessary level of detail—this issue stresses the importance of an adequate planning of the tasks to be performed both in the socio-economic and ecological analyses.

Usually, the main constraints encountered in socio-economic analyses are related to difficulties in getting reliable data, mainly at a local and regional level, for relevant variables. This may be especially the case for the economic activities or specific sector(s) related with the conflict. The conflicting economic activities

**Fig. 1** Presentation and discussion of first results about regional economy of the human-otter conflict in Germany. *Photo:* Jaroslav Myšiak

may only represent part of an economic sector for which data are available in official statistical data tables or data may only be available at higher, more aggregated levels, such as the state or national level. It is relevant to avoid an unbalanced analysis of the different variables due to a bias introduced by data availability and time constraints in fieldwork. It can easily drift towards a mainly descriptive analysis regarding key aspects (e.g., the economic impact of the conflict) and quite detailed for less relevant variables, but where more information is available (e.g., employment).

It is also important to balance the effort of collecting and analyzing socio-economic data, with its potential contribution to an improved understanding of the conflict and of the constraints to the implementation of effective policy instruments. Sometimes, the use of too many indicators does not bring a significant added value to the analysis. The collection of opinions from key stakeholders that can stress specific aspects, which are relevant for understanding the situation, may be more useful (Fig. 1). It is suggested that the data collection, including published information or fieldwork, should only start after a careful identification of the key variables, with the support of relevant stakeholders.

The analysis of the socio-economic context is very important to explain differences in the policy approach, in stakeholders' attitudes, or in the performance of past and present instruments and measures. In many situations there is no need, or no data available, to apply sophisticated methods for the regional economic analysis (e.g., input–output analysis, econometric analysis). A broad and sound characterization and analysis of the regional economic context, namely the economic sectors in conflict with the target species, is an important contribution to the assessment phase and provides relevant knowledge for conflict reconciliation.

## 2.2 Policy Analysis

The preliminary policy analysis is developed in this module, with the objectives to identify aspects that require or justify further detailed analysis and to provide relevant information to improve existing instruments or to develop new ones in the reconciliation and implementation stages. In general, it describes the policy instruments addressing the conflict, including more or less substantial data collection on their regional implementation. In this context, all the past and present policy instruments relevant to the conflict should be identified and studied. The contribution from relevant authorities is of particular importance for this task. The information from module 4 (Legal and institutional framework) is relevant, mainly for the selection of key entities. The classification of identified instruments should follow the general categories presented in module 4, namely command and control or regulatory instruments, economic instruments, and action directed at civil society.

Prior to starting data collection, the key criteria for policy evaluation have to be decided upon, both in terms of the preliminary and qualitative evaluation as part of this module and to prepare the detailed evaluation for the improvement or design of new policy instruments in module 9.

For each instrument a set of aspects should be addressed, according to their relevance to the conflict, including:

1. descriptive aspects: e.g., type of instrument, functioning, objectives (direct or indirect contribution to conflict mitigation), starting date, authority responsible for the initiative, target group and geographical scope, existing payments/financial transfers according to adequate spatial and temporal scales, beneficiaries, allocation of resources over time, enforcement and monitoring;
2. present implementation status and foreseen evolution: e.g., currently applied, in the pipeline, in study, abolished, reintroduced, to be substituted;
3. information sources;
4. preliminary and qualitative evaluation, e.g., has the instrument contributed to conflict mitigation? What are the main limitations/constraints? Are there different options for its application?

A minimum approach corresponds here to data collection necessary for the qualitative/quantitative preliminary evaluation of the instruments according to different basic criteria. These could be: effectiveness (success in terms of goal attainment); efficiency [in terms of cost-effectiveness (Wätzold and Schwerdtner 2005)];

**Table 1** Minimum, standard, and advanced approaches in module 5

| Tasks | Approaches | | |
|---|---|---|---|
| | Minimum | Standard | Advanced |
| Socio-economic characterization of the area | Published statistical data on standard socio-economic indicators | Other literature and reports<br>Interviews with key actors<br>Integration of specific indicators relevant for the study area | Fieldwork for data collection |
| Economic damage assessment | Expert guess<br>Study available literature | Integration of available knowledge from ecology with market values | Detailed ecological field data (densities, diet, behavior…) integrated with detailed information on supply and demand sides |
| Policy analysis | Definition of key evaluation criteria for policy evaluation to inform data collection<br>Published data<br>Qualitative description of past and present implementation of a broad range of instruments | Refinement of evaluation criteria<br>Interviews with key stakeholders<br>Expert knowledge<br>Add quantitative description of past and present implementation of instruments | Detailed data collection regarding spatial and temporal characteristics of policy instruments<br>Adjust data collection to needs for integration with ecological studies<br>Collection of most relevant quantitative data to validate key or controversial aspects of the conflict |
| Recommendations for policy development | Identification of instruments with higher potential for conflict mitigation | Develop general recommendations for the development of new instruments or for improvements in existing policies | Develop specific recommendations for in-depth evaluation and development of policy instruments |

management and implementation requirements (e.g., enforcement needs, administrative requirements, skills and competencies of actors involved), involvement of stakeholders (e.g., participation in the formulation or design, implementation, and evaluation of instruments), perception and fairness, and acceptance among stakeholders. The data to support this preliminary analysis should preferably be obtained from published reports. Recommendations for policy development are primarily of a qualitative nature.

A standard approach covers further evaluation criteria such as dynamic efficiency and innovation (e.g., incentives to the development of new solutions and continuous improvement) or legal compatibility and articulation of instruments with existing

**Box 1** Questions to be Answered

In the development of the socio-economic and policy analyses, the following questions should be considered:

- Are the economic activities likely to intensify or reduce the conflict? May the conflict shift to other activities, resulting in a different regional socio-economic relevance?
- Has financial support been provided to the people suffering damage? Has this support contributed to increase or decrease the conflict?
- Is the importance of the conflict or of the conflicting sectors different at the international, national, regional, and local levels?
- What is the experience in dealing with the conflict, or similar conflicts? What impact does it have in the adopted policy approaches?
- Is it possible to establish a clear link between the socio-economic relevance of the conflict and the existing policy instruments?
- Are economic incentives applied? Are the financial and compensation mechanisms based on damage assessment or ecological criteria? Are they intended to give an incentive to the provision of ecological services, promoting activities that otherwise would not be competitive? Are there funding problems?
- Do the instruments have a solid legal and institutional background? Were they specially designed to mitigate the conflict? Are existing institutions expected to be able to contain and manage the conflict? Are the instruments adapted to the conflict characteristics and the socio-economic context?

The following questions are relevant for informing modules 9 and 10:

- Is it possible to obtain some improvements with minor adjustments to existing instruments?
- Is there a policy mix, following an integrated approach, to deal with multiple objectives and interests? Is the role of each instrument clear? Were they planned to work together? Are there overlapping instruments?
- What barriers and wrong signals, which may be increasing the conflict or counteracting/hindering existing instruments, should be assessed and eliminated, before applying new instruments?
- How can policy integration be improved across different spatial scales, from the EU to the local level?
- Is there any recommendation regarding the characteristics of enforcement requirements? Is progressive implementation a relevant issue?
- Have the perceptions and views of the different actors been taken into account in policy formulation and implementation? If not, why? What has failed?
- Are there relevant data failures, justifying the change of reporting strategies and the implementation of new *ex ante* and *ex post* evaluation mechanisms?

policy frameworks at different levels (regional, national, and international). It requires the collection of quantitative data on the spatial and temporal implementation of the instruments, obtained from published data and complemented by interviews with the involved administrative entities or informal meetings with stakeholders. An in-depth data collection should be assured in an advanced approach, considering for instance the needs for integration with ecological studies and the assessment of key aspects of the conflict.

Table 1 provides guidelines for the regional economic and policy analysis for the different approaches: minimum, standard, and advanced. Box 1 contains a checklist of questions that can be used to guide the socio-economic and policy analysis tasks.

In any case, the policy analysis should provide a preliminary evaluation of the instruments and support the selection of those, which have the potential to be presented as best practices or to be improved and adopted as part of the policy mix in the reconciliation action plan.

As data on policy instruments may not always be available in published statistics, and obtaining them directly from relevant authorities or actors is time consuming, it is important to carefully plan this activity and to allocate enough effort to it. This is another reason why early contact with authorities is important. Statistical data are often biased (sometimes for strategic reasons), particularly at the local level; the support from different stakeholders is important to avoid or mitigate this information bias.

The collection of information regarding the use of instruments in similar conflicts with other species or about experiences with instruments in other regions or countries can also be helpful in the assessment phase and for the development of instruments in the reconciliation phase.

## References

Barbier EB, Burgess JC, Folke C (1994) Paradise lost? The ecological economics of biodiversity. Earthscan, London

Development Bank of Southern Africa (2001) Guidelines to regional socio-economic analysis, Development Paper 145, Development Information Business Unit, ISBN 1-919692-58-4

Gutman P (ed) (2003) From goodwill to payments for environmental services. WWF Macroeconomics Program Office, Washington

Karanth KU, Gopal R (2005) An ecology-based policy framework for human-tiger coexistence in India. In: Woodroffe R, Thirgood S, Rabinowitz A (eds) People and wildlife: conflict or coexistence?. Cambridge University Press, Cambridge, pp 373–387

McNeely JA (1988) Economics and biological diversity: developing and using economic incentives to conserve biological resources. IUCN, Gland

Nyhus PF, Osofsky SA, Ferraro P, Madden F, Fischer H (2005) Bearing the costs of human-wildlife conflict: the challenges of compensation schemes. In: Woodroffe R, Thirgood S, Rabinowitz A (eds) People and wildlife: conflict or coexistence?. Cambridge University Press, Cambridge, pp 107–121

Santos R, Madruga L, Antunes P (eds) (2005) Framework reconciliation action plan: Module regional economics and policy analysis. FRAP project report. Deliverable 8. ECOMAN Centre, Lisbon

Wätzold F, Schwerdtner K (2005) Why be wasteful when preserving a valuable resource? A review article on the cost-effectiveness of European biodiversity conservation policy. Biol Conserv 123:327–338

Western D, Waithaka J (2005) Policies for reducing human-wildlife conflict: a Kenya case study. In: Woodroffe R, Thirgood S, Rabinowitz A (eds) People and wildlife: conflict or coexistence?. Cambridge University Press, Cambridge, pp 357–372

# Module 6: Stakeholder Analysis

**Douglas Clyde Kongshøj Wilson**

**Abstract** Reconciliation Action Plans are political instruments that are implemented in political contexts. Getting a good grasp of who the relevant stakeholders are is critical to their effectiveness. A discourse analysis is a method for identifying stakeholders, understanding where they are in conflict, and identifying potential compromises among them. It investigates how facts, values and interests are linked together into coherent themes. These themes are drawn upon in various ways by different stakeholder groups as they express their understandings of the issues that the RAP must address. A discourse analysis always begins with in-depth qualitative interviews with stakeholders through which the themes are initially identified. Several more advanced methods are available, including quantitative ones that can be used to understand more complex discourses.

## 1 Rationale and Objective

This module describes who the local stakeholders in the conflicts are, what they believe, what they want, and what they might be willing to give up. This is critical if the development of policy instruments, the participatory decision strategies, and the framework of action plans are to respond meaningful to the political and social realities in which mitigation options are created and implemented. The "local" level always begins with the individual and community levels where behaviors take place, but what constitutes the upper bound of the "local" will vary among different conflicts.

D. C. Kongshøj Wilson (✉)
Department of Development and Planning, Innovative Fisheries Management,
An Aalborg University Research Centre, Nybrogade 14, 9000, Aalborg, Denmark
e-mail: dw@ifm.aau.dk

R. A. Klenke et al. (eds.), *Human-Wildlife Conflicts in Europe*,
Environmental Science and Engineering, DOI: 10.1007/978-3-540-34789-7_16,

**Fig. 1** Scientists interviewing a fisherman at the bank of a pond about damages caused by Eurasian otters and great cormorants. *Photo:* Irene Ring

A meaningful response will be best facilitated if both the identities and desires of the local stakeholders are understood in the terms and categories they themselves are using. The group implementing a reconciliation strategy or reconciliation action plan (RAP) should not assume that they, or even the key informants they interviewed during the screening (module 1), know who all the stakeholders are; let alone how the stakeholders see themselves and their problems. Above all, they should not make assumptions about which groups do or do not have a legitimate stake.

This module should be thought of as the "listening module" (Fig. 1). It is the time when those who wish to facilitate the creation of a RAP listen to the stakeholders and learn how they see themselves and their problems. It informs those responsible for facilitating the RAP about the best ways to approach the different stakeholder groups, the kinds of information they will be interested in, and where there is a potential for compromise on contentious issues. The main product of the module is a systematic description of the facts, values, and interests that the different stakeholder groups bring to the process of RAP creation.

# 2 Methods and Approaches

## *2.1 Theoretical Background*

The scientific approach used in this module is the discourse analysis (Hajer 1995; Phillips and Hardy 2002). Discourse analysis is more effective than alternative methods for facilitating a participatory process because, rather than focusing on reconstructing beliefs and attitudes that are believed to exist in the actors' heads, they focus on what actors are saying to one another. Discourse examines the shared and divergent understandings of social reality that form the background of both verbal and physical behavior. Most alternative socio-logical approaches focus on competition between either individuals or groups. While both are critically important for understanding the political context of the RAP, these theoretical perspectives achieve their insights by overemphasizing the competitive aspects of society and underemphasizing the cooperative aspects.

Discourse analysis is well suited for research in support of the development of a RAP or reconciliation strategies in general. Because of the sheer number of human interactions, most methods focused on interactions, e.g., network analysis, are very time consuming. Discourse analysis can be done very thoroughly, but it can also be done simply by holding a series of interviews and then intuitively arranging the results into themes. This gives only a limited picture of the social and political realities but for most RAP situations this will provide important information from a relatively small investment of time and resources.

Discourse analysis focuses on existing and potential communications among stakeholders. In particular, it is interested in the intersection of facts, values, and interests in the things that people say about the conflict. Understanding how people link these three things makes a number of contributions. On the side of facts, the discourse analysis helps to reveal where the presentation of validated scientific information may aid in reconciliation. It may also identify real disagreements that may benefit from, or even require, further research that will lead to potential solutions within the RAP. On the side of values, the discourse analysis helps to uncover those areas where compromise may be blocked by strong beliefs. On the side of interests, it helps to illuminate where compromises and/or *win–win* outcomes are possible.

The product of the discourse analysis is a description of the discursive themes. Themes are repeated patterns in which facts, values, and interests are linked in the same way by participants in the discourse (see Box 1. Discursive themes).

**Box 1** Discursive Themes

An example of a discursive theme that emerged in the Swedish seal-fishery conflict case analyzed by the FRAP project (Bruckmeier et al. 2012) was that "compensation payment for damage to gear or catch is a component of the Swedish mitigation method". Stakeholders would express this theme in as a valuing statement, such as "compensation payment for damage to gear and catch is an efficient form of mitigation because it helps the fishermen to continue their profession". Such statements combine assertions of fact "fishers are economically threatened" with assertions of interests "help the fishermen" and values "efficiency".

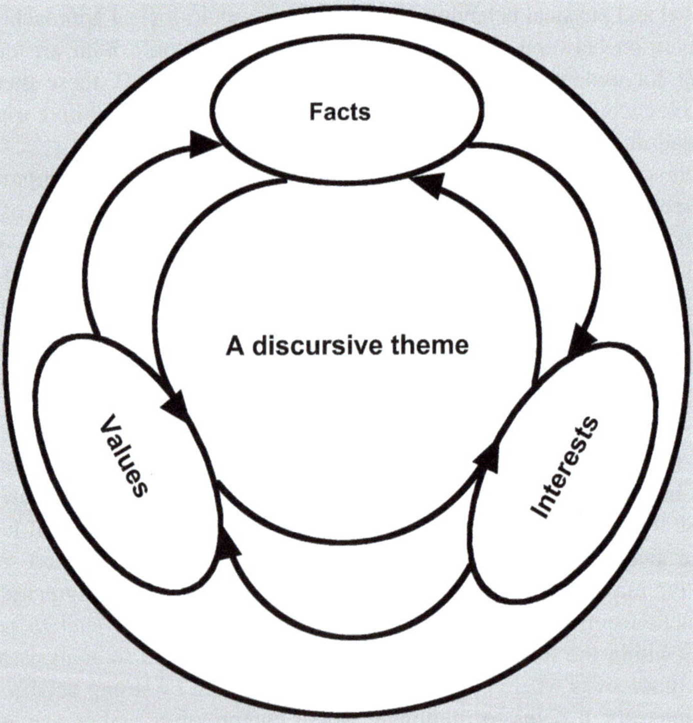


**Fig. 2** The parts of a discursive theme

Themes are interpretations of the interplay among facts, values and interests (Fig. 2). Themes cannot be given precise boundaries and different ways of describing themes can be valid. The only meaningful test is the degree to which participants see the themes as a coherent picture of their discussions. Some groups will agree with particular themes, some groups will disagree with the themes, but all the groups should be familiar with the themes as recurring parts of the discourse. Themes that are drawn on by only one group reflect world views and may be very limited in the degree to which they reflect shared understandings, while other themes that are drawn on by many groups may reflect shared understandings that can be the basis of conflict management and compromise.

## 2.2 The Definition of a Stakeholder

A stakeholder is any group that can influence the content or effectiveness of reconciliation strategies or an action plan. This definition includes government agencies at various levels, environmental groups, resource users, and local businesses, such as those related to tourism, etc. Stakeholders can wield such influence either by participating in the creation of the RAP or by helping or hindering its implementation. In other words, if an environmental group in a distant city has the possibility of having influence on the content of the RAP then that group is a stakeholder. Furthermore, if user groups have a possibility of violating the rules of a RAP then they are stakeholders because they can have a negative impact on its effective implementation, even if they did not have any influence on its content.

## 2.3 Minimum Required Steps

### 2.3.1 Regional Level Interviews

The module begins with interviewing people working in the region where the conflict is found. A set of initial interviews has already been done in the screening process in module 1. If those interviews have already mapped the main stakeholder groups in the region, then this module can move directly to the "local" level. In a more complex conflict further interviews on a regional level may be advisable to ensure that no necessary stakeholder group is being overlooked. These interviews provide more information about the nature of the conflict. They document the positions, of the various agencies and NGOs. Most importantly they identify the primary local stakeholder groups.

The main types of stakeholders in ecological conflicts are usually:

1. Conservationists;
2. Consumptive user groups and related industries;
3. Non-consumptive user groups and related industries;
4. Government officials from various agencies.

However, these categories can hide many differences. Conservation groups have many different priorities. Very large differences in interests can exist between consumptive user groups who use one technique versus those who use another. Different government agencies have different mandates. It is important that the investigator identifies the actual critical stakeholder groups at this more detailed level. The number of such groups will depend on the actual situation and may vary considerably between different issues. The first job of the investigator after the regional interviews is to decide which different groups have to be involved in the discussions around the RAP. This decision must remain open to changes and redefinition during the entire module as more will be learned about differences during the interviews.

The second job of the investigators at this point is to define what is meant by "local". The purpose of this module is to understand the local level political realities that will have an impact on the success of the RAP. "Local" should not be taken in this discussion to mean a predefined geographical or administrative area. What constitutes the upper bound of "local" in respect to a particular RAP should be defined by both the ecological, geographical, and social landscapes over which the conflict is taking place. The local is the space where the members of the stakeholder groups are operating and interacting with one another. An important question in determining what is local would be to whom the stakeholders complain about the conflict. Another will be the area over which members of a particular stakeholder group meet with one another, for example a chapter of a concerned conservation organization or business group. However, it is important to realize that the lower bound of the local is always the most local community level because it is the actual behavior of individuals that determines the effectiveness and relevance of any policy.

### 2.3.2 Local Level Interviews

Enough interviews should be carried out at the local level to get a complete picture of the conflict. The best practice is to keep interviewing until you start hearing mainly things you have heard before. A good initial goal would be three interviews for each of the main types of stakeholders (1 through 4 above). It is important here that they are people with quite different roles and at least one of these three should not be a "leader". The meaning of "leadership" in each group can be very different and this will be a critical question throughout the RAP process. The assessment of leadership in the interviews should still be looked at as very tentative as the reconciliation process continues.

Interview goal 1: *Understanding perceptions of interests*

This step begins with gathering information on perceived costs and benefits and their distribution. The most important part is collating information on how the local stakeholders see these costs and benefits and their distribution. The products of this work are of use not only to understand stakeholder interests in relation to the discourse analysis, but can also provide input into a regional level analysis of the economic context of the conflict (module 5).

In particular the following issues need to be covered:

1. The perceived economic costs of the conflict (e.g., reduced fish catch due to predation);
2. The perceived economic benefits of both consumptive and non-consumptive economic activities (e.g., commercial fishing, wildlife-based tourism);
3. The perceived social costs and benefits of conservation (e.g., education, pride, more diverse environment) and the social role played by both the consumptive and non-consumptive economic activities in the life of the community (e.g., traditions, local networks, and community organization);

4. Appropriate potential mitigation strategies for the RAP (module 7).

For each of these four items the analyst needs to discover how perceptions differ among local stakeholders, how extensive these social and economic costs and benefits are from the perspective of the overall community, and how these costs and benefits are distributed among stakeholder groups.

Interview goal 2: *Understanding perceptions of facts*

The second goal is approached in concert with the natural science team and begins with the prioritization of the factual basis for a potential RAP. The emphasis here is on biological and ecological facts (modules 2 and 3), but economic facts (module 5) may also play a role. The interviews involve:

1. Examining how respondents see causal processes. A good way to do this is to ask what they see as the most important changes in the ecosystem over some appropriate time period and why they think it happened. Asking them to draw maps is a useful technique where landscape factors are important (module 2);
2. Examining which facts they believe to be the most relevant to the conflict itself. What they see as the most important "science factors" in the conflict. What they believe has caused these factors and what they think can be done about it;
3. Finally, after having examined their "undiluted" perceptions, the interview introduces the biological factors that the RAP team sees as important and asks for reactions and assessments.

Interview goal 3: *Understanding stakeholders' values*

The final goal is to understand the values that stakeholders attach to the issues involved in the conflict. One part of this is uncovering important group identities for both the respondent and for how the respondent sees other stakeholders. It would not be effective to directly ask "what are your values," that would lead to an abstract discussion of the meaning of a "value". It would be better to use indirect approaches to this question such as the following:

1. Ask who the respondent sees as main players in the conflict and what they are after;
2. Ask how the respondent became involved personally in the conflict;
3. Ask the ways the respondent sees the conflict being resolved and what their preference would be;
4. Ask who the respondent thinks has the primary responsibility for creating the problem and why. Then ask who has primary responsibility for resolving the problem.

A number of other ways to explore the value issues will come up in the interviews. The important thing is to be alert to this issue and be willing to ask follow up questions.

# 3 Analysis

Analysis begins with writing up the notes from the interviews. From these notes a team should work together to identify the themes using the following steps:

1. On a set of cards write statements made by respondents that connect at least two of facts, interests, and values. Choose statements that are seen as relevant to the RAP and that meet one of two other criteria: (1) they are statements that are repeated by three or more respondents or (2) they are statements that the respondent felt strongly about. Continue until there are no more statements that fit these criteria.
2. As a team, group these cards intuitively into thematic categories. At first, do not worry about the number of categories or the number of cards in the categories. When you have finished the grouping then you should merge the cards until you have no groups with less than three cards.
3. Give each group a descriptive name that is an "assertion", meaning a statement that can be agreed or disagreed with. The reason is to make sure that the themes are comparable with each other and can be analyzed the same way. It is important to keep in mind that a theme is much larger than just its name.
4. Write paragraphs describing each theme. The name is a summary, it should be the most important, representative, and central assertion within a theme, but it is not the entire theme. The entire theme is the whole set of facts, values, and interests linked together and this should be described in the paragraph.

## *3.1 Standard Steps*

In some circumstances module 6 may be carried out just to provide background information for module 9 and 10 because some specific program or institution is going to carry out module 10 without involving stakeholders. This would most often be the case if there is already a functioning forum where the stakeholders are addressing these issues and the RAP would be most effectively picked up by this group rather than by trying to organize its own stakeholder group. In this case module 6 information may simply be handed over or presented to this group. If this is the case then the minimum methods described are enough and the information in module 6 can be taken directly to the resolution and implementation phase (modules 9 and 10).

# 4 Confirmation Meeting

Once the themes have been identified, they should be presented to the stakeholders you have interviewed for confirmation and reflection. As mentioned above, the only meaningful test of the validity of the themes is whether or not the stakeholders see them as a coherent representation of their discussion. Presenting them to the stakeholders makes it possible to ask this question. It also begins to put the results of the module to work, providing a way for the stakeholders to reflect on the nature of their disagreements and begin to discuss possible directions for the RAP.

One helpful way to make this presentation is as a graphic design that illuminates three important dimensions of the discourse analysis. This involves scoring all the themes you have identified along three dimensions. On each of these dimensions the scoring will have to be done intuitively, first by the team and then later by stakeholders. The dimensions can be represented by an X axis and Y axis and by colors as follows:

1. The Y axis: Frequency of Use. The Y axis is the frequency with which the theme is drawn upon.
2. The X axis: Importance for the RAP. If all stakeholders see the theme as a minor one, it is scored low, if everyone sees it as important it is scored high, if stakeholders are divided on this point it is scored in the middle.
3. The Color Code: Degree of Consensus. The third dimension, represented by colors, is the degree to which the theme is contested.

This graphic can be used to present the discourse analysis to the stakeholders for reaction. This can be done individually or with groups. Such sessions would elicit overall reactions to the discourse analysis, if they see any important issues and positions as being left out, and if they would score any of the themes differently.

## *4.1 Advanced Steps*

In some situations, particularly those involving strongly felt conflicts the RAP may not be facilitated by pulling together stakeholders early in the process. Q-sorting (Brown 1986) would then be an alternative method for confirming the discourse analysis (Brown 1986). In this technique you ask the stakeholders in individual interviews to rate the statements in terms of agreement following a forced normal distribution. These ratings are then subjected to an analysis that reveals the positions of various groups and how strongly they feel.

A number of software packages are available that can be used to identify and name discursive themes from the transcripts of the interviews. If time and resources are available these packages should be considered. The software facilitates coding the information to identify the themes.

The most thorough approach of all would be a formal survey of attitudes toward the conflict aimed at all the people living in the area or a sub-population. This is an expensive option that must be carried out by survey professionals. This might include seeing a need for a broad educational campaign or the need to introduce policies that will only succeed with public support.

**Table 1** Steps for implementing module 6

| Tasks | Approaches | | |
|---|---|---|---|
| | Minimum | Standard | Advanced |
| Data gathering | Regional level interviews | Local level interviews | Formal survey |
| Analysis | | Confirmation meeting | Q sorts<br>Textual analysis software<br>Survey analysis |

# 5 Summary and Conclusion

Table 1 summarizes the methods to be used in module 6. The minimum approach is to identify, listen to, and understand the people who are able to have an influence on the development and/or the implementation of the measures identified in reconciliation strategies or the RAP. In the standard approach this understanding is described in a clear form and then presented to the stakeholders for reflection and feedback. More precise methods of data gathering and analysis are also available at the advanced level to address particularly complex or sensitive situations. These can be implemented where need arises.

# References

Brown SR (1986) Q technique and method: principles and procedures. In: Berry WD, Lewis-Beck MS (eds) New tools for social scientists: advances and applications in research methods. Sage Publications, Beverly Hills, pp 57–76 (Simple introduction to Q sorting)

Bruckmeier K, Westerberg H, Varjopuro R (2012) Reconciliation in practice: the seal conflict and its mitigation in Sweden and Finland. In: Klenke R, Ring I, Kranz A, Jepsen N, Rauschmayer F, Henle K (eds) Human-wildlife conflicts in Europe—Fisheries and fish-eating vertebrates as a model case. Springer, Heidelberg. doi:10.1007/978-3-540-34789-7_3

Hajer MA (1995) The politics of environmental discourse: ecological modernization and the policy process. Oxford University Press, Oxford. (*A particularly good example of a discourse analysis of an environmental issue*)

Phillips N, Hardy C (2002) Discourse analysis: investigating processes of social construction. Sage Publications, London. (*This is a guide to methods*)

# Free and Commercial Software for Textual Analysis

QSR Software makes N6 and Nvivo –the leading packages for qualitative textual analysis. http://www.qsrinternational.com. Accessed 28 August 2006

Descriptions of several alternative software products for Q sorting, including freeware. http://www.lrz-muenchen.de/~schmolck/qmethod/. Accessed 30 May 2011

# Module 7: Ecological Mitigation

**Andreas Kranz**

**Abstract** Ecological mitigation comprises removal of specimens, changing their behavior or manipulating their habitat. In the case of protected species, such measures may not be applicable or may be restricted in the one or the other way. This chapter describes existing approaches and discusses their limitations in context of protected conflict species and in context of efficacy and cost-effectiveness. It outlines minimum, standard and advanced approaches and identifies their risks. The conclusions put ecological mitigation in the wider context and stress amongst others that a thorough understanding of the biology and ecology of the conflict is essential, that traditional ecological knowledge may be misleading, that human dimensions and emotions are essential aspects and costs of different methods are rather of secondary nature.

## 1 Rationale and Objectives

The aims of this module are to help assessing the suitability, advantages, and disadvantages of ecological mitigation methods in respect to protected wildlife and in particular to identify those parameters, specific for the conflict species and resource, which determine the applicability of a given method. This implies both, the efficacy to prevent the damage and the acceptability in order not to threaten the concerned wildlife population.

A. Kranz (✉)
Institute of Wildlife Biology and Game Management,
University of Natural Resources and Life Sciences,
Gregor Mendel Street 33, 1180, Vienna, Austria
e-mail: andreas.kranz@aon.at

R. A. Klenke et al. (eds.), *Human-Wildlife Conflicts in Europe*,
Environmental Science and Engineering, DOI: 10.1007/978-3-540-34789-7_17,

Ecological mitigation strategies depend on the characteristics of the conflict species (module 2) and the specific characteristics of the resource in relation to e.g. the diet of the species (module 3). Their implementation is the result of the existing legal framework (module 4) and the actual use of available policies (module 5), as well as the traditional knowledge of the people facing the threat of their resource (modules 5 and 6). Classic ecological mitigation strategies in wildlife management are reproductive and lethal control, wildlife translocation, fear-provoking stimuli, chemical repellents, diversion, exclusion, and habitat and resource manipulation (Conover 2002).

In the case of protected large vertebrates the straight forward application of one or the combination of several of these mitigation measures often is either not acceptable for some stakeholders, it is legally prohibited or restricted in its use, or it does work only under very restrictive conditions. In many, if not most cases, typical conflict species tend to be highly adaptable, skilful, and able to some degree of learning, thereby counteracting the efficacy of the chosen mitigation strategy. The tricky goal is a mitigation technique, which has a minimum effect on the population of the conflict species and at the same time has a maximum effect in damage reduction and thus is acceptable for most stakeholders.

## 2 Methods and Approaches

Crucial parameters for choosing an appropriate mitigation measure are status and trend of the conflict species (increasing/decreasing/constant), the fraction of the population living in the conflict region, the global abundance, and the location of the conflict area relative to the distribution of the conflict species (e.g., core area or population border, essential stepping stone). In addition, biological parameters matter for the choice of suitable mitigation measures. Among them are the morphology of the wildlife species (e.g., size, distinctness of sexes, difference to similar species), its reproductive traits, and its feeding strategy (e.g., dietary preferences like mentioned in module 3, scavenger versus predator or herbivore). Likewise, wildlife behavior patterns (physical capabilities, adaptability, and habituation), its home range, and movement patterns influence the effectiveness of measures.

Ecological mitigation measures can be grouped according to their underlying principle as follows:

1. Removal of specimens, which implies reproductive and lethal control or translocation or killing of "problem" individuals;
2. Changing the behavior of individuals by fear provoking stimuli, taste aversion, or lures to distract them from (vulnerable) resources;
3. Habitat manipulation, which includes all forms of exclusion of wildlife from the resource (e.g., fencing of ponds, construction of burglarproof traps) and

changes in habitat elements to affect the quality of the habitat for the conflict species (e.g., modification of roosting or resting sites, reduction of cover);
4. Resource manipulation, which refers to changes in the management of the resource to reduce the risk of predation or damage.

## *2.1 Removal of Specimens*

The efficacy of population or problem individuum management as a mitigation technique depends on basic ecological and biological traits of the concerned species, such as reproductive rate, breeding strategy (colonial or single), mobility and movement patterns (e.g., migration of birds), activity patterns (nocturnal or diurnal), social organization (territoriality in the relevant time of the year), and hunting strategy (solitary or cooperative). It also depends on behavioral patterns, if certain "problem" individuals cause the damage (e.g., individual seals with a special knowledge for entering or destroying fishing gear).

## *2.2 Reproductive Control*

Reproductive control comprises contraception and oiling, addling, or puncturing of eggs, but also destroying reproduction sites and disturbance. Contraception for large protected vertebrates, such as elephants (*Elephas maximus* and *Loxodonta africana*), lions (*Panthera leo*), or beavers (*Castor spp.*), is still in the stage of testing and development (Conover 2002). This method faces several critical obstacles, such as affecting non-target species and a prolongation of the annual breeding season with adverse effects on energy expenditure for males and females, but also with the ultimate consequence that young are born out of their optimal season. Up to now it was successfully used for controlling some classical pest species, such as pigeons (*Columba livia f. domestica*), voles (*Microtus spp.*), and rabbits (*Oryctolagus cuniculus*).

In contrast to contraceptive methods, killing of embryos in eggs is a well-tested method. Its success depends mainly on the access to the eggs and whether the species is breeding in colonies or not. However, reduced recruitment may be compensated for by an enhanced survival of the recruited individuals by density dependency of survival (Caughley and Sinclair 1994).

Destroying breeding sites or disturbance aiming at preventing successful reproduction is an old method which is still applied by hunters and indigenous people. One of the limitations of that method are difficulties to find breeding sites, which depend on the involved species. In North America, it is applied in modern wildlife management for wolf (*Canis lupus*), when a pack settles down for breeding near live stock holders, which frequently result in considerable losses of domestic animals (Bangs et al. 2005). Module 8 provides guidelines how to assess the effects of fertility control on the population level of the wildlife species.

## *2.3 Wildlife Translocation*

Wildlife translocation may be an option in the case of problem individuals that learn to ignore deterrents and have developed a preference for the competed resources, especially in the case of a particularly high value of the "problem" individual (critically endangered species). Public approval of this method may be high; however, translocation faces many severe difficulties. It may only be a success beyond a placebo effect, if very few individuals cause problems. Grizzly bears (*Ursus arctos horribilis*), polar bears (*Ursus maritimus*), and African wild dogs (*Lycaon pictus*) are typical examples for which such mitigation techniques have been applied successfully. For species like Eurasian otters (*Lutra lutra*), cormorants (*Phalacrocorax spp.*), or seals, any translocation would only show a short-term effect until the individual is substituted by another attracted to the resource.

## *2.4 Lethal Control*

Lethal control implies shooting, trapping, or poisoning. It may be an option, if it is possible to exclude negative effects upon the viability of the wildlife population (conservation status). A certain degree of selectivity (age, sex) in killing will be a requirement in most cases and it has to be carried out under controlled conditions (e.g., compliance with animal welfare regulations). For example, unless it can be guaranteed that females with dependent cubs are spared of the killing effort, continuous breeding capability, as in otters, may be exclusion criteria for lethal control on ethical and legal (module 4) grounds.

*Shooting* may meet these requirements, if sufficient sex and age dimorphism exist and if it is feasible to identify this under field conditions (large distance, often inadequate light conditions). *Life trapping* may be a suitable technique, e.g., in box traps or snares, with killing only specimens of known age and sex. *Poisoning* can hardly provide selectivity and control in terms of affected individuals.

Lethal control must not create sink habitats in the sense of Pulliam and Danielson (1991). Sink habitats are areas, where the population survival may only be guaranteed by immigration of individuals from other areas. Negative effects on viability highly depend upon population size, distribution, and status of the conflict species (ecological scale), the intensity of killing and the area across which it is applied.

Each of these aspects as well as a combination of several of them may become relevant in deciding for or against lethal control. If a population is so small that every individual counts, lethal control cannot be an option and the same is true, if a critical stepping stone within the population is concerned or if the population is already decreasing. Module 8 provides advice how the effects of lethal control can be assessed in terms of viability of the wildlife species and how its effectiveness to reduce population size can be monitored.

## 2.5 Behavioral Manipulation of the Conflict Species

The behavioral repertoire of large vertebrates offers three principally different approaches to reduce their interest in the resource: fear-provoking, taste aversion, or diversion to other resources. Fear-provoking stimuli are not related to the resource, but make use of anti-predator behavior. In contrast, taste aversion and diversion work only in context with the resource. Taste aversion leads to a marked decrease of interest in feeding on the resource by reducing the palatability without harming the resource and the conflict species. Diversion offers another resource, which is more readily available, more nutritious, or more palatable.

*Fear-provoking stimuli* are widely used by most resource holders. They comprise some of the few methods that are legally allowed for protected species: visual, auditory (bangers, unfamiliar sounds, distress, and alarm calls), and olfactory stimuli. However, their degree of application contrasts significantly with their effectiveness, since for many wildlife species effectiveness decreases due to habituation. To overcome this problem, it is necessary to provide a re-enforcement of the negative stimuli, such as actual killing or attacks. In order to not create undesirable population effects, it is necessary to limit killings, e.g., only a small percentage of gregarious problem birds (e.g., cormorants) may be actually shot (compare module 8). The survivors—the vast majority—will take the message. The brown bear may serve as an example for a solitary problem species: after repeated use, they ignore bangers and in such cases it is necessary to shoot them with rubber bullets, which do not cause any injuries, but aim at counteracting habituation.

*Chemical repellents* may shape food preferences. Many plants and animals use this strategy naturally to avoid being predated or consumed. Efforts to make use of such effects to prevent damage of wildlife to human resources usually face serious difficulties. A frequent problem is that a repellent makes the resource also unpalatable or poisonous for human consumption. Conditioned food aversions based on deception may be a strategy for carnivores, but in several cases it turned out not to show the expected result as it was the case with coyotes (*Canis latrans*) predating on lambs and raccoons (*Procyon lotor*) depredating nest of sea turtles (Conover 2002). Furthermore, most predators depend on live prey and then this method is not applicable at all.

A quite promising strategy to reduce losses due to wildlife is *diversion*. Food has to be offered in a highly attractive way and this will distract the animal from the concerned resource. This mitigation technique takes advantage of optimal foraging strategies (Begon et al. 1996). Critical for the long-term success of this approach is that it does not increase the carrying capacity of the habitat, resulting in a higher density of the concerned conflict species with finally no positive effects in terms of damage reduction. As it became clear in the case of the otter in fish farms, such an approach works well, if applied for relatively short periods only (Poledníková et al. 2013). This in turn implies that it does not work as an all year round approach, but should be applied during the most critical time of the year, such as in wintertime.

## 2.6 Habitat Manipulation

Mechanical obstacles hindering the access to the resource are one way of habitat manipulation. The other approach is any intentional modification of habitat, which leads to a lower probability of presence, foraging success, or density of the conflict species at locations of vulnerable resources.

Excluding the predator from the resource is a long established and well tested method to reduce losses caused by wildlife, but it does not help in all cases for a given conflict species. It depends on the specific conditions of the resource, the abilities of the concerned wildlife species to access the resources, and the exclusion device used. For example fences will work only, if the vulnerable resource is limited in space and if the fences can be made wildlife proofed. Examples are fences around small artificial ponds used for wintering fish stocks or the development of seal proofed fish traps (Bruckmeier et al. 2013). A novel approach in exclusion is to provide save refuge for the prey. In the case of cormorants, this may be effective for certain fish farms, but it does not work with otters (Fig. 1).

*Habitat modification* is another sophisticated method to reduce losses due to wildlife. This concept is based on key habitat factors necessary for the conflict species. Modifications, which reduce the availability of these factors, can prevent the presence or at least the extent of stay of the species in the conflict area. One example is the reduction of suitable roosting places and breeding colonies for cormorants. Another example is the location of predation-sensitive estuarine fish farms distant to fresh water, because access to fresh water is essential for otters (Santos-Reis et al. 2013).

*Resource manipulation* implies all types of resource management practices that reduce resource vulnerability to wildlife. Temporal and spatial management of resource utilization need to be considered in combination with information about landscape factors promoting or hindering the conflict species (module 2) to identify management options that incur low versus high risks of wildlife damage. Being aware of landscape or habitat factors, which favor the probability of damage, it may be possible to select those areas, where least damages may be

**Fig. 1** Use of wide-meshed nets to exclude great cormorants from small ponds with fish offspring in the Po delta, Italy. *Photo:* Reinhard Klenke

expected. Taking into account food or size preferences (module 3) and aspects of food availability (e.g., stocking density) may help to avoid or at least reduce damages. Finally, some resources may be more valuable than others, e.g., breeding stocks or seed production fields. Valuable resources should be maintained in the safest place and less valuable stock or crop may be left in those areas, where predation and damage are more likely to occur.

Examples for a temporally increased vulnerability of a resource are fish ponds stocked in winter, when fish is less mobile and thus more vulnerable to otter predation. Farmers with ponds in the core area of otters have to consider a production scheme, which reduces the probability of otter—fish contact, e.g., sell the fish already in autumn instead of doing so in the following spring or constructing wintering ponds such that they can be effectively fenced-off. Similar changes in management practices avoiding critical seasons or life cycles will be possible also for many kinds of terrestrial live stock in the presence of medium-sized or large carnivores. An example could be the transfer of mother individuals to carnivore proof enclosures.

## 3 Efficacy and Cost-Effectiveness of Mitigation Methods

Efficacy highly depends on the specific local peculiarities of the involved resource, conflict species, and area. It is not possible to give an overall rating or recommendation. In some species, deterring by shooting can be highly effective; in others it may be even counterproductive (chasing the animals over larger areas, increasing their feeding demands). Reproduction control may work perfectly in colonial breeding birds or mammals, but will fail in solitary carnivores. Methods aiming at a change in behavior of the conflict species may be more effective in herbivorous wildlife than in omnivorous, piscivorous, or carnivorous species, but even this is little more than a rule of thumb. Exclusion may work when the habitat allows a perfect application, but it also depends greatly upon the involved conflict species and resource. Changes in management practices that aim at reducing damage again may be the option to go for in one case, but not in another case.

The financial costs of ecological mitigation measures are of primary importance to any professional resource user trying to avoid and reduce damage. Additionally, this is an important input to the integrated evaluation of policy instruments addressing technical mitigation measures (module 9). In many cases it is already highly complicated to properly assess damage caused by the conflict species in material terms (module 3). Economic damage further depends on changing market conditions. Any business will compare economic damage to the costs of mitigation techniques that the firm has to bear while considering available subsidies—and the latter represent costs to overall society (modules 5 and 9). So there is a need to assess costs at the individual firm level that finally is decisive for the use of certain mitigation measures by a professional resource user. However, aggregated costs to society should not be neglected, including subsidies to technical mitigation and activities that are directly carried out by public authorities.

Comparing costs of mitigation measures is meaningful, if similar techniques with comparable effectiveness are looked at (e.g., different types of fences). A comparison of costs of different mitigation methods may be complicated and even meaningless in some cases due to the incommensurability of different effects of the respective measures, e.g., when comparing costs of shooting with those of other methods. Furthermore, leisure activities and habits will have an influence. The willingness of people to carry out shooting for free may vary among regions and wildlife species and may change with time. Consequently, costs to be born by the various stakeholder groups or the general public could also change. They depend on the culture, hunting practices, general ethics of the people, policy instruments implemented, and local circumstances.

Wildlife species usually have the chance to shift to other resource holders. Therefore, the costs needed to resolve the conflict may have to be multiplied by the number of resource holders until the entire conflict area is appeased. In many cases mitigation measures work only as long as the animal has the option to shift to other resources. As a consequence, they may fail, if they are applied at every resource location, because they may have adverse effects on the population level, which is not desirable in the case of protected species. It is also important to keep in mind that the costs of a given mitigation measure do not necessarily need to be less expensive than the amount of damage to justify its application. In many cases the society or at least certain stakeholders are willing to accept costs of mitigation to be higher for the sake of a non-lethal or more humane method.

Nevertheless it is worthwhile and recommended to estimate costs of different mitigation measures for a given conflict, especially, but not only, of those techniques, which stakeholders ask to implement. It may give the decision makers an idea of whether a measure is within a realistic scope. Such financial figures may also serve as an important argument to convince stakeholders to refrain from the one or the other demand.

## 4 Minimum, Standard, and Advanced Approaches

According to the result of the screening process (module 1), it has to be decided, if a minimum, a standard, or an advanced approach is appropriate in order to identify the best fitting ecological mitigation strategy.

### *4.1 Minimum Approach*

An in-depth literature survey, including grey literature and reports from administrative bodies of other areas, is a minimum requirement. Those findings have to be compared with the locally applied methods as indicated by interviews of concerned resource holders (module 6). This will offer the possibility to learn from other cases. Such information may allow the introduction of mitigation strategies new to the conflict area, it may be of informal use in participatory decision processes (module 10), or it may go straight into policy development (module 9). The minimum approach is only justified, if good literature exists and if the type of conflict is well described for similar conditions. However, it carries the risk that crucial local circumstances are overlooked.

In this context it has to be noted that the traditional knowledge and managerial experience of resource holders may vary significantly both within and among given areas. Furthermore, it may be difficult for the conflict manager to assess which parts of their "experience" are based on true experience and which just represent strategic behavior.

### *4.2 Standard Approach*

A standard approach asks for additional investigation of the concrete and specific conditions of the conflict in question. Experiences of related conflicts in other areas or even continents with similar resources and wildlife species have to be taken into account and have to be adapted and tested under the local conditions. Since ecological mitigation is a key element in any reconciliation action plan, due to its role in the past and the expectations of various stakeholders, a standard approach should be the goal in most, if not all, situations.

### *4.3 Advanced Approach*

An advanced approach implies developing completely new mitigation techniques. This will be necessary, if it becomes obvious that existing strategies do not work or cannot be applied, because of specific reasons, e.g., legal requirements (module 4), high costs (module 5), or rejections by some stakeholder groups (module 6). A precondition for such an approach is a thorough understanding of predator–prey relationships before designing new techniques and testing them experimentally. If this is neglected, a failure is to be expected.

## 5 Recommendations and Conclusions

Usually there is no panacea for a generally acceptable mitigation measure. This is little surprising since this lack is one of the core reasons of human–wildlife conflicts. Nevertheless, this fact needs to be stressed, because all too often it is neglected in practice.

1. Usually, a combination of several mitigation measures is required to provide satisfactory effectiveness.
2. Wildlife usually ignores boundaries created by humans, such as administrative bodies or hunting grounds. Therefore, the coordination of mitigation activities in time and space and regionally differentiated approaches are crucial (wildlife ecological spatial planning: Reimoser 2002).
3. A full understanding of the ecology, status and distribution of the conflict species and the resource is indispensable for choosing the appropriate strategy.
4. Know-how about factors favoring and reducing the risk of predation and damage has to be distributed to resource owners in order to enable them to make full use of their leeway within their management practices.
5. The traditional ecological knowledge and managerial experience of resource holders of the efficacy of a given strategy may be correct only in some cases, but not in others. Notwithstanding, it must not be ignored and has to be considered in the conflict management process.
6. Any purely cost-based balancing between the financial costs of the mitigation measure and the monetary value of the saved resource can be problematic, although it is the favored approach from a business perspective and a prerequisite for profitability assessment at the individual firm level. Note that there

may be an individual or social willingness to pay more for some mitigation techniques e.g., to avoid lethal control.

7. Any cost balancing between alternative mitigation measures is tricky and complicated. Comparisons depend on many factors and can be evaluated only when incorporated into a multi-criteria decision process as outlined in module 10. However, a cost estimate of competing means may be useful even when the figures are not accurate. They will still indicate which means should be taken into closer consideration and which are out of scope.
8. It is not only a mix of mitigation measures, which brings the success, but their integration in policy instruments (module 9) and participatory decision-making processes (module 10).

## References

Bangs EE, Fontaine JA, Jimenez MD, Meier TJ, Bradley EH, Niemeyer CC, Smith DW, Mack CM, Asher V, Oakleaf JK (2005) Managing wolf-human conflict in the northwestern United States. In: Woodroffe R, Thirgood S, Rabinowitz A (eds) People and wildlife: conflict or coexistence. Cambridge University Press, Cambridge

Begon M, Harper JL, Townsend CR (1996) Ecology: individuals, populations, and communities, 3rd edn. Blackwell Science, Cambridge

Bruckmeier K, Westerberg H, Varjopuru R (2013) Reconciliation in practice: the seal conflict and its mitigation in Sweden and Finland. In: Klenke R, Ring I, Kranz A, Jepsen N, Rauschmayer F, Henle K (eds) Human-wildlife conflicts in Europe—fisheries and fish-eating vertebrates as a model case. Springer, Heidelberg

Caughley G, Sinclair ARE (1994) Wildlife ecology and management. Blackwell Science, Cambridge

Conover M (2002) Resolving human—wildlife conflicts: the science of wildlife damage management. Lewis Publishers (CRC Press LLC), Boca Raton

Poledníková K, Kranz A, Poledník L, Myšiak J (2013) Otters causing conflicts: the fish farming case of the Czech Republic. In: Klenke R, Ring I, Kranz A, Jepsen N, Rauschmayer F, Henle K (eds) Human-wildlife conflicts in Europe—fisheries and fish-eating vertebrates as a model case. Springer, Heidelberg

Pulliam HR, Danielson BJ (1991) Sources, sinks, and habitat selection: a landscape perspective on population dynamics. Am Nat 137:51–66

Reimoser F (2002) Wildlife ecological spatial planning (WESP): an instrument for integrating wildlife into comprehensive land management. In: 24th Congress of the international union of game biologists, Thessaloniki, 20 sept 1999, pp. 176–185

Santos-Reis M, Santos R, Antunes P, Sales-Luís T, Gomes J, Freitas D, Madruga L (2013) Reconciliation of the conflict between otters and fish farmers: lessons from Sado estuary in Portugal. In: Klenke RA, Ring I, Kranz A, Jepsen N, Rauschmayer F, Henle K (eds) Human-wildlife conflicts in Europe—fisheries and fish-eating vertebrates as a model case. Springer, Heidelberg

# Module 8: Management and Viability of Target Species: Modeling and Monitoring

**Karin Frank, Peter M. Alberti, Klaus Henle and Reinhard A. Klenke**

**Abstract** One way to mitigate conflicts between species protection and economic damage is population management, i.e. active control of the target species. Any strategy of population control, however, has to meet the constraint that the population stays viable. This chapter presents a framework for modelling and monitoring the viability of populations under active control and for assessing chances and risks of management scenarios already in advance. It accounts for the demographic parameters in the population and the management measure of interest. It highlights that active control is only recommended in cases where population growth rate is above 1.0. Management ought to be adaptive (activity only above a certain threshold) and differentiate between breeders and non-breeders. This requires an appropriate monitoring. Mean age was found to be a useful indicator for population monitoring. The chapter also provides a list of free and commercial software for population viability analysis and monitoring.

K. Frank (✉) · P. M. Alberti
Department of Ecological Modelling,
UFZ—Helmholtz Centre for Environmental Research,
Permoserstraße 15, 04318, Leipzig, Germany
e-mail: karin.frank@ufz.de

P. M. Alberti
e-mail: alberti@fh-erfurt.de

K. Henle · R. A. Klenke
Department of Conservation Biology,
UFZ—Helmholtz Centre for Environmental Research,
Permoserstraße 15, 04318, Leipzig, Germany
e-mail: klaus.henle@ufz.de

R. A. Klenke
e-mail: reinhard.klenke@ufz.de

R. A. Klenke et al. (eds.), *Human-Wildlife Conflicts in Europe*,
Environmental Science and Engineering, DOI: 10.1007/978-3-540-34789-7_18,

# 1 Rationale and Objectives

This module becomes relevant if wildlife management aiming at active control of the target species is considered (module 7). Population management has to meet two conditions to become appropriate for conflict reconciliation: Firstly, it has to be safe, i.e. it has to secure the viability of the target species. Secondly, it has to ensure a significant and rapid reduction of population size and thus of damage. Such management is inappropriate unless the target species has an average growth rate markedly above 1.0. Otherwise, active control ought to be avoided and limited to other, less intervening conflict reconciliation strategies, such as technical mitigation (module 7) or compensation payment (module 9).

Population management can be classified according to whether it manipulates (1) the population size itself, (2) the growth rate of the population, or (3) the carrying capacity of the habitat. Population management should be implemented in an adaptive way, i.e., it should continuously take into account information about the state of the population. The following questions have to be answered as an indispensable input for the development of policy instruments and participatory decision processes (see modules 9 and 10):

- What type and intensity of population management is appropriate for conflict reconciliation?
- Which option is the best, regarding ecological effectiveness and considering costs, risks, and social acceptance?
- How can we monitor the success of a selected management option and detect early signs of critical decline?

To answer these questions requires predicting and analyzing the dynamics of the population under the influence of management scenarios and monitoring the population. Modeling is powerful in this context. Models allow all relevant details to be integrated and logically combined. On this basis, long-term and/or large-scale effects on the dynamics of the population can be consistently extrapolated, quantified, and analyzed, and key factors of the effects can be determined. Last but not least, models allow alternative mitigation scenarios (input from modules 6 and 7) to be simulated, assessed, compared, and ranked in terms of their effect. In this way the best scenario as well as general qualitative trends can be determined. This makes models useful for prediction, understanding, and decision-support.

Models are always abstractions. Clearly, the more details are included, the higher is the predictive power of the model for the particular situation. In many cases, information about all the details is not available. Often, however, it is helpful to obtain a rough orientation to set correct management priorities. In this case, so-called generic models that focus on the essential factors and ignore specific details (such as a sketch in comparison with a naturalistic painting) are useful. Because of their generic nature, these models are applicable to a broad range of situations.

Modeling needs to be combined with a monitoring program since there is never a full insight into the demographic processes, the size and structure of the population, or environmental factors influencing viability of the managed population. Hence, long-term observations are needed for detecting critical changes in time, so regulation can be stopped and an irreversible collapse of the population can be prevented. Furthermore, monitoring managed and unmanaged populations with an appropriate experimental design allows to test whether the management has the expected and desired effects on the population and to assess the strength of this effect.

# 2 Methods and Approaches

## *2.1 Modeling*

There are different approaches to assess the effect of population regulation on population viability by models. In principle, any model designed for population viability analysis (PVA; Brook et al. 2000a, b; Possingham et al. 2001; Beissinger and McCullough 2002) can be used. The variety of PVA models, however, is huge. The models vary in their level of detail (individual-, cohort- versus population-based; non-spatial versus spatial, deterministic versus stochastic). Hence, it is not useful for practical applications to refer to the variety of existing PVA models and to leave the practitioner alone with it.

We focus on presenting a generic model for assessing the dynamics and viability of populations under population management. It is applicable to target species with age-dependent demographic parameters. This model takes the following demographic parameters and data of the target species into account (provided by module 2):

- survival rates and fecundities, depending on age-class and social status;
- breeding and resource capacity of the habitat;
- type of density regulation; (we make a distinction between "breeders" and "non-breeders" in the population—a situation that is common in nature).

These data are included in a so-called matrix model (Caswell 2001). This model describes the development of the population in the course of time and has the number of individuals for the different age-classes (cohorts), the number of breeders and non-breeders, as well as the risk of extinction after 100 years (measure for population viability) as outcome. The model allows different types of population management to be assessed: (1) population reduction (e.g., protective hunting), (2) manipulation of the survival rates or fecundities, and (3) manipulation of the breeding capacity (e.g., destruction of breeding places). Moreover, it can account for other impacts (e.g., natural mortality, individual loss caused by by-catch or road kill) on population viability. Such analyses of combined effects are especially important for ensuring that population management is safe, i.e. does not make the population vulnerable to accidental individual loss. Typical outcomes of the described model

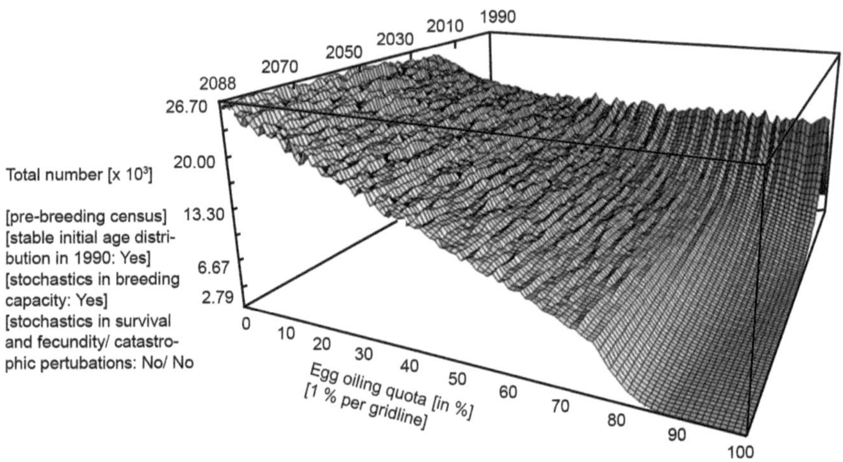


**Fig. 1** Example of a viability analysis made with the generic model developed in FRAP for the great cormorant (*Phalacrocorax carbo*) population of the Ringkøbing Fjord, Denmark. The development of its size is plotted from 1990 to 2089 as a function of egg oiling quota after start of management (in %). Evidently, there is a critical quota of 75 % above which the population rapidly collapses

are the critical average age in the stock, appropriate mixtures of regulation types and their intensities, as well as census data and the risk of extinction after 100 years (Fig. 1).

Applying a model to a particular situation means to parameterize it accordingly. This draws the attention to the problem of uncertainty in the model parameters. Clearly, an uncertain input increases the risk of an uncertain output. However, the degree of uncertainty in the outcome depends on the entire model structure. It primarily affects the absolute results, i.e. the viability status of the population under a particular regulation scenario. Qualitative trends and relative results (e.g., ranking orders, relative importance of a certain factor) are much more robust against uncertainty.

Performing a population viability analysis requires experience with using and parameterizing models as well as interpreting model results. This includes the understanding of the potential and limitations of using models. Hence, the work with module 8 is to some extent knowledge demanding. As far as the required skills are concerned, the standard approach in module 8 is comparable with advanced approaches in other modules.

## 3 Monitoring

The aim of monitoring in wildlife management is to answer questions about the size and status of the managed population and the effectiveness of the management. This requires a carefully designed sampling scheme and appropriate field methods.

Ideally, a monitoring scheme should be experimental and adaptive. Experimental means a comparison of managed populations with unmanaged control populations using an experimental design, i.e., populations are randomly selected and as similar to each other as possible except of the management (Caughley and Sinclair 1994). Only then it can be tested whether the management of the population has the expected and desired effects or whether any observed change in abundance is due to natural fluctuations. Adaptive means that monitoring methods are revised according to weaknesses identified during data analysis, and monitoring goals are revised according to new needs or more precisely defined questions. Various tools exist (see Table 1 at the end of this chapter) for analyzing monitoring data; however, it is important to be familiar with some basic monitoring principles before using any existing tool (Fig. 2). The BioMAT tool provides general guidelines for the planning and analysis of monitoring data. The software TRIM has been particularly devised for the analysis of monitoring data that covers larger regions and may have many missing values due to incomplete coverage of the monitoring scheme.

Since the abundance of wildlife often is a critical element in conflicts, it is a basic parameter to be monitored. Determining abundance usually relies on indices of abundance or, in an advanced approach, on the estimation of population size (module 2). The processes underlying change in abundance can be disentangled by monitoring demographic parameters (advanced approaches of module 2). For this, two main types of data are used: individual follow-up (e.g., capture-mark-recapture) and age (or size) structure of populations. Individual follow-up is to be preferred, whenever possible, because it makes less simplifying assumptions for parameter estimation, inferences are more robust, and variations in detection probability can be accounted for Demographic monitoring, however, is usually time and resource consuming and can be implemented for a restricted number of sites only.

Monitoring data should have four components attached to them: the measure, the site, the year, but also uncertainty. That is, the sampling design should allow the estimation of measurement error, e.g., whether detection probability differs between managed and unmanaged populations, due to different wariness of the individuals (module 2). In the latter case, statistical corrections for detection probability is required.

Ideally, measures are taken every year at every site to simplify statistical analysis. This is usually far from being achievable in schemes that rely on a large number of observers, sites, and years. As a consequence, in most cases, some sites are not sampled in some years. In technical words, monitoring data include missing points. Linear modeling, with appropriate distribution of data, link-function, and parameterization of sites and year effects, can account for the main problems faced when analyzing temporal series of monitoring data: heterogeneity among sites, among observers, through time, and in precision. The computer program TRIM is a widely used tool designed for such analyses.

Three basic properties are worth highlighting. First, the computation of temporal trends does not require complete time series; missing counts can be accounted for by extrapolation. Second, including site effects in the statistical model largely accounts

**Fig. 2** Modelers and ecologists discussing approaches, data needs and possible outcomes of a population model for the great cormorant. *Photo:* Oliver Zwirner

for differences in detection probability among observers and habitats. For this reason, it is strongly recommended that each site is monitored by the same observer, as long as he/she is involved in the monitoring scheme. Third, the trend in abundance is not the simple difference between the first and the last year of the monitoring, but instead data of all years equally contribute to the trend.

Several software programs are available to calculate trends in count data (e.g., Gerrodette 1987, 1993; Ter Braak et al. 1994; van Strien et al. 2004), some are freely available on the net, e.g., TRIM3 and TRENDS (see for Tools below). A comprehensive manual for monitoring wildlife populations (field work and sampling design) is provided by Cooperrider et al. (1986), and one for sampling design and statistical analysis by Pollock et al. (2002). A recent special issue on biodiversity monitoring (Schmeller 2008) provides various guidelines and examples for the design and analysis of biodiversity monitoring, including the integration of different monitoring schemes. The BioMAT tool developed by the European project EuMon (EU-wide monitoring methods and systems of surveillance for species and habitats of Community interest—http://eumon.ckff.si) provides online guidance for planning and analyzing monitoring schemes, presents background information on monitoring species, and allows to contact existing schemes that monitor particular taxonomic groups.

## *3.1 Minimum Approach*

The minimum approach of modeling depends on whether we are interested in relative or absolute results. If relative results (e.g., ranking of different management options) are the aim, then it is sufficient to know the range of parameters that are relevant for population viability. In this case, a parameter variation within the biologically plausible range has to be carried out and the robustness of the relative results against this variation has to be tested. In this way, the best/worst scenario can be determined even if detailed demographic data for the target species are not available. If absolute results are required, then at least an estimation of the demographic parameters of the target species is needed (module 2).

A minimum approach in monitoring collects demographic information for the regulated part of the population using an index of abundance (module 2). Since managed populations usually exchange individuals among each other, the monitoring usually should be extended to a wider region with the size of the region depending on the dispersal ability of the target species. If fecundity is regulated, reproductive output of the managed population also needs to be monitored. The minimum approach allows detection of undesired changes in the population but does not provide information whether any observed change is due to management or other factors.

### *3.2 Standard Approach*

As far as modeling is concerned, one should take demographic data for the target species from the literature or field studies (module 2).

As a standard approach in monitoring, one should compare the abundance for managed and unmanaged populations (module 2), using an experimental design (Caughley and Sinclair 1994). When planning a monitoring program to assess trends in the abundance of a regulated population, one should consider particularly type II errors instead of the usual type I errors. Type II errors are erroneous acceptance of the hypothesis of "no change", when the population in reality declined, whereas type I errors incorrectly conclude a change, when in reality no change has taken place. The former is more serious in wildlife management but is often ignored.

### *3.3 Advanced Approach*

In population viability analysis, this approach becomes important if (1) predictions for specific case studies are the aim, (2) a management alternative is suggested that is not covered by a generic model, or (3) factors become important that are not covered by a generic model (e.g., complex individual behavior, dispersal, spatial heterogeneity, interactions with other species). In the first case, an appropriate field study is mandatory, as predictions for specific case studies require specific data.

The second and third case requires more detailed and complex models that need to be developed by a modeler and targeted directly to the relevant situation.

Regarding monitoring, an advanced approach will combine modeling with the monitoring program. For example, it would use the model results to identify threshold values for key parameters of viability above which regulation management can be continued, but below which it has to be stopped for not jeopardizing the viability of the managed wildlife population. In addition, in an advanced monitoring program, data collection is designed such that the data analysis allows reduction of uncertainty in the values of the key parameters of viability identified. Advanced monitoring further collects data for testing of corollary predictions of the models (patterns, such as sex ratio or age structure). The test results can be used then to refine the models and adjust the monitoring to new questions that arise.

# 4 Recommendations and Conclusions

## *4.1 Methodological Implications of the Results From the Model Cases of FRAP*

In the FRAP project several model results of certain generality with major implications for the methodology of conflict reconciliation were obtained:

1. It is indispensable to differentiate between "breeders" and "non-breeders" when analyzing the dynamics of managed populations (cf. the cormorant case described by Jepsen and Olesen 2013 in this book). Otherwise, there is a high risk of underestimating the effort required for population regulation or overestimating population viability. Both effects are counterproductive from the point of view of conflict reconciliation. They result from the fact that "non-breeders" serve as a reservoir, i.e., they can compensate losses among the "breeders".
2. Regardless of the type of regulation, constant management with fixed rules, quota or rates was found to be either too ineffective in terms of population reduction or too risky for population viability. Hence, an adaptive regulation management with rules taking the current state of the population into account is essential as it is both effective and safe because of the chance to stop before the population is critically reduced.
3. Adaptive regulation management requires appropriate indicators for the state of the population. Usually, population size (only breeders) is used as indicator. This quantity, however, is often difficult to determine in the field. Our studies revealed an alternative: the age structure of the population was found to be an indicator for its viability. There is a threshold value for the mean age, above which regulation management can be continued, but below which it has to be stopped. To use the age structure as indicator has one advantage: It can be determined from samples of dead individuals, provided the sample is large

enough and representative for the structure of the population. The threshold value for the mean age depends on the demographic attributes of the target species and the type of regulation planned.

4. Regardless of the type of population management, the effect of the management should be monitored using an experimental design, type II erros should be given more weight than type I errors, and the management should be adjusted according to the monitoring outcome.

## 4.2 Consequences for the Reconciliation Process

Our findings have consequences for different parts of the reconciliation process:

1. Since population management ought to be adaptive, it has to encompass monitoring. Monitoring is indispensable for detecting changes before critical thresholds are reached so that regulation can be stopped and an irreversible collapse of the population can be prevented. Monitoring should consider the whole population (not only the regulated part, e.g., breeders) and compare the managed population with control populations to test whether the management has the expected biological outcome.
2. To know whether a certain type of population management (e.g., hunting) is suitable for a certain target species or not is important for the design of appropriate policy instruments (module 9). The finding will influence whether a considered species has to remain under protected status or can be actively controlled. The opportunity to analyze and compare the suitability of centralized and decentralized forms of population management is of particular importance in this context. The results indicate what level of governance (local, regional, national, pan-European) is appropriate.
3. Ranking orders among alternative regulation scenarios are usually quite robust against uncertainty in the input parameters. As was mentioned above, however, the threshold values for the average age of the population and the regulation intensity do not only depend on the type of population regulation but also on the demographic attributes of the target species. This shows that the two threshold values are vulnerable to uncertainty in the age-dependent survival rates and fecundities. The model, however, allows upper and lower values for these threshold values to be determined, depending on the degree of uncertainty in the input data. This allows the development of conservative strategies.

**Table 1** Free and commercial software for population viability analysis and monitoring

| Title/Organization | URL |
|---|---|
| BioMAT—Biodiversity Monitoring Assessment Tool (2011): | http://eumon.ckff.si/biomat/. Accessed 30 May 2011 |
| Gerrodette T, Brandon J (2006): Trends—the software. | http://swfsc.noaa.gov/textblock.aspx?Division=PRD&ParentMenuId=228&id=4740 Accessed 26 September 2008 |
| Lenth RV (2006). Java Applets for Power and Sample Size | http://www.stat.uiowa.edu/~rlenth/Power Accessed 25 September 2008 |
| Statistics Online Computational Resource (SOCR) | http://socr.ucla.edu/SOCR.html Accessed 26 September 2008 |
| van Strien A, Pannekoek J, Hagemeijer W, Verstrael T (2006) Indices and trends (TRIM). Version 3.53, released 8 November 2006 | http://www.cbs.nl/en-GB/menu/themas/natuur-milieu/methoden/trim/default.html Accessed 25 September 2008 |
| Zucchini W, Borchers DL, Erdelmeier M, Rexstad E, Bishop J. (2007) WiSP 1.2.4. Institut fur Statistik und Okonometrie, Georg-August-Universitat Gottingen, Platz der Gottinger Seiben 5, Gottingen, Germany. | http://www.ruwpa.st-and.ac.uk/estimating.abundance/WiSP/index.html Accessed 25 September 2008 |

## References


Beissinger SR, McCullough DR (eds) (2002) Population viability analysis. University of Chicago Press, Chicago

Brook BW, Burgman MA, Frankham R (2000a) Differences and congruencies between PVA packages: the importance of sex ratio for predictions of extinction risk. Conservation Ecology 4(1):6; http://www.consecol.org/vol4/iss1/art6/. Accessed 26 September 2008

Brook BW, O'grady JJ, Chapman AP, Burgman MA, Akcakaya HR, Frankham R (2000b) Predictive accuracy of population viability analysis in conservation biology. Nature 404:385–387. doi:10.1038/35006050 Accessed 23 March 2000

Caswell H (2001) Matrix population models: construction, analysis, and interpretation, 2nd edn. Sinauer Associates, Sunderland

Caughley G, Sinclair ARE (1994) Wildlife ecology and management. Blackwell Science, Cambridge

Cooperrider AY, Boyd RJ, Stuart HR (1986) Inventory and monitoring of wildlife habitat. US Department of the Interior, Bureau of Land Management, Service Centre, Denver

Gerrodette T (1987) A power analysis for detecting trends. Ecology 68:1364–1372

Gerrodette T (1993) Trends: software for a power analysis of linear regression. Wildl Soc Bull 21:515–516

Jepsen N, Olesen T (2012) Cormorants in Denmark—reinforced management and scientific evidence. In: Klenke R, Ring I, Kranz A, Jepsen N, Rauschmayer F, Henle K (eds) Human-wildlife conflicts in Europe—fisheries and fish-eating vertebrates as a model case. Springer, Heidelberg

Pollock KH, Nichols JD, Simons TR, Farnsworth GL, Bailey LL, Sauer JR (2002) Large scale wildlife monitoring studies: statistical methods for design and analysis. Environmetrics 13:105–119

Possingham HP, Lindenmayer DB, McCarthy MA (2001) Population viability analysis. Encyclopedia of Biodiversity Vol 4. Academic Press, New York, pp. 831–843

Schmeller D (2008) Biodiversity Monitoring. Biodivers Conserv, 17 (Special Issue)
Ter Braak CJF, van Strien AJ, Meijer R, Verstrael TJ (1994) Analysis of monitoring data with many missing values: which method? In: Hagemeijer EJM, Verstrael TJ (eds) Bird numbers 1992. distribution, monitoring and ecological aspects. In: Proceedings of the 12th international conference of IBCC and EOAC, Noordwijkerhout, The Netherlands, Statistics Netherlands, Voorburg/Heerlen & SOVON, Beek-Ubbergen, pp 663–673
van Strien AJ, Pannekoek J, Hagemeijer W, Verstrael TJ (2004) A loglinear Poisson regression method analyse bird monitoring data. In: Anselin A (ed) Bird numbers 1995. In: Proceedings of the international conference 13th meeting of the European bird census council, Pärnu, Estonia. Bird Census News 13 (2000):33–39

# Module 9: Development of Policy Instruments

**Irene Ring, Kathleen Schwerdtner Máñez and Rui Santos**

**Abstract** In phase 3, natural and social scientific results from phase 2, modules 2–6 (assessment and analysis of the conflict), are integrated in order to design and implement conflict resolution strategies. Module 9 deals with the evaluation of policy instruments in terms of their effectiveness, their cost-effectiveness, their perception by stakeholders, and their potential for including relevant actors in wildlife decision making. If none or few instruments are present, new policies addressing the human-wildlife conflict need to be developed. Wildlife conflict resolution policies ideally are a policy mix consisting of species conservation and derogation measures, economic instruments, and action directed at civil society.

I. Ring (✉) · K. Schwerdtner Máñez
Department of Economics, UFZ—Helmholtz Centre for Environmental Research, Permoserstraße 15, 04318 Leipzig, Germany
e-mail: irene.ring@ufz.de

K. Schwerdtner Máñez
Department of Social Sciences, Leibniz Center for Tropical Marine Ecology, Fahrenheitstrasse 6, 28359 Bremen, Germany
e-mail: kathleen.schwerdtner@zmt-bremen.de

R. Santos
Centre for Environmental and Sustainability Research, Faculdade de Ciências e Tecnologia, Campus de Caparica, Universidade Nova de Lisboa, 2829-516 Caparica, Portugal
e-mail: rfs@fct.unl.pt

R. A. Klenke et al. (eds.), *Human-Wildlife Conflicts in Europe*,
Environmental Science and Engineering, DOI: 10.1007/978-3-540-34789-7_19,

# 1 Rationale and Objectives

This module integrates all findings from the assessment phase: the ecological results regarding abundance and habitat of the wildlife species (module 2) and ecological damage assessment (module 3). It further considers the specific situation of economic sectors in conflict with wildlife (module 5), builds on the legal framework and policy analysis (modules 4 and 5), and integrates the perception of the stakeholder groups (module 6) (Ring 2009).

The module involves an evaluation of existing policy instruments aiming at improving single instruments and better coordinating the different policy instruments in place. If no or very few specific instruments addressing the conflict exist, although the conflict is salient, the development of new policy instruments may be necessary. If policy instruments were in place but abolished in the past, it may be helpful to identify the reasons for this and the potential for reintroducing these instruments.

The final aim of this module is to propose a conflict-reducing mix of policy instruments that considers the criteria of effectiveness regarding conflict reconciliation, cost-effectiveness, as well as the perceptions of relevant stakeholders while considering ecological aspects. General guidelines provide a starting point. However, due to the different contexts of conflicts, policies must be adapted to the specific situation in place.

# 2 Methods and Approaches

## *2.1 Selection of Instruments to be Evaluated*

The first step consists in selecting instruments to be evaluated. This decision must ultimately be based on the relevance of instruments for conflict reconciliation, but should also consider ecological and socio-economic knowledge from other modules, and whether instruments are controversial or illegal action is taking place.

A minimum approach with just one or two instruments in focus may suffice when basic conflict management is already in place, and the improvement of one or two specific instruments is the major aim (Table 1). A standard approach will cover a mix of policy instruments, with regulatory and economic instruments, as well as action directed at civil society (see module 4). Regulation sets the frame for all other measures, economic instruments help to equally distribute the benefits and costs among various stakeholder groups and civil-society action is an essential ingredient of socially acceptable conflict management. Sometimes not all categories are relevant due to the existing legal framework or cultural traditions. An advanced approach looking at several instruments within each of the categories should be adopted for high priority species and/or salient conflicts. It is highly recommended to use the advanced approach when a species-specific management plan (e.g., at state, national, or international level) is revised or newly adopted [see the case studies by Bruckmeier et al. (2013) and Santos-Reis et al. (2013)].

## 2.2 Data Collection

For all approaches (minimum, standard, and advanced), ecological and socio-economic results of the assessment phase should be used to identify further needs for specific information. Close cooperation with authorities is recommended for all approaches, but will vary according to depth and detail of data needed.

## 2.3 Evaluation of Policy Instruments

All approaches include an evaluation of the selected single instruments according to three basic evaluation criteria: effectiveness, cost-effectiveness, and perception by stakeholders (Table 1). A minimum approach may add further evaluation criteria, such as management requirements or innovation potential, that are necessary for their improvement. A standard approach looks at the interplay between instruments to suggest an improved policy-mix for conflict management. The advanced approach covers all these steps and builds on continuous interaction between policy development and the other modules of the resolution and implementation phase. The advanced approach may also create a feedback loop to phase 2, if more detailed data and assessments are needed to improve single instruments or coordinate various policy instruments (see Box 1).

### 2.3.1 Effectiveness Regarding Conflict Reconciliation

Policy instruments should strive for reducing damage caused by wildlife. In many cases, damage by wildlife can only be reduced to a certain level but not completely avoided. Therefore, another important feature of policies is to facilitate consensus among the various stakeholders involved regarding the important aspects of the conflict. Even though instruments might have been implemented in the first place for other reasons, e.g., species protection regulation, they may need to be selected for further analysis due to their role in conflict mitigation. Does the instrument intensify or ease the conflict? What aspects of the conflict are mitigated and to what extent? Often, it will be difficult to quantify effectiveness. It may only be possible to describe a positive or negative direction of the effect.

How effective can an instrument be for conflict mitigation? As a first answer, one can say: the more an instrument mitigates a conflict, the better. However, there are limits to conflict mitigation, especially if we aspire to mitigate conflicts in the framework of sustainable development and consider ecological, economic, and social aspects. Given species protection and biodiversity conservation as a generally accepted societal aim, instruments should reduce damage as far as possible, but not endanger the viability of the target species population (Ring 2009). This is especially important for all instruments aiming at population management.

Here, there is an important link to and necessary interaction with module 8 "Management and viability of target species".

Concerning economic aspects, wildlife conservation should not endanger the viability of economic activities. The costs of wildlife conservation in the form of damage caused by certain species should not only be loaded on the shoulders of specific economic sectors or stakeholder groups. As legal standards for biodiversity conservation are in the public interest and set by society, there may be a need to compensate persons or groups. Such compensation can take on different forms: the public sector itself can adopt measures that decrease damage, e.g., by population management; more common is financial support to the persons suffering damage by financially supporting technical mitigation measures or partial compensation.

A special case is given when the species themselves are the economic resource to be harvested. Here, population viability of the target species is the first condition to fulfill, whereas the viability of the relevant economic activities comes second. The conservation and sustainable use of resources is a prerequisite for long-term social welfare, and economic activities should not destroy the natural basis of this.

### 2.3.2 Improving the Cost-Effectiveness of Policy Instruments

Policy instruments should be as cost-effective as possible to save public and private resources. For this purpose, direct costs of an instrument should be analyzed, such as the actual payments for damage compensations to resource holders. Furthermore, the so-called transaction costs of an instrument must not be neglected involving an assessment of the information, management, and bargaining costs (Birner and Wittmer 2004). Information costs relate to the costs of informing stakeholders about existing policies and how to apply. Management costs consist of the regular procedural costs of managing the instrument, e.g., the administrative costs associated with assessing the amount of compensation payments that should be paid for a particular claim. Bargaining costs involve, among others, a potential dispute between authorities and other stakeholders about the correctness or fairness of certain instruments.

Analysis of cost-effectiveness can be understood in terms of two principles: the minimum principle asks for the potential of an instrument to reach the same effect with less costs, the maximum principle asks for the potential to reach a better effect with the same budget. Improving cost-effectiveness in conflict reconciliation can be reached in three ways:

- The cost-effectiveness of existing single instruments can be improved. For example, damage compensation payments should not be higher than the estimated maximum damage that can be caused according to the estimated numbers and diet of the wildlife species (see Box 1).
- The overall cost-effectiveness can be improved by a better coordination of the existing mix of policy instruments. For example, in the long run it might be

more cost-effective to support technological mitigation measures that prevent damage rather than paying large sums of damage compensation for many years.
- The overall cost-effectiveness can be improved by introducing new instruments. For example, if interview results indicate that there is lack of information about the conflict among stakeholders, an information campaign can increase cost-effectiveness.

**Box 1** Improving the cost-effectiveness of compensation payments

In Saxony, a comparison of the allocation of compensation payments for otter damages with the actual occurrence of the species on pond level has shown some discrepancies. A model was developed that used various landscape factors, such as habitat types and visiting rates of otters in order to predict damage. Comparing predicted damage with actual claims by fish farmers revealed that high damage claims did not correspond with predictions based on the landscape model. In order to improve the cost-effectiveness of compensation payments in terms of actually compensating damage where it occurs, it was suggested to base the damage compensation scheme on biological data and landscape factors. Obtaining biological data requires species monitoring, whereas landscape factors may be extracted from GIS-data.

Cost-effectiveness analysis of damage compensation schemes must also consider transaction costs that are closely related to the spatial and temporal distribution of damages. Transaction cost analysis can help deciding whether *ex post* compensation or damage compensation in advance is more cost-effective (Schwerdtner and Gruber 2007).

The analysis of the Saxon case has also highlighted the need to combine technical mitigation measures with various types of compensation payments. Damage compensation alone is a passive strategy, thus providing no incentive to prevent damage. Where applicable, payments for ecological services provided by land users should be preferred instead of damage compensation. In Saxony, this is covered by a specific measure as part of an agri-environmental program aiming at compensating fish farmers for providing a feeding habitat for protected species, such as the otter.

The analysis of cost-effectiveness of a policy instrument is closely linked to the analysis of its effectiveness. The costs of an instrument are finally evaluated with respect to the original objective of the instrument. In this context it has to be checked whether the design and implementation of the instrument is still consistent with the most recent knowledge regarding the abundance and diet of the species and the process of damage causation. It may well be that the existing workflow in administrations, possibly combined with outdated methods of damage assessment, lead to inefficiency of an instrument. Furthermore, outdated or incorrect abundance estimates can easily cause over- or under-compensation. Therefore, close interaction with modules 2 and 8 in terms of species abundance and monitoring is necessary (see Box 1). However, the uncertainties and costs involved in getting both regular and exact estimates for the abundance of certain species and the actual damage caused (compare modules 2, 3 and 8) require a certain trade-off between scientific exactness, administrative feasibility, and stakeholder consensus.

#### 2.3.3 Perception and Involvement of Stakeholders

The results of stakeholder analysis (module 6) should be thoroughly consulted with respect to the various conflict mitigation measures and policy instruments. If policy instruments are exclusively designed as part of a top–down process, they can easily miss the needs of stakeholders, making them less effective in terms of conflict reconciliation. Whenever possible, relevant stakeholder groups should meet and commonly contribute to the design and development of policy instruments. This is important for critical aspects of the conflict, such as decisions on hunting quotas or the number of species causing damage. Close interaction with module 10 "Designing participatory decision strategies" is recommended, especially when new instruments have to be designed.

#### 2.3.4 Further Evaluation Criteria

It is often important to consider management and implementation requirements of different instruments. A standard evaluation should also include a basic consideration of the interaction between policy instruments. Decision-makers could additionally consider the dynamic efficiency of policy instruments. This relates to the capacity of the instrument to trigger innovative solutions for conflict mitigation. For example, it might pay off in the long run to financially support research into effective damage mitigation technologies that are able to keep wildlife away from the biological resources they feed on.

### *2.4 Development of New Policy Instruments*

A specific case exists if none or few policy instruments are present. It creates the need for the development of new policies addressing the conflict (Table 1). In this case it is important to obtain information about similar conflicts in other areas. This could relate to experiences with similar conflicts regarding other protected species or the same protected species in other areas. It is important to consider the transfer potential of a range of instruments adapted to the specific case in place.

It is recommended to combine the development of new instruments with a participatory decision-making process involving relevant stakeholders as described in module 10. The Portuguese case study addressing the conflict between otters and fish

**Table 1** Minimum, standard, and advanced approaches for module 9

| Steps/approach | Minimum | Standard | Advanced |
|---|---|---|---|
| Selection of instruments | Improve one or two specific instruments | Evaluate existing conflict mitigation policies | Improvement and coordination of various instruments |
| | Prerequisite: conflict management with policy instruments in place | Consider regulatory and economic instruments, and civil-society action | Species with high priority or salient conflicts |
| | | | Prepare or improve a species-specific management or reconciliation action plan |
| Evaluation of instruments | Consider effectiveness, cost-effectiveness, and perception by stakeholders | Assess single instruments as well as coordination among them | Collect more specific data (feedback loop to phase 2) |
| | Add further criteria according to specific needs | | Feedback between modules of phase 3 |
| Development of new instruments | No minimum approach applicable | Cooperation with and involvement of relevant authorities and other stakeholder groups | Close interaction with authorities and between ecological and socio-economic experts in policy design |
| | | Consider policy mix | Participatory process involving relevant stakeholders |
| | | Consider basic evaluation criteria | |

farmers (Santos-Reis et al. 2013) provides an example for such a process. In some conflict cases, it may be possible to develop incentives for alternative economic activities that may even build on the presence of the conflicting species, e.g., for eco-based tourism. Here, the initial economic losses are made up for by new economic gains. For example, wildlife farming in African countries is based on this principle (Henle and Luiselli 2003). Initially, farmers suffered economic loss due to wildlife foraging but by developing eco-tourism (wildlife watching) they developed a new economic activity that may have a better return on investment than the original one.

## 3 Recommendations and Conclusions

Based on the results regarding the characteristics and ecology of the respective species and the social sciences results regarding the institutional frame and scale of conflict, policy instruments have to be developed at the adequate level of government. Conflict management strategies and their implementation involving relevant stakeholders are best placed at decentralized levels of government. However, the spatial distribution and mobility of the species influence both the level of conflict and the required conflict management strategies. Hence, they may range from local and regional solutions for less mobile species (e.g., otter in Saxony) up to national solutions and bilateral or international coordination for highly mobile species. This can be the case for relatively fast moving species or migrating birds (e.g., Baltic grey seal *Halichoerus grypus* and great cormorant *Phalacrocorax carbo sinensis*). In the latter cases, coordination is especially important regarding population management measures (Similä et al. 2006; Bruckmeier et al. 2013; Rauschmayer and Weiss 2013).

Nevertheless, wildlife conservation and derogation regulation is correctly placed at centralized levels of decision-making, be it the European level or the national level outside Europe. The conservation and sustainable use of biodiversity is a global public good and associated with large-scale benefits. However, the costs of conservation, in our case the damage caused by wildlife, have to be borne mainly at the local level and by specific economic sectors. Here, the primary economic sector, including agriculture, forestry, and fisheries, is especially affected. Therefore, supporting financial funds for environmentally sound agriculture (including pond farming and certain types of aquaculture) and fisheries are available at the European level. They should be fully utilized for regional conflict mitigation policies, but this requires special programs in the EU member states (compare module 4). The analysis of our model conflicts showed that European agri-environmental funds are already much more widely used than fisheries funds in terms of mitigating human-wildlife conflicts. So there is potential for better use of European fisheries funds for the various conflicts.

Relevant economic instruments include payments to land users for ecological services, financial support schemes to develop or buy technical mitigation measures that reduce damage to valuable resources, publicly supported insurance schemes for valuable technical gear that may be destroyed by the conflict species (Bruckmeier et al. 2013), and damage compensation schemes. The latter are widely used, though not all governments support this type of instrument. Before damage compensation schemes are adopted, decision-makers should always strive to implement support schemes for land users that provide ecological services. This turns wildlife conservation into something positive instead of focusing on damage and creating a negative or rather passive attitude for continuous claims. Secondly, damage prevention measures should be used to the extent possible.

By the use of economic instruments, the costs of wildlife conservation are more equally shared among the various members of society. However, there will always remain a certain benefit of wildlife conservation at decentralized levels. Therefore, all European funds require co-funding by public resources at national or state levels of government. Furthermore, land users benefit from the measures supported and are expected to provide matching funds, i.e. have to bear part of the damage. Where financial resources at the local level are too low for generating co-funding, certification schemes for environmentally sound management practices provide a promising option (Santos-Reis et al. 2013).

Policy instruments aiming at conflict mitigation ideally represent a policy-mix (Ring 2009). This includes binding regulations, such as species protection regulation and derogation measures, economic instruments to adjust private costs and social benefits of species protection, and last but not least action directed at civil society. Civil-society action is absolutely essential in biodiversity conflict management, though sometimes still neglected. Early participation of all relevant stakeholders in decision-making and policy design is recommended. Successful strategies for biodiversity conflict management across Europe include institutionalized stakeholder *fora*, bringing together the conflicting parties to find consensus on critical issues. In this way, public and private actors cooperate actively, aiming both at sustaining viable populations of protected species and viable economic activities based on natural resource use.

## References

Birner R, Wittmer H (2004) On the efficient boundaries of the state—the contribution of transaction costs economics to the analysis of decentralisation and devolution in natural resource management. Environ Plan C—Gov Policy 22(5):667–685

Bruckmeier K, Westerberg H, Varjopuro R (2013) Reconciliation in practice: the seal conflict and its mitigation in Sweden and Finland. In: Klenke RA, Ring I, Kranz A, Jepsen N, Rauschmayer F, Henle K (eds) Human-wildlife conflicts in Europe—fisheries and fish-eating vertebrates as a model case. Springer, Heidelberg. doi: 10.1007/978-3-540-34789-7_3

Henle K, Luiselli LM (2003) Progress with conservation strategies of selected threatened animals. In: Gherardi F, Corti C, Gualtieri M (eds) Biodiversity conservation and habitat management. Encyclopedia of life support systems (EOLSS), developed under the auspices of the UNESCO. EOLSS Publishers, Oxford. http://www.eolss.net. Accessed 19 Sept 2010

Rauschmayer F, Weiss V (2013) Screening the cormorant conflict on the European level. In: Klenke RA, Ring I, Kranz A, Jepsen N, Rauschmayer F, Henle K (eds) Human-wildlife conflicts in Europe—fisheries and fish-eating vertebrates as a model case. Springer, Heidelberg. doi: 10.1007/978-3-540-34789-7_9

Ring I (2009) Toward a framework for integrating human dimensions in wildlife management. In: Manfredo MJ, Vaske JJ, Brown PJ, Decker DJ, Duke EA (eds) Wildlife and society: the science of human dimensions. Island Press, Washington, pp 90–103

Santos-Reis M, Santos R, Antunes P, Sales-Luís T, Gomes J, Freitas D, Madruga L (2013) Reconciliation of the conflict between otters and fish farmers: lessons from Sado estuary in Portugal. In: Klenke RA, Ring I, Kranz A, Jepsen N, Rauschmayer F, Henle K (eds) Human-wildlife conflicts in Europe—fisheries and fish-eating vertebrates as a model case. Springer, Heidelberg. doi: 10.1007/978-3-540-34789-7_4

Schwerdtner K, Gruber B (2007) A conceptual framework for damage compensation schemes. Biol Conserv 134:354–360

Similä J, Thum R, Varjopuro R, Ring I (2006) Protected species in conflict with fisheries: the interplay between European and national regulation. J Eur Environ Plan Law 3(5):432–445

## Further Reading

Barbier EB, Burgess JC, Folke C (1994) Paradise lost? The ecological economics of biodiversity. Earthscan, London

Bräuer I, Müssner R, Marsden K, Oosterhuis F, Rayment M, Miller C, Dodoková A (2006) The use of market incentives to preserve biodiversity. Final report, project under framework contract for economic analysis ENV.G.1/FRA/2004/0081, Berlin

Bulte EH, Rondeau D (2005) Why compensating wildlife damages may be bad for conservation. J Wildl Manag 69(1):14–19

McNeely JA (1988) Economics and biological diversity: developing and using economic incentives to conserve biological resources. IUCN, Gland

TEEB (The Economics of Ecosystems and Biodiversity) (2009) The economics of ecosystems and biodiversity for national and international policy makers. http://www.teebweb.org. Accessed 10 Jan 2010

Woodroffe R, Thirgood S, Rabinowitz A (2005) People and wildlife. Conflict or coexistence?. Cambridge University Press, Cambridge

# Module 10: Designing Participatory Decision Strategies

**Felix Rauschmayer**

**Abstract** Human-wildlife conflicts are necessarily linked to stakeholders. Resolving these conflicts requires multilateral interaction. Participatory developed decisions promise to have a better information base, include all actors with relevant knowledge and implementation power, be based on mutual trust and understanding and, finally, reduce overall costs. This chapter gives some general guidelines on what to consider for participatory processes in human-wildlife conflicts in order to live up to some of these promises. While the minimum approach consists of characterizing the conflict and of informing the stakeholders, the advanced approach eventually consists of employing a fully fledged participatory method.

## 1 Rationale and Objectives

Human-wildlife conflicts are by nature conflicts between different stakeholders (Conley and Moote 2003; Schusler and Decker 2003). Assessing whether participatory processes are useful for decision-making is particularly valuable when observing the following changes: a shift from species conservation to species management, the emergence of new actors in the conflict, or the escalation of the conflict due to environmental change or changing human and/or animal behavior.

Participatory processes (see Box 1) can lead to a better inclusion and processing of information. They can create or support positive social dynamics between

F. Rauschmayer (✉)
Department of Environmental Politics,
UFZ—Helmholtz Centre for Environmental Research,
Permoserstraße 15, 04318 Leipzig, Germany
e-mail: felix.rauschmayer@ufz.de

R. A. Klenke et al. (eds.), *Human-Wildlife Conflicts in Europe*,
Environmental Science and Engineering, DOI: 10.1007/978-3-540-34789-7_20,

stakeholders and public administration, create decision options such as new or adapted policy instruments with higher legitimacy and, thus, with higher acceptance. Finally, they can lead to options with lower costs. In biodiversity conflicts, participation will mainly be played on the stakeholder level and not involve the general public.

Whereas mere disposal of credible information might be enough in low-intensity conflicts about facts, highly elaborate and costly processes may be the adequate means to deal with high-intensity conflicts about values and interests. The results of discourse analysis (module 6) will indicate the intensity of the conflict.

**Box 1** Participatory Processes

"Numerous participatory processes have been designed, implemented, and analyzed in various contexts. In the field of environment and sustainable development, they are manifold. They include: focus groups, citizens' juries, consensus conferences, cooperative discourse, dialogue groups, stakeholders' workshops, participatory expert workshops, reflection forums, deliberative interviews, voluntary agreements, eco-audits, policy simulation exercises, deliberative foresights, concerted environmental management, mediation, regulatory negotiation, consultative forums, deliberative conflict resolution processes, environmental negotiations, etc." (van den Hove 2006, p. 11)

Most of the legal and institutional settings will have been analyzed in previous steps of analysis (modules 4 and 5). Negotiation-oriented processes aiming at decision-making are in the centre of this module. Analyses of different participatory methods and processes can be found in the literature (e.g., Renn et al. 1995; Dietz and Stern 2008).

This module aims at selecting a method appropriate to the specific case according to the main characteristics of the situation. The initial conflict characterization in module 1 has been redefined during the further analysis of the conflict (modules 2–6). In severe conflicts, data collection and scientific work will have to be embedded in a participatory process in order to achieve a general acceptance of the data. Therefore, particular attention should be paid to information management: Whereas scientific knowledge often is the first base of reconciliation action plans, local knowledge frequently does not find its way into the decision processes on action plans. This one-sidedness, apart from substantial shortcomings, can have severe implications on the legitimacy of the plan and, hence, on its implementation.

Designing participatory decision strategies always involves the cooperation with government agencies and may include new policy instruments; a close cooperation with module 9 (policy instruments) is therefore indicated. Close cooperation with modules 7 and 8 about ecological mitigation of the conflict and population modeling will enrich the quality of the discussion as well.

## 2 Methods and Approaches

Before really conducting the participatory decision process, one has to understand the conflict to select an appropriate method, and to identify the appropriate social organization or person for leading the process. One should therefore differentiate between the following steps when deciding about the participatory process to undertake:

1. Characterizing the conflict
2. Selecting the method
3. Leading the process
4. Conducting the process

(1) The conflict characterization mainly builds on the information obtained in the stakeholder analysis (module 6): it is the judgment of the stakeholders, which provides the necessary information on the type of conflict. One can roughly differentiate between conflicts on facts, on interests, and on values. In praxis, though, elements of all three categories are often intermingled. Furthermore, conflicts can be dormant, slowly evolving, or erupting. The institutional analysis (modules 4 and 5) provides information on existing and potential ways of conducting participatory processes. The importance of the conflict for society is another criterion for choosing the proper intensity of the participatory process. Attention must be paid to differentiate between different spatial and political levels of a conflict: local, regional, national, or supra-national. Each level usually evokes different attitudes of the stakeholder, has different institutional settings, and the human-wildlife conflict at hand has different relative importance. All this information is then brought together in order to define the situational requirements for a participatory method (Table 1). (1) Conflicts mainly caused by differences of belief in facts ask for better information management; (2) conflicts caused by the exclusion of one or several stakeholder groups demand an improved legitimacy of processes and decisions for their resolution; (3) ongoing conflicts with changing ecological and socio-economic behaviors require enhanced social dynamics; finally (4) costs of the decisions, the process, and of decision failures will play a role in all conflicts. Using analytical methods allowing for the simultaneous evaluation of different options along clearly outlined decision criteria, such as the methods of multi-criteria analysis (Salminen et al. 1998; Bouyssou et al. 2000; Stirling 2001), should help in most cases (NRC 1999).

(2) The aim of this step is to select a participatory process appropriate to the conflict situation. The categories described in step (1) (Table 1) can be used to evaluate different participatory methods (for an overview see Box 2; Rauschmayer and Wittmer 2006). These evaluations help the analyst to select the most appropriate method. They are based on the translation of the results of the discourse analysis into a conflict characterization. Participatory methods can only be evaluated in a standardized version and have to be adapted to the specifics of the situation. In practice, the deployment of the selected method also largely depends on the skills of the moderator. Selecting the appropriate process is thus dependent on the

**Table 1** Evaluation criteria for participatory processes

| Metacriteria | Criteria |
|---|---|
| Information management | Integrating different types of information |
| | Coping with uncertainty |
| | Coping with complexity |
| Legitimacy | Legal compatibility |
| | Inclusion/representation |
| | Transparency of rules and assumptions to insiders and outsiders |
| | Accountability |
| Social dynamics | Changing behavior, changing perspectives/learning |
| | Agency/empowerment |
| | Respect/relationship |
| | Facilitation of convergence or illustration of diversity |
| Costs | Cost-effectiveness |
| | Costs of the method |
| | Decision failure costs |

*Source*: Wittmer et al. (2006)

conflict characterization, builds on the institutional setting and demands from the analyst to specifically consider the possibilities and willingness of the authorities. Traditions of participation, legal options for participation, and factual openness to participation, especially of the administration, play an important role in selecting suitable participatory methods. Traditions and legal possibilities of participation differ between countries, between policy fields, and between policy levels. It has been shown that the commitment of the authorities to really consider the results of a participatory process is a key element to its success.

(3) It must be decided who leads the process, but often it is not even clear, which is the relevant organization for authorizing the request for a biodiversity reconciliation action plan. Usually, scientists have no mandate to initiate or to conduct a participatory process, and they may be perceived as partial or incompetent. Often, the authorities are dominated either by the users or by the protectors of nature. In many countries, though, legal or semi-legal participatory bodies already exist, which may or may not use external facilitators to conduct the process. Having external facilitators conducting the process improves impartiality and, therefore, legitimacy. Their external perspective can help to include new or currently excluded stakeholders (with their knowledge, interests, and values) in the process. At first sight, external facilitators raise the costs of the process, but through a potentially shorter process considering a fuller range of options, they can contribute to lower process and follow-up costs. Most often, a conflict has already been acknowledged, and activities started to resolve it. Adapting one's own participatory process to the existing calendar of events and to the policy cycle improves the chances of having enough participation and of influencing policy by the process results.

**Fig. 1** Scientists demonstrating the involvement of stakeholders in a participatory process. *Photo:* Norma Neuheiser

(4) The first step in conducting participatory decision processes is to identify its participants. Further steps may include elements of decision analysis, i.e. to elaborate clear decision options, to identify the criteria relevant to the decision, and to evaluate the options, which then leads to the discussion of the final recommendation. The selection of participants depends on the prior existence of participatory bodies, on the authorities responsible for and involved in the process, and, of course, on the diversity of the stakeholders, which becomes apparent through the discourse analysis (module 6). It might be difficult to introduce new stakeholders into an already existing participatory body for the creation of a reconciliation action plan. One should consider whether their participation could be channeled in a different way than through their direct and full participation (Fig. 1).

Different options, partly coming from module 9 and all using information from modules 1–8, are developed further and decided upon according to the agreed criteria. There are several benefits of making options, criteria, and the formal decision procedure transparent. The decision process should make sure, firstly, that the considered options cover the whole range of possibilities and, secondly, that the criteria cover the whole range of values and interests of people concerned. Together with a clear and comprehensive evaluation of the options, completeness and clarity will increase the legitimacy of the process and of the decisions made. The options will be elaborated based on the results of natural and social analyses while considering improved or newly emerging policy options (module 9), new mitigation techniques (module 7), and the result of population viability analysis (module 8). At the same time, the results of these other modules will provide the foundation for the evaluations. The first ideas on the relevant decision criteria will stem from the discourse analysis (module 6).

# 3 Recommendations: Selecting the Appropriate Approach

Following the four steps outlined above, the characterization (step 1) constitutes the minimum approach. Giving advice to the authorities on how to conduct an appropriate participatory process can be considered as the standard approach. The appropriate way of conducting participatory processes, and of advising authorities in the standard approach, depends on the financial and institutional possibilities and on the intensity of the conflict. Simple participatory methods require only little money and effort (for example an embedment in existing well-functioning institutional settings), while highly elaborated methods ask for highly skilled facilitators, profound elaboration of decision options, and a balanced consideration of stakeholders, administration, and the public (compare Webler 1995; Renn 1999; Dietz and Stern 2008).

## *3.1 Minimum Requirement: Conflict Characterization and Information*

The characterization of the conflict is the basis for deciding whether to undertake a participatory process. It indicates type, severity, and intensity of the conflict. It mainly builds on the screening phase (module 1), enriched by stakeholder and discourse analysis (module 6), the institutional setting (modules 4 and 5) and on first ideas of potential policy options (modules 1–9). Stakeholders should be informed and perhaps educated about the main conflict characteristics in an appropriate form.

## *3.2 Standard: Process Selection and Advice*

If the characterization of the conflict indicates a need for a participatory decision process, the conflict manager should select appropriate methods (see Box 2) and check whether those can be embedded in the existing institutional setting, such as stakeholder *fora* or advisory councils. The resulting advice is meant to enrich and complement the existing setting in order to help relevant authorities and further stakeholders to identify better ways of making joint decisions.

## *3.3 Advanced: Conducting a Process*

If there is a window of opportunity for a new and specific participatory process as identified in the standard approach, i.e.

- if the authorities are committed to the process,
- if the relevant stakeholders are willing to participate,
- if a selected facilitator is judged neutral by all participants,
- if there is enough time, and
- if there are enough finances for the selected method,

then such a process can be conducted. Box 2 gives a short example of how to conduct such a process.

**Box 2** Participatory Decision Process in the Portuguese Sado Estuary Nature Reserve (see Santos-Reis et al. 2012)

In the Sado Estuary Nature Reserve a participatory process for the development of policy instruments was undertaken within the FRAP project in order to reconcile the conflict between otter conservation and fish farming. Assessment showed that a growing conflict between the fish-farmers and the nature reserve administration was motivated by opposing interests and a communication gap, without instruments in place to address it.
Rui Santos and colleagues initiated a participatory conflict reconciliation process, combining the use of formal participation techniques—consultation workshops—with an informal approach of information gathering and building of trust, based on individual meetings. This has resulted in the development of a collaborative decision option, involving fish-farmers, the reserve administration and the municipality. A first result of this agreement is the development of a fish certification scheme, promoting an ecologically sustainable production and the sector's economic competitiveness.

## References


Bouyssou D, Marchant T, Pirlot M, Perny P, Tsoukias A, Vincke P (2000) Evaluation and decision models: a critical perspective. Kluwer, Dordrecht

Conley A, Moote MA (2003) Evaluating collaborative natural resource management. Soc Nat Resour 16:371–386

Dietz, T, Stern, PC (eds) (2008) Public participation in environmental assessment and decision making. National Research Council, Panel on Public Participation in Environmental Assessment and Decision Making. National Academies Press

NRC (National Research Council) (1999) Perspectives on biodiversity: valing its role in an ever changing world. National Academy Press, Washington

Rauschmayer F, Wittmer H (2006) Evaluating deliberative and analytical methods for the resolution of environmental conflicts. Land Use Policy 23(1):108–122

Renn O, Webler T, Wiedemann P (1995) Fairness and competence in citizen participation. Kluwer, Dordrecht

Renn O (1999) A model for an analytic-deliberative process in risk management. Environ Sci Technol 33(18):3049–3055

Salminen P, Hokkanen J, Lahdelma R (1998) Comparing multicriteria methods in the context of environmental problems. Eur J Operational Res 104:485–496

Santos-Reis M, Santos R, Antunes P, Sales-Luís T, Gomes J, Freitas D, Madruga L (2012) Reconciliation of the conflict between otters and fish farmers: Lessons from Sado estuary in Portugal. In: Klenke R, Ring I, Kranz A, Jepsen N, Rauschmayer F, Henle K (eds) Human-wildlife conflicts in Europe—Fisheries and fish-eating vertebrates as a model case. Springer, Heidelberg, pp xxx–xxx

Schusler TM, Decker DJ (2003) Social learning for collaborative natural resource management. Soc Nat Resour 15:309–328

Stirling A (2001) A novel approach to the appraisal of technological risk: a multicriteria mapping study of a genetically modified crop. Environ Plan C, Gov Policy 19:529–555

van den Hove S (2006) Between consensus and compromise: acknowledging the negotiation dimension in participatory approaches. Land Use Policy 23(1):10–17

Webler T (1995) "Right" discourse in citizen participation: an evaluative yardstick. In: Renn O, Webler T, Wiedemann P (eds) Fairness and competence in citizen participation. Kluwer, Dordrecht, pp 35–86

Wittmer H, Rauschmayer F, Klauer B (2006) How to select instruments for the resolution of environmental conflicts? Land Use Policy 23(1):1–9

# Editors Biography

Reinhard A. Klenke is biologist and obtained his PhD in Ecology at the University of Greifswald, Germany. He has managed the EU FRAP project, the main inspiration for this book, at the Department of Conservation Biology of the Helmholtz Centre for Environmental Research in Leipzig. His research focuses on population ecology and conservation of larger vertebrates (birds and mammals) and how they are influenced by changes in landscape structure and land use activities.

Irene Ring is an environmental scientist and obtained her PhD in Economics at the University of Bayreuth, Germany. She is deputy head of the Department of Economics at Helmholtz Centre for Environmental Research in Leipzig and head of the Social Sciences Working Group on Nature Conservation and Biodiversity. Together with Klaus Henle, she coordinated the EU project FRAP. Her current research focuses on biodiversity conflict reconciliation, environmental federalism, conservation economics and environmental policy instruments.

R. A. Klenke et al. (eds.), *Human-Wildlife Conflicts in Europe*,
Environmental Science and Engineering, DOI: 10.1007/978-3-540-34789-7,

Andreas Kranz is wildlife ecologist and obtained his PhD at the University of Natural Resources and Life Sciences in Vienna, Austria. His main fields of interest are conservation research and conflict management. In terms of habitats he is active from the Alps down to the Danube Delta. His focal species are the Eurasian otter and the European mink. However, in everyday life he has to cope with conflicts arising from many species, in particular large carnivores, ungulates, mustelids, corvids and piscivorous birds.

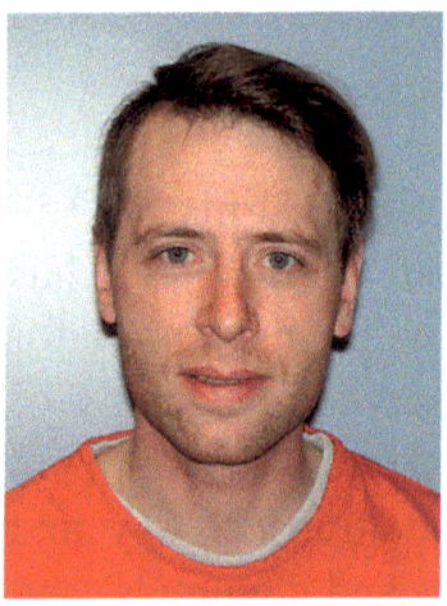

Niels Jepsen is senior researcher, working for DTU Aqua, where his research has focused mainly on behavior of freshwater fish and management of fisheries, including investigations of the effect of cormorant predation. He obtained his PhD degree from University of Aalborg in 1999. Since then he has worked on a wide variety of projects, mainly cross-disciplinary, international (EU) projects. He is expert member of the Danish National Cormorant Management Group.

Felix Rauschmayer received his PhD in economics in 1999 at the University of Heidelberg and subsequently worked as assistant professor in Philosophy in Leipzig until 2001. He now is deputy head of the Social Sciences Working Group on Governance and Institutions at the Helmholtz Centre for Environmental Research in Leipzig. His research focuses on theory, practice, and evaluation of decision processes in environmental governance, on science-society interfaces in nature conservation policy, and on needs-based sustainability conceptualisation.

Klaus Henle obtained his PhD in Zoology at the Australian National University, Canberra. He is head of the Department of Conservation Biology at the Helmholtz Centre for Environmental Research, Germany. Together with Irene Ring, he coordinated the EU project FRAP. His research focuses on biodiversity conservation with particular emphasis on habitat fragmentation, vertebrate population biology, and methods to analyse and monitor changes in biodiversity.

# Index



R. A. Klenke et al. (eds.), *Human-Wildlife Conflicts in Europe*, Environmental Science and Engineering, DOI: 10.1007/978-3-540-34789-7,

**I** (*cont.*)



**K**



**L**

MIX
Papier aus verantwortungsvollen Quellen
Paper from responsible sources
FSC® C105338

If you have any concerns about our products,
you can contact us on
**ProductSafety@springernature.com**

In case Publisher is established outside the EU,
the EU authorized representative is:
**Springer Nature Customer Service Center GmbH**
**Europaplatz 3, 69115 Heidelberg, Germany**

Printed by Libri Plureos GmbH
in Hamburg, Germany